THE NORMAL MICROBIAL
FLORA OF MAN

THE SOCIETY FOR APPLIED BACTERIOLOGY
SYMPOSIUM SERIES NO. 3

THE NORMAL MICROBIAL FLORA OF MAN

Edited by

F. A. SKINNER

AND

J. G. CARR

1974

ACADEMIC PRESS · LONDON · NEW YORK

ACADEMIC PRESS INC.(LONDON)LTD
24-28 OVAL ROAD
LONDON N.W.1

U.S. Edition published by
ACADEMIC PRESS INC.
111 FIFTH AVENUE
NEW YORK, NEW YORK 10003

Copyright © 1974 By the Society for Applied Bacteriology

ALL RIGHTS RESERVED

NO PART OF THIS BOOK MAY BE REPRODUCED IN ANY FORM BY PHOTOSTAT, MICROFILM, OR ANY OTHER MEANS, WITHOUT WRITTEN PERMISSION FROM THE PUBLISHERS

Library of Congress Catalog Card Number: LCCCN 73-19019
ISBN: 0-12-648040-0

Printed in Great Britain by
The Whitefriars Press Ltd., London and Tonbridge, England

Contributors

CHARLOTTE M. ANDERSON, *The University of Birmingham Institute of Child Health, The Nuffield Building, Francis Road, Birmingham B16 8ET, England*

D. C. BLENDEN, *Department of Veterinary Microbiology, School of Veterinary Medicine, University of Missouri, Columbia, Missouri 65201, U.S.A.*

G. H. BOWDEN, *Dental Bacteriology Laboratory and MRC Dental Epidemiology Unit, The London Hospital Medical College, Turner Street, London E1 2AD, England*

GLENNA C. BURTON, *Department of Veterinary Microbiology, School of Veterinary Medicine, University of Missouri, Columbia, Missouri 65201, U.S.A.*

D. N. CHALLACOMBE, *The University of Birmingham Institute of Child Health, The Nuffield Building, Francis Road, Birmingham B16 8ET, England*

J. G. COLLEE, *University Medical School, Edinburgh EH8 9AG, Scotland*

R. M. DAVIES, *Department of Oral Medicine, University of Manchester Dental Hospital, Bridgeford Street, Manchester M15 6FA, England*

B. S. DRASAR, *Department of Bacteriology, St. Mary's Hospital Medical School, London W2 1PG, England*

S. M. FINEGOLD, *Wadsworth VA Center Hospital, Los Angeles and the Department of Medicine, UCLA School of Medicine, Los Angeles, California, U.S.A.*

D. A. M. GEDDES, *Department of Oral Physiology, University of Newcastle upon Tyne, Markham Laboratories, Upper Claremont Street, Newcastle upon Tyne, NE2 4AJ, England*

J. M. HARDIE, *Dental Bacteriology Laboratory and MRC Dental Epidemiology Unit, The London Hospital Medical College, Turner Street, London E1 2AD, England*

D. C. HIRSH, *Department of Veterinary Microbiology, School of Veterinary Medicine, University of Missouri, Columbia, Missouri 65201, U.S.A.*

ROSALINDE HURLEY, *Queen Charlotte's Hospital for Women and the Institute of Obstetrics and Gynaecology, University of London, Goldhawk Road, London W6 0XG, England*

G. N. JENKINS, *Department of Oral Physiology, University of Newcastle upon Tyne, Markham Laboratories, Upper Claremont Street, Newcastle upon Tyne, NE2 4AJ, England*

BARBARA G. S. LEASK, *Queen Charlotte's Hospital for Women and the Institute of Obstetrics and Gynaecology, University of London, Goldhawk Road, London W6 0XG, England*

O. M. LIDWELL, *Central Public Health Laboratory, Colindale Avenue, London NW9 5HT, England*

CONTRIBUTORS

J. DE LOUVOIS, *Queen Charlotte's Hospital for Women and the Institute of Obstetrics and Gynaecology, University of London, Goldhawk Road, London W6 0XG, England*

J. H. McCOY, *Public Health Laboratory, Hull Royal Infirmary, Anlaby Road, Kingston upon Hull, England*

MARY J. MARPLES, *1 Vanburgh Close, Old Woodstock, Oxford OX7 1YB, England*

R. R. MARPLES, *Central Public Health Laboratory, Colindale Avenue, London NW9 5HT, England*

H. N. NEWMAN, *MRC Dental Unit, The Dental School, Lower Maudlin Street, Bristol BS1 2LY, England*

D. F. G. POOLE, *MRC Dental Unit, The Dental School, Lower Maudlin Street, Bristol BS1 2LY, England*

SYLVIA E. REED, *Clinical Research Centre, Watford Road, Harrow, Middlesex HA1 3UJ, England*

JUDITH M. RICHARDSON, *The University of Birmingham Institute of Child Health, The Nuffield Building, Francis Road, Birmingham B16 8ET, England*

D. A. SHAW, *Unilever Research, Isleworth Laboratory, Unilever Limited, 455 London Road, Isleworth, Middlesex, England*

VALERIE C. STANLEY, *Queen Charlotte's Hospital for Women and the Institute of Obstetrics and Gynaecology, University of London, Goldhawk Road, London W6 0XG, England*

VERA L. SUTTER, *Wadsworth VA Center Hospital, Los Angeles and the Department of Medicine, UCLA School of Medicine, Los Angeles, California, U.S.A.*

D. A. J. TYRRELL, *Clinical Research Centre, Watford Road, Harrow, Middlesex HA1 3UJ, England*

SIR GRAHAM WILSON, *London School of Hygiene and Tropical Medicine, Keppel Street, London WC1, England*

R. C. S. WOODROFFE, *Unilever Research, Isleworth Laboratory, Unilever Limited, 455 London Road, Isleworth, Middlesex, England*

JUDY L. ZEIGLER, *Department of Veterinary Microbiology, School of Veterinary Medicine, University of Missouri, Columbia, Missouri 65201, U.S.A.*

Preface

A SYMPOSIUM on the topic 'The Normal Microbial Flora of Man' was held during the Summer Conference of the Society for Applied Bacteriology at the University College of Wales, Aberystwyth, in July 1973. The papers, which were given by leading specialists in their fields, are published in this, the third volume in the Symposium Series of the Society's publications.

In its early days bacteriology was, for obvious reasons, concerned largely with abnormal microbial development in man. Today, now that many clinical problems of infection have been solved, and because of the current interest in microbial ecology, there is both opportunity and inclination to study the normal microbial flora of man and to elucidate the reasons for its establishment and persistence.

In the following pages three microbial habitats, skin, mouth and intestine, receive especial consideration. For each of these systems contributors have tried to characterize the normal microflora and to determine changes in its composition in response to intrinsic regulatory mechanisms or to antimicrobial agents. Other topics such as the microflora of the vagina, dental plaque, dispersal of micro-organisms from the respiratory tract and viruses in the healthy subject, are covered by single contributions.

F. A. SKINNER
Rothamsted Experimental Station
Harpenden AL5 2JQ
Hertfordshire
England

J. G. CARR
University of Bristol
Research Station
Long Ashton
Bristol BS18 9AF
England

February 1974

Contents

LIST OF CONTRIBUTORS v

PREFACE vii

The Normal Flora of Man: Introduction, General Considerations and Importance
SIR GRAHAM WILSON
    Introduction 1
    Origin, localization, interference and protection 2
    Qualitative and quantitative assessment of the normal flora . . 3

The Normal Microbial Flora of the Skin
MARY J. MARPLES
    Introduction 7
    Members of the cutaneous biocenose 7
    Location of the cutaneous flora 8
    Distribution of the bacterial flora 8
    Composition of cutaneous populations 9
    Dispersal of the cutaneous flora 10
    Coactions among the resident flora 10
    References 11

Natural Control and Ecology of Microbial Populations on Skin and Hair
R. C. S. WOODROFFE and D. A. SHAW
    Introduction 13
    Microflora of the hair 14
    Microflora of the skin 15
        (a) Transient and resident flora 15
        (b) The normal flora 17
        (c) Density of microbial populations at various skin sites . . 18
    Factors in the microbial ecology of skin 18
        (a) Water availability 19
        (b) Microbial nutrients and inhibitors on the skin . . . 20
            (i) Apocrine sweat 21
            (ii) Eccrine sweat 21
            (iii) *Stratum corneum* 22
            (iv) Lysozyme 22
            (v) Skin surface lipids 22
            (vi) Bacteriocins and other inhibitors 25
        (c) Population-density control of resident micro-organisms 29

Conclusions 29
References 30

Effects of Soaps, Germicides and Disinfectants on the Skin Flora
R. R. MARPLES
Introduction 35
Effects of systemic antibiotics 36
   (a) Oral demethylchlortetracycline and the flora of the axilla . 36
   (b) Other systemic antibiotics 36
Effects of simple washing 38
Effects of antibacterial agents in soaps 39
Occlusion effects 40
   (a) Testing topical antibacterial agents 41
   (b) Prolonged occlusion and pyoderma 43
Experimental infections and antimicrobial agents . . . 43
Clinical trials 44
Discussion 45
References 45

The Normal Microbial Flora of the Mouth
J. M. HARDIE and G. H. BOWDEN
Introduction 47
The oral environment 48
The development of the oral microflora 49
The microbial flora of different parts of the mouth . . . 50
   (a) Tongue 51
   (b) Saliva 52
   (c) Dental plaque 53
   (d) Gingival crevice 59
The oral flora of primitive man 61
Some ecological determinants 62
   (a) Anaerobiosis 62
   (b) Adherence and aggregation 64
   (c) Microbial inter-relationships 64
The principal groups of micro-organisms found in the mouth . . 67
   (a) Gram positive cocci 67
   (b) Gram negative cocci 70
   (c) Gram positive rods and filaments 71
   (d) Gram negative rods and filaments 72
   (e) Other micro-organisms 73
References 74

Intrinsic and Extrinsic Factors Influencing the Flora of the Mouth
D. A. M. GEDDES and G. N. JENKINS

Introduction	85
Establishment of flora	85
Physical effects of saliva	86
Mechanical removal of plaque	87
Composition of saliva	89
Plaque acid production from exogenous sugar	91
Microbial interactions	92
Antibacterial systems of non-bacterial origin	92
Dietary factors	93
Factors inferred from presence of caries	94
Effect of fluoride	95
Conclusions	96
References	96

Control of Oral Flora by Hibitane and Other Antibacterial Agents
R. M. DAVIES

Introduction	101
Antibacterial agents	102
(a) Theoretical considerations	102
(b) Antibiotics	102
(c) Synthetic antibacterial agents	103
(i) Saliva	104
(ii) Gingiva and tooth surface	104
References	108

Structural and Ecological Aspects of Dental Plaque
H. N. NEWMAN and D. F. G. POOLE

Introduction	111
Plaque formation	112
The acquired pellicle	113
The attachment of plaque to the tooth	114
The plaque flora	115
Plaque structure	116
Interbacterial contacts	118
Filament formation and plaque	118
The dynamics of plaque formation	120
Plaque and other natural microbial films	120
Plaque and disease	124
References	127

Aerial Dispersal of Micro-organisms from the Human Respiratory Tract
O. M. LIDWELL
- Introduction 135
- Droplet dispersal from the respiratory tract 135
- Salivary streptococci in the air 140
- Some examples of the dispersal of specific pathogens . . . 143
    - (a) β-haemolytic streptococci 143
    - (b) Mycobacteria 144
    - (c) *Staphylococcus aureus* 146
    - (d) Respiratory viruses 149
- Conclusion 150
- Summary 152
- Acknowledgements 152
- References 152

Microflora of the Vagina During Pregnancy
ROSALINDE HURLEY, VALERIE, C. STANLEY, BARBARA G. S. LEASK and J. DE LOUVOIS
- Introduction 155
- Anatomy of the female external genitalia and the vagina . . 156
- The internal genitalia 156
- Physiology of the vagina 158
- Relationship of flora to vaginal pH and glycogen . . . 158
- Experimental procedures 159
    - (a) General ecological survey of the vaginal flora of pregnant women 159
    - (b) Studies on yeast flora 161
- Results 161
    - (a) General ecological survey 161
    - (b) Yeast flora 164
    - (c) Clinicopathological correlations 165
- Discussion 172
- Acknowledgements 182
- References 182

Some Factors Associated with Geographical Variations in the Intestinal Microflora
B. S. DRASAR
- Introduction 187
- Mechanisms controlling the microflora 187
    - (a) The small intestine 187
    - (b) The large intestine 188

Geographical variations in the microflora 190
   (a) The small intestine 190
   (b) The large intestine (faeces) 191
Factors influencing mechanisms controlling the microflora . . 192
   (a) Race 192
   (b) Tropical residence 194
   (c) Diet 194
Conclusion 194
References 195

The Bacterial Flora of the Upper Gastrointestinal Tract in Children both in Health and Disease
CHARLOTTE M. ANDERSON, D. N. CHALLACOMBE and
JUDITH M. RICHARDSON

Introduction 197
The normal intestinal microflora in infants and children . . . 198
Alterations in small-intestinal microflora 199
Relationship of *E. coli* in the upper small intestine to diarrhoeal symptoms 201
References 202

Clostridium perfringens (Cl. welchii) in the Human Gastro-intestinal Tract
J. G. COLLEE

Introduction 205
Occurrence in human faeces 206
Distribution in the human gut 207
The ingestion of *Cl. perfringens* 208
Growth control mechanisms in the gastro-intestinal tract . . 208
Host protective factors 209
Endogenous pathogenic potential in the gut 209
Skin contamination with faecal clostridia 210
Specific intestinal infections 210
Clostridium perfringens food poisoning 211
The borderland of cases and carriers 212
Toxigenicity and sporulation 214
Challenges 215
Summary 215
Acknowledgement 216
References 216

Enteric and Salmonella Infection: the Carrier State
J. H. McCOY

Introduction	221
Development of the carrier state	223
The chronic carrier	223
(a) Frequency of excretion	223
(b) Numbers of bacilli excreted	223
Prevention of typhoid fever	225
Salmonella infection	225
References	227

The Effect of Antimicrobial Agents on Human Faecal Flora: Studies with Cephalexin, Cyclacillin and Clindamycin
VERA L. SUTTER and S. M. FINEGOLD

Introduction	229
Experimental	230
(a) Subject material	230
(b) Collection and processing of faecal specimens	230
(c) Bacteriological studies	230
Results	232
(a) Cephalexin	232
(b) Cyclacillin	232
(c) Clindamycin	234
Discussion	236
Acknowledgements	239
References	239

The Effects of Tetracycline on the Establishment of *Escherichia coli* of Animal Origin, and *in vivo* Transfer of Antibiotic Resistance, in the Intestinal Tract of Man
GLENNA C. BURTON, D. C. HIRSH, D. C. BLENDEN and JUDY L. ZEIGLER

Introduction	241
Materials and Methods	243
(a) Volunteers	243
(b) Properties of the ingested organism	243
(c) Antibiotic and X-314 administration	244
(d) Handling of faecal specimens	245
(e) Statistical tests	245
Results	245
Discussion	249
Acknowledgements	252
References	252

Viruses Associated with the Healthy Individual
   Sylvia E. Reed and D. A. J. Tyrrell
      Introduction 255
      Viruses of healthy children 255
      References 257

Index 259

The Normal Flora of Man: Introduction, General Considerations and Importance

SIR GRAHAM WILSON

*London School of Hygiene and Tropical Medicine,
Keppel Street, London W.C.1., England*

CONTENTS

1. Introduction 1
2. Origin, localization, interference and protection 2
3. Qualitative and quantitative assessment of the normal flora 3

1. Introduction

THE SUBJECT for our discussion this year is one that has been particularly well chosen. It is of wide interest and presents a number of unsolved problems, both technical and general. The many ways in which it is of interest to the microbiologist is evident from the list of papers before us. The interest it presents to workers in other fields is perhaps not so evident.

To the ecologist it presents a fascinating kaleidoscope of variations and fluctuations among a population that is never constant yet is always tending to modulate about a mean—a population too in which there is continuous competition for possession and maintenance of favoured sites.

For the physiologist and biochemist it provides a field in which a host of biochemical reactions of different kinds is being carried out, many of which besides being of academic interest contribute to the support of the body and the promotion of health.

The normal must be known before the abnormal; hence its appeal to the pathologist. Just as anatomy and physiology must be learnt before the study of clinical medicine, so must the normal flora of the body be known before the pathogenic role of any unknown organism can be assessed.

The part played by the normal flora in preventing the access of pathogenic bacteria to the tissues is of interest to the immunologist; as is also the possession by certain members of the normal flora of antigens common to pathogenic organisms which, by stimulating the production of so-called natural antibodies, renders more difficult the interpretation of serological findings in diagnostic medicine.

The effect on the composition of the normal flora of changes in diet, in climate, in various combinations of atmospheric temperature and humidity, in insanitary environmental conditions, and in personal hygiene is of interest to the epidemiologist whose function it is to unravel the tangled web of aetiological factors that govern the occurrence and outcome of so many different diseases.

Next I must mention the practitioners of clinical medicine—the general physicians and surgeons, and the more specialized obstetricians and gynaecologists, the paediatricians, the dermatologists, and those engaged in dentistry and in surgical treatment of diseases of the genito-urinary system. All these representatives of clinical medicine are affected in one way or another by the nature and frequency of the organisms that help to make up the natural flora of any particular part of the body.

Last among those I mention—and there are others that I do not—must be included the manufacturers of disinfectants and of chemotherapeutic and antibiotic drugs which so often disturb the balance of the normal flora and do more harm than good. Against those whose lives have been saved by these products must be set the great number that have died as the result of displacement of the susceptible intestinal flora by resistant pathogenic bacteria and fungi.

2. Origin, Localization, Interference and Protection

No one can study the normal flora of the body for long without realizing how many problems are posed by its establishment and composition. The origin of the different organisms is the immediate environment of the infant. During the process of delivery the skin, the nose, the mouth and the conjunctiva all become contaminated with organisms from the genital passages of the mother; and within a few hours of birth micro-organisms are proliferating within the alimentary tract.

Observations on the established normal flora have shown that certain organisms are adapted to certain sites and entrench themselves so firmly that it is difficult except by drastic means to displace them.

What is it that determines the selective localization of these different organisms? Why, among the aerobes, do the Gram positive cocci choose the skin for their permanent abode? Why do coliform organisms, with which hands so frequently become contaminated, not form part of the resident cutaneous flora? And why do these organisms flourish so abundantly in the intestine? Again, why among the coliform organisms is it only *Escherichia coli* that assumes dominance in the gut; and why is this organism confined to certain parts only of the gut? How much of the selective localization is due to the presence in the tissues of substances that act as specific nutrients for some organisms and give them an advantage over other organisms that cannot use these nutrients so readily; or how far is it due to the absence of certain nutrients that are essential for other organisms?

How much does it depend on other conditions provided by the host, such as the blood supply, the degree of moisture, the temperature of the part concerned, the pH or E_h values, the production of fatty acids and of various secretions having a stimulating or inhibitory effect on different organisms, the lysozymes of mucosal surfaces, the gastric and intestinal juices of the alimentary tract, the acidity of the urine, and the various hormones of the body?

Now comes another question. What part does the normal flora play in the defence of the host? In healthy persons potentially pathogenic organisms, when they do gain lodgement in the tissues, seem to be kept in check and prevented from free multiplication. How long are they able to maintain their hold? Does *Streptococcus pyogenes,* for example, which is found in the pharynx of a small proportion of children, reside there continuously, or is it displaced after a time and fail to get a footing again? Similarly with *Staphylococcus aureus* in the nose; why do phage types of the same organism interfere with each other and why are fresh organisms so difficult to implant in the nose or the intestine? The replacement of potentially pathogenic by non-pathogenic organisms, similar to the control of plant pests by biological rather than by chemical means, might be of great advantage to the host, but so far little success has been reported from attempts in this direction.

By what means is this interference of the resident flora with the establishment of fresh organisms brought about? Is it by the formation of bacteriocines, antibiotic substances, acid, hydrogen peroxide, reducing agents, or some other inimical product? Or is it more associated with the ability to make use of the nutrient substances locally available? These and numerous other questions are raised by the existence and behaviour of the normal flora, but I must leave it to those who follow me to discuss and elaborate them and to draw such conclusions as the evidence allows.

3. Qualitative and Quantitative Assessment of the Normal Flora

The technical problem of determining the qualitative composition and the quantitative distribution of the organisms making up the normal flora is almost insuperable, at any rate as far as certain sites are concerned. For instance, all sorts of different methods have been employed in studying the flora of the skin but none of them so far has proved entirely satisfactory. With organisms situated on the surface and at different depths beneath, in hair follicles and in sebaceous glands, it is almost impossible to devise a technique suitable for practical purposes that will yield trustworthy and reproducible results. Again, in the intestine, withdrawal of the contents at various levels by a magnetically guided tube affords a sample only of the organisms free in the lumen; those attached to the villi or other parts of the surface, which there is reason to believe are often present in large numbers, remain untouched. For the nose, nasopharynx, throat and other mucosae neither the swab nor the rinse method can be expected to yield accurate results; the most one can hope for is a qualitative assessment with no more than a rough estimate of the proportions in which the different organisms are present.

The treatment of the sample once it is obtained must necessarily be laborious. Microscopical examination with various stains and sometimes with fluorescent antibody should indicate broadly what sort of organisms are to be looked for, including spirilla, spirochaetes, yeasts and fungi and, in the mouth and faeces, protozoa. To culture these organisms is the next step, and to make reasonably certain that all the different types noted under the microscope are represented. For this purpose a great variety of selective media may be necessary, including tissue cultures for viruses and sometimes egg and animal inoculation. The cultural conditions must comprise the presence and absence of oxygen and carbon dioxide, together with incubation at different temperatures for varying lengths of time. For the strict anaerobic non-sporing organisms, such as *Bifidobacterium, Bacteroides* and some members of the *Fusobacterium* group, a technique may have to be employed that protects them against contact with oxygen from the time they are sampled to the time they are cultured and, of course, during the process of subculture.

Once satisfactory cultural requirements have been worked out for each species of organism an attempt may be made to ascertain their numbers. This is a formidable task. The organisms are often present in clumps which may have to be freed from tissue cells and debris. Methods for breaking up the clumps themselves vary but it should be remembered that violent methods may lead to destruction of many of the organisms. Selective media, which are unavoidable in any attempt to count mixed cultures, nearly always have some inhibitory action on the organisms they are designed to favour. Provided this is not too great, comparison of the cultural results with the microscopic picture may afford a rough guide to the proportion of living and dead bacteria, but this is not always practicable.

As in the counting of all mixed cultures, the results obtained should be expressed in terms of colony count and not of viable organisms. It is perhaps elementary to say so, but a true viable count can be realized only when a single strain of organism is cultured in a fluid medium under conditions in which every reproducible organism is capable of growing and in which the individual bacteria are separated from one another so that each colony formed can be taken as arising from a single organism. Such conditions exist only in precise experiments in the laboratory. In the type of enumeration that is carried out on the normal flora the relation between the colony count and the true viable count cannot be more than guess-work. In practice, however, accuracy is seldom required. It suffices to determine what sort of organisms are present and approximately in what proportions.

Finally, interpretation of the results is often difficult. The flora of any given area as determined by sampling is subject to such large variations, not only between individual subjects but from day to day in the same subject, that the effect of extraneous factors, such as washing, disinfection, changes in the diet, and so on, is often impossible to evaluate.

There are many more things that might be mentioned in an introductory address of this kind, but I hope I have said enough to indicate the wide field for observation, discussion, conjecture and conclusion that lies open to us during this symposium.

The Normal Microbial Flora of the Skin

MARY J. MARPLES

*1 Vanbrugh Close, Old Woodstock,
Oxford OX7 1YB, England*

CONTENTS

1. Introduction 7
2. Members of the cutaneous biocenose 7
3. Location of the cutaneous flora 8
4. Distribution of the bacterial flora 8
5. Composition of cutaneous populations 9
6. Dispersal of the cutaneous flora 10
7. Coactions among the resident flora 10
8. References 11

1. Introduction

TEN YEARS AGO it would have been difficult, if not impossible, to organize a two-day symposium on the normal microbial flora of man. Although the existence of such a flora was recognized, it had received very little investigation and attention had been directed primarily towards the activities of pathogenic species and to the response they evoked in the host. In spite of its accessibility the cutaneous community had been studied in much the same piece-meal way, to a great extent in relation to surface disinfection. In recent years, however, the skin has been considered as an ecosystem and attention has been directed to the harmless cutaneous residents rather than to transient pathogens. It is still not always easy to distinguish between these two groups, nor between non-pathogenic and potentially pathogenic species. In general, however, it is now realized that human skin provides an inhospitable habitat, so that except in the very young, the very old and the diseased, the true resident community is restricted to a very few microbial species, always present in relatively large numbers and, in ordinary circumstances, exciting no pathological changes in the substrate.

2. Members of the Cutaneous Biocenose

Only one animal, *Demodex folliculorum,* can be regarded as a true cutaneous resident. This very small mite lives in the hair follicles and sebaceous glands chiefly on the face. It can be found in the skin of most adults if an intensive search is made and, in some individuals, dense populations may develop.

The fungi are represented by a small number of species. Two lipophilic yeasts, *Pityrosporum ovale* and *P. orbiculare,* are carried by the majority of adults. Both are found in many cutaneous areas but Roberts (1969) has shown that *P. ovale* predominates in the scalp while *P. orbiculare* occurs more frequently on the chest and back. This yeast is a potential pathogen and appears to be causally related to the superficial infection *tinea versicolor*. Non-lipophilic yeasts, for example *Torulopsis glabrata,* and nonpathogenic species of *Candida,* frequently form sparse populations in the interdigital spaces of the foot and in tropical countries, may be inhabitants of the general skin surface (Marples & Somerville, 1968). The filamentous fungi associated with foot ringworm can be regarded as true residents of many human feet because they can be isolated from skin which shows minimal pathological changes. It is difficult to determine the extent to which viruses can be regarded as cutaneous residents. *Herpesvirus hominis,* the cause of 'cold sores', appears to remain in the skin in a latent phase during interim periods, and can, therefore, be classified as a member of the normal flora. It is now known that many cutaneous bacteria are lysogenic. Their virus partners may be important members of the flora, as they influence the structure of individual cutaneous communities.

The bacterial flora of the skin has received the greatest attention from investigators, and the results of many studies indicate that bacteria dominate the cutaneous biocenose. The number of species living on the skin is limited. Coagulase negative, Gram positive cocci and diphtheroids predominate. Gram negative species may be present in small numbers and mycobacteria and aerobic spore-formers can be found in some, if not all, individuals. The following paragraphs are mainly concerned with discussion of recent investigations of the bacterial communities living on the skin of young adults.

3. Location of the Cutaneous Flora

Most authorities believe that the vast majority of micro-organisms are living in the most superficial layers of the epidermal *stratum corneum* and in the upper parts of the hair follicles. Montez & Willborn (1969) have demonstrated this location in a series of beautiful electron-microscope pictures of skin sections. A proportion of skin bacteria, however, are so deep in the follicular canals that up to 20% are beyond the reach of ordinary disinfection procedures (Selwyn & Ellis, 1972). The presence of this reservoir permits the rapid re-establishment of the surface flora after it has been removed by artificial means.

4. Distribution of the Bacterial Flora

It has been repeatedly shown that different areas of skin support populations which differ in density, and to a less extent, in composition. There seems to be no doubt that available water is the most important environmental factor influencing the size of cutaneous populations. Marples (1965) has shown that simple occlusion of the relatively dry skin of the forearm results in a rapid build-up of the bacterial population from an initial colony count of $3 \times 10^3/cm^2$ to a level of 3.8×10^7 by the fourth day of occlusion. The high relative humidity of the groin and axilla is an important environmental feature influencing the survival of dense populations in these areas. There is good evidence that humidity is largely responsible for the presence of dense populations of diphtheroids in the axilla. In the occlusion experiments already quoted the increase in population reflected primarily an enormous increase in lipophilic diphtheroids.

5. Composition of Cutaneous Populations

Most authorities agree that although great variations in the size of cutaneous populations occur in the same individual from day to day, differences between individuals are recognizable and valid. Differences in composition of the communities living on different hosts can also be demonstrated. In an interesting study Marples & Williamson (1969) found that they could separate their subjects into 'coccal' and 'diphtheroid' men, on the basis of the community structure present in the axilla. Gram positive cocci and diphtheroids formed 87% of the organisms isolated from this habitat, but in the axillae of 'coccal' subjects, coagulase negative cocci formed >50% of the colony count/unit area, while diphtheroids predominated in the axillae of the other group. 'Coccal' and 'diphtheroid' subjects had other recognizable differences such as population density, the diversity of species carried and response to antibiotic treatment. The aerobic population on the foreheads of both groups was predominantly coccal, but the colony count/unit area of skin was significantly lower in 'diphtheroid' than in 'coccal' men.

Both the cocci and the diphtheroids of the skin can be divided into subgroups which differ in their distribution on the skin. Marples (1969) classified 422 isolates of cutaneous cocci into the groups in Baird-Parker's (1965) classification. He found that *Staphylococcus* S II predominated on the head and was less common in the axilla. *Micrococcus* M 3 was the type most frequently isolated from the forehead while M 2 had a predilection for the axilla.

Studies on the incidence and distribution of *Staphylococcus aureus* on the skin are too numerous to summarize. It is clear that this pathogen can be isolated from almost any area of skin of almost any individual at some time or other. But it is also apparent that *Staph. aureus* is truly resident only in the nostrils and perhaps the perineal skin of a proportion of hosts, and on healthy skin elswhere it should be regarded as a transient member of the community.

The classification of the diphtheroids continues to be unsatisfactory, but certain groups can be recognized. Lipophilic diphtheroids are extremely common in the axilla (Marples & Williamson, 1969), while non-lipophilic strains are found more commonly on the glabrous skin. The species *Corynebacterium minutissimum*, associated with the disease erythrasma, can readily be found on apparently normal skin (Somerville *et al.*, 1970).

The anaerobic diphtheroid *C. acnes* has been divided into 2 phage types (Voss, 1970) which may represent 2 species of *Propionibacterium* (Whiteside & Voss, 1973). The contribution which these anaerobes make to the community on the skin has been a subject of conflict. Recently Somerville & Murphy (1973) have made quantitative estimations of *C. acnes* and of aerobic bacteria at 18 skin sites on 22 individuals. They found a geometric mean density of $4 \times 10^2/\text{cm}^2$ of *C. acnes* on the forehead, 7.2×10^4 presternally and 9.1×10^4 on the back. Corresponding values for aerobes were 1.1×10^4, 2.1×10^3 and 1.1×10^3, respectively. Aerobes outnumbered the anaerobes on the arms and legs. One surprising finding was a mean population of $280/\text{cm}^2$ of *C. acnes* on the palm. It is difficult to believe that these bacteria were truly resident on the palm. It is more probable that they were transients derived from the face and head as a result of the constant handling of these areas.

6. Dispersal of the Cutaneous Flora

The dispersal of cutaneous micro-organisms has been extensively studied in recent years, primarily in attempts to prevent cross infections in hospitals. It has also had an important bearing on problems involved in space travel. It is now known that every movement an individual makes dislodges epidermal fragments which act as 'rafts', carrying micro-organisms to new habitats. Certain normal individuals, and others suffering from skin diseases, can be identified as 'dispersers' who shed vast numbers of bacteria into their surroundings. Smith & Bruch (1969), in a carefully planned study, identified 4 healthy individuals who, while exercising naked for 30 min, each dispersed 2–6 million viable organisms. It is valuable to find such dispersers and to use them as subjects in tests of methods designed to reduce inter-host transfer of potential pathogens.

7. Coactions among the Resident Flora

The coactions of the cutaneous bacteria are little understood but are being intensively studied. It is well known that the free fatty acids found on the skin surface are inhibitory to many potential pathogens and there seems little doubt that these substances are products of microbial metabolism. Although the Gram positive cocci have been shown to have active lipases (Freinkel, 1968), Marples *et al.*, (1970) have demonstrated that *C. acnes* is the most important lipolytic organism in the skin. It is probable that these products of fat metabolism are involved in the inhibitory effect exerted by the Gram positive over Gram negative organisms on the skin. It has been shown repeatedly that suppression of the Gram positive residents may be followed by an overgrowth of Gram negative strains, sometimes with disastrous results (Taplin, 1972; Amonette & Rosenberg, 1973). The inhibition appears to depend on complex factors, for Selwyn & Ellis (1972) failed to demonstrate *in vitro* inhibition of Gram negative bacteria by cutaneous isolates, though several of their coagulase negative coccal strains inhibited closely related species. Their results indicate that members of the resident cutaneous flora possess 'micrococcines' or 'epidermins' which influence the growth of related organisms.

Selwyn & Ellis (1972) also reported *in vitro* growth enhancement or satellism between isolates from healthy skin. This finding is of very great interest and illustrates the complex relationships to be found among many individual species which compose any living community. All these studies indicate that the composition of the communities living on the skin depends not only on the attributes of the substrate but also on the type and nature of its inhabitants.

Many fascinating problems of the cutaneous ecosystem remain to be elucidated. The exciting progress made in recent years shows that the field is a fruitful one, and supports the view that the study of the skin as an ecosystem is of more than limited and academic value.

8. References

AMONETTE, R. R. & ROSENBERG, E. W. (1973). Infection of toe webs by Gram negative bacteria. *Arch. Derm.* **107**, 71.

BAIRD-PARKER, A. C. (1965). The classification of staphylococci and micrococci from world-wide sources. *J. gen. Microbiol.* **38**, 363.

FREINKEL, R. K. (1968). The origin of free fatty acids in sebum. I. Role of coagulase negative staphylococci. *J. Invest. Derm.* **50**, 186.

MARPLES, M. J. & SOMERVILLE, D. A. (1968). The oral and cutaneous distribution of *Candida albicans* and other yeasts in Rarotonga, Cook Islands. *Trans. Roy. Soc. Trop. Med. Hyg.* **62**, 256.

MARPLES, R. R. (1965). The effects of hydration on the bacterial flora of the skin. In *Skin Bacteria and their Role in Infection.* Eds H. I. Maibach & G. Hildick-Smith. New York: McGraw-Hill Co.

MARPLES, R. R. (1969). The resident coccal flora of human skin. *J. Invest. Derm.* **52**, 397.

MARPLES, R. R., KLIGMAN, A. M., LANTIS, L. A. & DOWNING, D. T. (1970). The role of the aerobic microflora in the genesis of fatty acids in human surface lipids. *J. Invest. Derm.* **55**, 173.

MARPLES, R. R. & WILLIAMSON, P. (1969). Effects of systemic dimethylchlortetracycline on human cutaneous microflora. *Appl. Microbiol.* **18**, 228.

MONTEZ, L. F. & WILLBORN, W. H. (1969). Location of bacterial skin flora. *Brit. J. Derm.* **81**, *Supplement I,* 23.

ROBERTS, S. O. B. (1969). *Pityrosporum orbiculare,* incidence and distribution on clinically normal skin. *Brit. J. Derm.* **81**, 523.

SELWYN, S. & ELLIS, H. (1972). Skin bacteria and skin disinfection reconsidered. *Brit. Med. J.* **1**, 36.

SMITH, F. W. & BRUCH, M. (1969). Reduction of microbiological shedding in clean rooms. *Devs ind. Microbiol.* **10**, 290.
SOMERVILLE, D. A. & MURPHY, C. T. (1973). Quantitation of *Corynebacterium acnes* on healthy human skin. *J. Invest. Derm.* **60**, 231.
SOMERVILLE, D. A., SEVILLE, R. H., CUNNINGHAM, R. C., NOBLE, W. C. & SAVIN, J. D. (1970). Erythrasma in a hospital for the mentally subnormal. *Brit. J. Derm.* **82**, 35.
TAPLIN, D. (1972). The use of antibiotics in dermatology. *Adv. Biol. Skin* **12**, 315.
VOSS, J. G. (1970). Differentiation of two groups of *Corynebacterium acnes. J. Bact.* **101**, 392.
WHITESIDE, J. A. & VOSS, J. G. (1973). Incidence and lipolytic activity of *Propionibacterium acnes* (*Corynebacterium acnes* Group I) and *P. granulosum* (*C. acnes* Group II) in acne and in normal skin. *J. Invest. Derm.* **60**, 94.

Natural Control and Ecology of Microbial Populations on Skin and Hair

R. C. S. WOODROFFE AND D. A. SHAW

Unilever Research, Isleworth Laboratory, Unilever Limited, 455 London Road, Isleworth, Middlesex, England

CONTENTS

1. Introduction 13
2. Microflora of the hair 14
3. Microflora of the skin 15
   (a) Transient and resident flora 15
   (b) The normal flora 17
   (c) Density of microbial populations at various skin sites 18
4. Factors in the microbial ecology of skin 18
   (a) Water availability 19
   (b) Microbial nutrients and inhibitors on the skin 20
     (i) Apocrine sweat 21
     (ii) Eccrine sweat 21
     (iii) Stratum corneum 22
     (iv) Lysozyme 22
     (v) Skin surface lipids 22
     (vi) Bacteriocins and other inhibitors 25
   (c) Population-density control of resident micro-organisms 29
5. Conclusions 29
6. References 30

1. Introduction

TO ATTEMPT, in the few pages allotted, more than a broad outline of the scope of the topic suggested by the title would be an impossible task. By far the major part of this review will be concerned with bacteria, mainly for the reason that they have been and still are the focus of interest both for research and in pathology. For information on viruses and microfauna the reader is referred to the unique treatise by Marples (1965) in which are considered all the microflora and microfauna known to reside both on and in normal and diseased human skin. Relevant aspects of the structure and physiology of skin are also described in that work. Other related books, reports, and reviews of the microbiology of skin we have found particularly useful include those of Rosebury (1962), Maibach & Hildick-Smith (1965), Lowbury (1969), Marples (1969), Noble (1969) and Somerville (1969 a, b; 1972).

In this review we shall only summarize briefly present knowledge of the nature and numbers of the species of micro-organisms that occur on hair and skin and then discuss in greater detail the local factors which determine the types and numbers of them at particular sites. Both the growth-promoting and inhibitory properties of skin secretions and excretions will be considered. The interactions of the resident micro-organisms on each other and on transient species, i.e. the ways in which the resident organisms limit, inhibit, or promote growth of each other and non-resident micro-organisms, will be described.

2. Microflora of the Hair

Very few studies have been made of the microflora of hair as distinct from skin. Summers, Lynch & Black (1965) examined the pathogenic bacteria on hair of groups of in-patients, out-patients and medical and nursing staff of a hospital. They found *Staphylococcus aureus* to be the most frequently encountered pathogenic organism, occurring in > 30% of individuals. *Escherichia coli* (nearly 20% occurrence) and *Streptococcus viridans* (>10% occurrence) were the next most frequently found pathogenic organisms. Less than 10% of those subjects who were both hair and nasal carriers of *Staph. aureus* yielded the same strain at both sites and the authors concluded that much of the carriage on hair was simple contamination from exogenous sources.

Noble (1966) extended these studies by comparing a group of subjects with no hospital contact to a group of in-patients with skin diseases, in view of the tendency of subjects in the latter group to be heavily colonized with *Staph. aureus*. He found that *Staph. aureus* was present on the hair of *c.* 10% of the people with no hospital contact and on that of *c.* 50% of the patients with skin diseases. Noble concluded that hair may form a considerable reservoir of organisms for cross-infection in such situations.

The ecology of the microflora of the scalp (hair and skin) is markedly different from that of other parts of the body surface and this is determined to a large extent by the presence of the hair which helps to maintain a higher temperature and humidity on the surface of the scalp skin. In addition, sebaceous gland activity is relatively high on the adult scalp (Nicolaides, 1965) and this significantly affects the microbial ecology. Numerous species of yeasts, moulds, actinomycetes and bacteria can be recovered frequently; moulds such as *Aspergillus awamori* and *Asp. fumigatus* are particularly prevalent but as these organisms are readily found in the environment they are likely to be transient rather than resident organisms. The yeasts *Pityrosporum ovale* and *Pityr. orbiculare* are found on the scalps of most adults and, being lipophilic, there is no doubt that these are residents well-adapted to their environment. The majority of the bacterial species recoverable from the scalp are Gram positive, with species of the genus *Bacillus* occurring most frequently, followed by species of Gram positive cocci (Roberts, 1969; Roia & VanderWyk, 1969).

The dermatophyte fungi found on human beings include species of the genera *Epidermophyton, Microsporum* and *Trichophyton*. These fungi are causative agents of *tinea pedis* (athlete's foot), *tinea capitis* (ringworm) and similar conditions in the pubic and axillary regions in which the keratin of *stratum corneum* and hair is destroyed. These organisms have been found on the skin, especially the feet, of a substantial proportion of the population which shows no signs of disease, and whether the onset of disease is due to endogenous or exogenous infection by dermatophytes has been a matter for dispute (Rosebury, 1962).

The causative fungus of ringworm is *Microsporum audouini*. Because adult infections are rare and the incidence of infection drops abruptly at puberty it was suggested that the growth of this fungus on the scalp is inhibited by the increased sebacous gland secretion which occurs at the onset of puberty and persists until old age (Rothman *et al.,* 1947). However, Weary (1968) demonstrated that *Pityr. ovale* has inhibitory action against dermatophyte fungi *in vitro* and this has led to the suggestion that *Pityr. ovale* may be responsible for the resistance of adult scalps to ringworm. It remains to be demonstrated that the numbers of *Pityr. ovale* on the scalp increase dramatically at puberty.

In the dermatophyte infections the hair cuticle and cortex are invaded, causing the hair to weaken and break off near the skin surface. The diphtheroid *Corynebacteriun tenuis* also breaks down the structure of hair and is the causative organism of *trichomycosis axillaris,* a condition in which axillary and occasionally pubic hair are involved. Yellowish or orange-coloured 'coatings' or 'sheaths' lying on the hair shafts are readily visible to the naked eye (Orfanos, Schloesser & Mahrle, 1971).

3. Microflora of the Skin

The 2 square metres area of skin of the human adult have been colourfully described by Kligman (1965) as a vast empire in which contrasts of terrain and climate are as varied as those of the earth itself. Yet in spite of this great diversity similar types of organisms of remarkably few genera only have succeeded in colonizing the various habitats. Where differences exist they are quantitative rather than qualitative. With a few exceptions the density of the microbial population on skin is fairly low, generally hundreds or thousands rather than millions/cm^2.

(a) *Transient and resident flora*

Considering that the skin is continually exposed to the environment it is often difficult to decide whether an organism found on the skin is a resident or a transient. In spite of an extensive literature on the microflora of skin little is known even in semi-quantitative terms of the numbers of any particular species found on various areas of the skin surface. Usually, only the prevalence, i.e. the percentage of persons studied from whom the organism was isolated, is assessed. It is implicit that a resident organism on skin is one that multiplies and not merely survives there. This is not readily demonstrable but we can infer that an organism is a resident if it is recovered repeatedly in large numbers. Discrimination between minor residents and transients is much more difficult, if at all possible, at least by methods based on performing viable counts on aqueous rinses of the skin.

Errors in deciding whether an organism is a resident or a transient can arise from a number of sources. First, the various methods of sampling can give viable counts that differ by orders of magnitude. Secondly, it is possible to obtain misleading results because of the complex nature of the terrain of the skin surface, e.g. soil under nails removed during sampling of the hands may contain a very high count of an organism yet it is obviously a transient. Thirdly, some species of skin microflora have special cultural requirements so that although the chosen method of sampling may remove adequately such a particular species from skin, a conventional culture medium will be incapable of demonstrating its presence when the sample is cultured *in vitro*. The lipophilic diphtheroids and *Pityrosporum* species are 2 examples of such microflora.

The method of sampling skin microflora which gives the highest and most reproducible viable count is that due to Williamson (1965). In this method the skin is scrubbed gently with 0.075 M phosphate buffer at pH 7.9 containing 0.1% of Triton X-100 (a nonionic detergent) by means of a blunt Teflon rod for 2 min. The solution is confined in a glass cylinder held against the skin. These conditions ensure optimal recoveries of bacteria from the skin in that the bacterial colonies on the squames released from the skin surface are dispersed as much as possible into colony-forming units consisting of single or only numerically small groups of cells.

Other methods of sampling skin bacteria e.g. tape-stripping, scraping the *stratum corneum*, pressing nutrient agar blocks against the skin, or rubbing with swabs soaked in water or nutrient medium have been and still are used but they are generally less efficient and less reliable than the method of Williamson.

Stuttard (1961) drew attention to the importance of hydration of skin in the release of bacteria from it. Smylie, Webster & Bruce (1959) and Smylie & Webster (1960) had recovered only relatively few bacteria from the dry skin of surgeons' finger-tips; Stuttard showed that prior addition of 1% peptone water to the surgeons' gloves to ensure that the fingers were hydrated markedly increased the number of bacteria that could be removed. Ulrich (1965) and other groups of workers (Anon., 1965) have subsequently confirmed this repeatedly, viz. that increased hydration of the *stratum corneum* by washing, bathing etc., results in greatly increased recoveries of bacteria obtainable from the skin surface.

A former, long-held belief that bacteria multiplied in the deep structures of the skin whence they were delivered to the surface has not been substantiated, with the exception that the strictly anaerobic *Corynebacterium acnes* has its headquarters in the sebaceous ducts. Lovell (1945) postulated that the majority of the skin bacteria were located in the openings of hair follicles. This was confirmed by electron microscopy by Montes & Wilborn (1969), who also observed bacteria in the superficial layers of the *stratum corneum;* these were always extracellular. From his observations, Kligman (1965) concluded that bacterial colonization of eccrine and apocrine sweat gland ducts in normal, healthy skin is negligible. The attempts by Price (1938) to sterilize surgeons' hands by scrubbing with soap and water is an often-quoted example used to illustrate the apparent fruitlessness of such methods for sterilizing skin. However, hands with their large reservoirs of bacteria under the nails and in the nail folds are the least typical as an example of the general skin surface. Even washing hands with soaps containing germicides does not remove bacteria *in toto* (Hurst, Stuttard & Woodroffe, 1960). Other areas of skin surface less topographically varied than the hands can be virtually sterilized by washing with bactericidal solutions, especially those areas without sebaceous glands, indicating that most bacteria at or near the surface are readily removed (Kligman, 1965; Selwyn & Ellis, 1972).

(b) *The normal flora*

The incidence and distribution of the various species of micro-organisms on normal human skin have been described in detail elsewhere (see Introduction) and also in this symposium (Marples, 1973). We shall therefore describe them only briefly.

The dominant resident flora on skin comprises certain aerobic genera of the family Micrococcaceae, in particular the 'aerobic staphylococci', and *Sarcina* spp. Next in importance as residents are the lipophilic and non-lipophilic diphtheroids. Occasionally, certain Gram negative rods e.g. *Mima* spp., *Herellea* spp., and *Alkaligenes* spp., may achieve resident status, especially in the more humid regions such as the axillae and toe-webs. Included in the heterogeneous group 'aerobic staphylococci' are numerous sub-groups of staphylococci and micrococci which can be distinguished according to the scheme of Baird-Parker (1963, 1965). The majority of the strains of skin staphylococci are coagulase negative. Coagulase positive strains are found quite frequently on the skin of infants without signs of disease but less often on the skin of adults. The normal flora of skin in different age groups was studied by Somerville (1969a, b) who found that *Sarcina* spp., occurred more frequently on children (3-12 years old), and the diphtheroids less frequently, than on adults (18-45 years old). Also, Gram negative rods were found with greater frequency on children. Streptococci were not infrequently recovered from the skin of infants (4-7 days old) and elderly people (>60 years old). Infants in general appeared to be more likely to carry pathogens or potential pathogens. Sarkany & Gaylarde (1967, 1968) investigated the skin flora of the newborn and during the first few days after birth. They found that even at the moment of delivery the normal adult skin bacteria viz. cocci and diphtheroids, could be recovered from the skin. The same species were also found in the birth canal. From their observations on some 1400 babies Evans, Akpata & Baki (1970) concluded that the umbilicus was more likely to be colonized by aerobic flora than the nares during the first 3 days of life.

(c) *Density of microbial populations at various skin sites*

Of the few studies that have been published on the variation of total microbial population with site of skin, none are directly comparable because of the wide variation in techniques used. Selwyn & Ellis (1972), using the method of Williamson (1965), found a mean count of $80,000/cm^2$ of aerobic bacteria on the axillary skin of 42 adult subjects. Considerably lower counts were obtained from skin at other sites such as abdomen, forearm and interscapular regions. Thus, the mean aerobic count for forearm skin was $2,300/cm^2$. Much higher counts were obtained by shaking excised specimens from 20 cadavers in buffered Triton X-100 for 2 min, e.g. $920,000$ aerobic organisms$/cm^2$ from axillary skin and $1,360,000/cm^2$ from scalp skin.

Noble (1968), using a tape-stripping method (which measures only the number of separate colonies originally removed from the skin), found significant differences in the microbial populations of male and female subjects at several skin sites, viz. thighs, shins, sternum, and peri-umbilicus, the female subjects tending to have the higher bacterial populations. He suggested that these differences might be accounted for by different washing habits i.e. women tend to wash these areas more frequently than men.

Ulrich (1965) studied microbial populations on skin by the contact-plate method. He found considerable variation both from site to site and from individual to individual. The head, axillae, groin, perineum, hands and feet supported the largest recoverable populations. A large increase in surface bacterial populations was observed immediately after either bathing or showering which was followed by a fall to normal levels 1-2 h later. This increased recovery of organisms from fully hydrated skin has already been described (v.s.). No seasonal variations in total population were evident in Ulrich's study. Nevertheless, Blank (1965) noted that a few subjects (studied at Boston, U.S.A.) showed seasonal variation in the microbial ecology of skin; a preponderance of micrococci during the winter months was succeeded by a preponderance of diphtheroids as the climate became warmer and more humid.

4. Factors in the Microbial Ecology of Skin

The phenomena determining the ecology and total numbers of bacteria on skin will be discussed below in separate sections. Apart from the prime importance of water, the relative importance of other factors in promoting or inhibiting bacterial growth on skin will depend on the species or even the strain under consideration so no generalisations can be made. In our view the skin is an environment unfriendly to most microbial species but sufficient nutrients are available to allow a small number to survive and multiply. Other species related to those normally resident on skin which might also be expected to survive and reproduce on skin because of nutritional needs similar to those of the resident flora, especially low requirements for water, are unable to do so. These species are sensitive either to the inhibitory action of free fatty acids of skin lipids or to antibacterial substances produced by the resident flora.

The early literature on the relative importance of water, free fatty acids, and pH on the survival and growth of skin bacteria both *in vivo* and *in vitro* was critically reviewed by Pillsbury & Kligman (1954).

(a) *Water availability*

It has long been accepted that the primary rate-limiting factor for the multiplication of bacteria on skin (and, in some cases, their survival) is the restricted supply of water. The most densely populated regions of the skin are those where humidity is highest, i.e. axillae, toe-webs. Most of the water on skin is derived from eccrine sweat. In the axillary region there is also an intermittent contribution of water from the secretion of apocrine sweat (Hurley & Shelley, 1960). Transpiration of water from the dermis through the *stratum corneum* is quite low in adult skin, of the order 0.2-$0.4 mg/cm^2/h$, although the rate will be considerably increased if the skin is damaged by contact with certain organic solvents, detergents, etc. (Scheuplein & Blank, 1971). Although no data on transpiration of water through the *stratum corneum* of children has been published it is likely that it is in fact considerably greater than in adult skin. Somerville (1969 *a, b,*) suggested that the greater incidence and numbers of Gram negative bacteria on children's skin compared with that of adults might be due to the higher humidity on the skin of children. Although the diphtheroids have an even greater requirement for water than Gram negative bacteria, Somerville found a relative absence of diphtheroids on the skin of children; their absence may be due to the low levels of skin lipids (see *Skin surface lipids*).

Blank & Dawes (1958) studied the effects of hydration of the *stratum corneum* on supporting bacterial growth *in vitro* using callus (from the soles of feet) whose water content was adjusted by equilibration over aqueous sulphuric acid of various concentrations. They found that Gram positive micrococci grew on callus with a lower degree of hydration (29%) than did the strain of diphtheroid studied, which grew only when hydration was 70%. A strain of Gram negative coliform bacteria grew on callus when the degree of hydration was 50%. The moisture requirement of a *Staph. aureus* was similar to that of the micrococci.

R. Marples (1965) observed the effects of occlusion of skin *in vitro* by wrapping the forearms of adult male volunteers in an inert, impermeable sheet of polyethylene and sealing it at the wrists and elbows. The occlusion resulted in a 100% hydration of the *stratum corneum* and the accumulation of liquid water. The total viable bacterial count, after sampling the skin by Williamson's method, increased several thousandfold, reaching a peak on about the fourth day. Equally dramatic was the change of composition of the microflora. Before occlusion the population consisted of *c.* 80% of Gram positive cocci (staphylococci, *Sarcina*), 8% of lipophilic diphtheroids and the remainder mainly non-lipophilic diphtheroids. After occlusion for several days the stable population consisted of mostly diphtheroids (both lipophilic and non-lipophilic) with only a few cocci and a small, but definite number of Gram negative rods. Neither pathogenic staphylococci nor streptococci became established as result of the occlusion. However, Marples (1965) stressed that due to the impermeability of the occlusive material the growth of facultative anaerobes and microaerophilic strains would be favoured. This could account for the lack of growth of *Sarcina* spp. Also, the pH value of the skin would have probably increased. Since diphtheroids prefer both a higher pH value (Pillsbury & Rebell, 1952) and a moister environment than micrococci (Blank & Dawes, 1958) their eventual dominance on occluded skin is not unexpected.

Duncan, McBride & Knox (1969) noted that increases in both the temperature and humidity of the environment were necessary before microbial populations on skin were increased significantly. No lasting qualitative change in the composition of the flora was observed. The large increase in skin bacteria of subjects *immediately* after bathing or showering which was observed by Ulrich (1965) and other groups of workers (Anon., 1965) obviously cannot be due to stimulation of bacterial growth by hydration of the *stratum corneum*. It is more likely that softening and loosening of the cornified layers by the water made more bacteria accessible for sampling.

(b) *Microbial nutrients and inhibitors on the skin*

Nutrients on the skin surface which may be utilized by the microflora are derived from 4 sources, apocrine and eccrine sweat, *stratum corneum* and the lipoidal secretions of the sebaceous glands which form the bulk of the skin surface lipids. Sebaceous glands can occur in various modified forms in specialized areas of skin e.g. the Meibomian glands of the eyelids and the glands of Zeis which open into the follicles of the eyelashes. The ceruminous glands of the external ear canal are specialized apocrine glands whose secretion is mainly lipoidal material (Strauss & Ebling, 1970). However, because the normal flora of these specialized areas has not received much attention (excepting the external ear canal) and the chemical natures of their secretions have not been studied in detail, they will not be considered further.

(i) *Apocrine sweat*

Apart from some localized apocrine glands to be found in the anogenital regions, the mammary areolae, and a few scattered about elsewhere in the skin over the body surface, the vast majority are concentrated in the axillary region. Like sebaceous glands, the functioning of the apocrine glands is dependent on the androgenic steroids elaborated and secreted by the gonads and adrenal cortex (Strauss & Ebling, 1970). Since the levels of endogenous androgens are very low in children, sebaceous and apocrine gland activity is minimal during childhood but dramatically increases at puberty.

Hurley & Shelley (1960) stated that apocrine gland secretion from the human axilla gave positive reactions to tests for ammonia, protein, carbohydrate and ferric iron. In some unpublished studies on the composition of a pool of apocrine sweat also obtained from the axillae of normal men, we have confirmed in our laboratory the presence of substantial amounts of protein ($>1\%$) and ninhydrin-reacting substances (*c.* 1%). Carbohydrate was present, too, but only a trace of lipid, probably as a contaminant from skin lipids, was detectable. The sweat was collected after stimulation of secretion of the apocrine glands by intradermal injection of 0.1% aqueous adrenaline (Woodroffe, Shaw, Flawn & Fox, unpublished).

In addition to the nutrients contained in apocrine sweat and the high humidity of the region, the pH value of axillary skin is 0.5-1 pH unit higher than that of other regions of the skin in both the adolescent and the adult, i.e. closer to 6 than 5 (Behrendt & Green, 1971). The pH value of freshly-secreted apocrine sweat is only slightly higher than that of eccrine sweat and the higher pH value on adult axillary skin may be due to alkaline metabolites, e.g. ammonia, amines, produced by bacterial metabolism of apocrine sweat: this bacterial metabolism is claimed to be mainly responsible for axillary odour (Hurley & Shelley, 1960). The higher pH value on the axillary skin surface and the increased humidity probably account for the relatively higher numbers of diphtheroids usually found in this region since diphtheroids prefer environments of higher pH value than those preferred by cocci (Pillsbury & Rebell, 1952) and also have a greater requirement for water than cocci (Blank & Dawes, 1958).

(ii) *Eccrine sweat*

Apart from its vital aqueous content, eccrine sweat may contain a total of up to 1% of other substances, the main constituent being sodium chloride which accounts for *c.* 90% of the osmotic activity. The acidity of eccrine sweat is due mainly to lactic acid (up to 0.1%). Urea is also an important constituent (up to 0.06%) (Weiner & Hellman, 1960). In addition many other substances are found in lesser amounts: most of the amino acids of plasma have been detected and together comprise *c.* 0.05% of the sweat (Coltman, Rowe & Atwell, 1966); proteins in trace amounts have been detected (Jirka & Kotas, 1957; Vanfraechem & Poortmans, 1968; Seutter *et al.*, 1970); traces of combined zinc, iron and copper have also been noted (Suetter & Sutorius, 1972). That eccrine sweat provides a medium suitable for the growth of each species of resident skin micro-organisms is a matter of deduction rather than direct proof. No comprehensive information is available comparing the relative growth rates of the skin residents in eccrine sweat (with or without added skin lipids) *in vitro.*

(iii) *Stratum corneum*

The outer layer of the skin consists mainly of enucleated keratinized cells and it contains both water-soluble and ether-soluble materials potentially useful as bacterial and fungal nutrients. The water-soluble fraction contains amino acids and peptides (Laden, 1965; Burke, Lee & Beuttner-Janusch, 1966), *c.* 2% sodium-2-pyrrolidone-5-carboxylate (Laden & Spitzer, 1967) and many other substances including breakdown products of nucleic acids (Laden, 1965). These water-soluble substances are responsible for the high water-binding capacity of the *stratum corneum*. The importance of the hydration of the *stratum corneum* for permitting bacterial growth in the absence of active sweating was described earlier. The ether-soluble material is epidermal and sebaceous lipid and will be considered separately *(Skin surface lipids)*. Traces of several hydrolytic enzymes, no doubt released from epidermal lysosomes during keratinization, viz. acid phosphatase, acid protease, ribonuclease, β-glucuronidase, as well as more substantial amounts of non-specific esterase (mainly bacterial in origin), are detectable on the surface of the skin (Steigleder, 1964). The non-specific esterase or lipase will be considered under *Skin surface lipids*.

(iv) *Lysozmye*

The antibacterial activity of this enzyme is well known. It is found in saliva, tears, milk, leucocytes and elsewhere in the body. Thus it will exert some effect on the ecology of the microflora of the nose, eyes and mouth. There are reports of the occurrence of lysozyme both in skin (Ogawa, Miyazaki & Kimura, 1971) and on the surface of the skin (Kleňha & Krs, 1967). It is not known if the amount on skin (other than those areas supplied with lysozyme-rich secretions) is sufficient to affect the microbial ecology. The source of the enzyme on skin is unknown. It may be synthesized there or transported from the blood. However, *c.* 10% of strains of *Staph. epidermidis* from human skin produce lysozyme (Heczko, Kasprowicz & Kucharczyk, 1971) and an even greater proportion of coagulase positive staphylococci secrete lysozyme (Jeljaszewicz, 1967). Thus it is not surprising that most of the resident cocci on skin are insensitive to the lytic activity of this enzyme (Woodroffe & Wilkinson, unpublished) but invasion by some potential pathogens is perhaps discouraged by the lysozyme on skin.

(v) *Skin surface lipids*

Except in the few regions where sebaceous glands are absent, viz. palms, soles and dorsal surface of the feet, the skin surface lipids (SSL) are quantitatively by far the most important class of substances occurring on adult human skin. Both the total production and composition of SSL vary considerably over the body surface and between individuals for corresponding sites. A typical composition for human male scalp SSL is shown in Table 1. The sterol and sterol ester fractions of SSL are derived from the epidermal cells (Nicolaides, 1965).

Table 1

Composition of Skin Surface Lipids (SSL) of adult human male scalp*

SSL Fraction	Relative amount present (%)
Saturated hydrocarbons	0.8
Squalene	12.8
Wax esters	20.2
Sterol esters	3.3
Sterols	2.4
Free fatty acids	29.6
Glycerides (tri, di and mono)	31.7

*Mean values for 23 males, aged 17-23 years. Skin sampled by irrigation with acetone (Lewis & Hayward, 1971).

The sebum as synthesized in the sebaceous gland contains little, if any, free fatty acid (Kellum, 1967). Therefore the free fatty acids of SSL arise by hydrolysis of the sebum triglycerides, the extent of the hydrolysis varying widely at different skin sites and between subjects. There is now no doubt that most, and possibly all, of the hydrolysis is due to bacterial lipases, in particular those of the lipophilic diphtheroids (Scheimann *et al.,* 1960; Freinkel & Shen, 1969; Marples, Downing & Kligman, 1971;1972).

The bacteriostatic and bactericidal effects of individual fatty acids *in vitro* on a wide range of Gram positive bacterial species have been known for a long time. The most active antibacterial saturated fatty acid is lauric acid (C_{12}) and its activity is generally greater at lower pH values (Hassinen, Durbin & Bernhart, 1951; Karabinos & Ferlin, 1954). A recent extensive study was made by Galbraith *et al.,* (1971) who attempted to define the relative antibacterial properties of saturated fatty acids in physico-chemical terms, in particular the inversely related properties of water-solubility and lipophilic activity which are optimum for lauric acid. Both SSL and the free fatty acid mixture isolated from it after saponification exerted bactericidal action on 9 strains of coagulase positive, and 2 strains of coagulase negative, staphylococci (Joiris, 1957). Ricketts, Squire & Topley (1951) demonstrated that *Strep. pyogenes, Staph. aureus* and the skin micrococci were all inhibited by the saturated free fatty acid fraction of SSL *in vitro,* whereas the Gram negative species *Pseudomonas aeruginosa* and *E.coli* were resistant. These authors noted that the self-sterilizing power of human skin *in vivo* was reduced after washing with acetone. Pillsbury & Rebell (1952) also noted the antibacterial activity of caprylic (C_8), capric (C_{10}), lauric (C_{12}), tridecanoic (C_{13}) and myristic (C_{14}) acids towards skin micrococci and coagulase positive staphylococci *in vitro*. Lipophilic diphtheroids and *Pityr. ovale* were sensitive to undecylenic (mono-unsaturated C_{11}) acid in the pH range 5-7. Scalp lipid extract was weakly bactericidal but forehead SSL was not. In a series of carefully designed experiments, Aly *et al.* (1972) have recently confirmed and extended the earlier observations of the antibacterial effects of human SSL. They showed that inocula of *Candida albicans* and of pathogenic strains of both *Staph. aureus* and *Strep. pyogenes,* applied under an occlusive device to the surface of forearm that had been washed with acetone, survived much longer than when similarly applied (also under an occlusive device) to unwashed skin. If, after removal of solvent, the residue from the extract was re-applied to the skin or if the acetone was simply applied to the skin and allowed to evaporate *in situ,* then no significant increase was found in the survival rate of the organisms applied to treated skin compared with untreated skin. The poor survival rates of *Ps. aeruginosa* and *E.coli* on unwashed skin were not increased significantly when the skin was washed. Full antibacterial activity of washed skin was restored after 5, but not 2, h. Because this treated skin still had a very low count of normal resident flora 5 h after the washing, the solvent effect was not due to removal of resident organisms which inhibited the growth of the pathogens. The authors suggested that when desiccation of the skin is prevented alternative mechanisms inhibitory to the growth of micro-organisms become operative, one of them being the antibacterial activity of SSL.

Pillsbury & Rebell (1952) concluded that the inhibitory effects of SSL on microbial growth are not due to their lowering of skin pH value. The growth of skin micrococci *in vitro* in media whose pH value was lowered by addition of either lactic or hydrochloric acids was not significantly decreased until a pH of 4.5 was reached which is about the lowest limit of normal skin pH. Viability was lost at a pH of *c.* 3.7. The aerobic diphtheroids grew only poorly at pH 5; the optimal pH value for their growth was 7. The scalp yeast *Pityr. ovale* is favoured by acid media, growing well in the pH range 4-6.

For some skin residents SSL may have a permissive rather than inhibitory role to play. Oleic acid (mono-unsaturated C_{18}) stimulates the growth *in vitro* of lipophilic strains of human cutaneous diphtheroids. Puhvel & Reisner (1970) found that although saturated fatty acids (C_8–C_{16}) inhibited the growth *in vitro* of 2 strains of *Coryn. acnes,* oleic acid stimulated their growth. This acid also stimulates the growth *in vitro* of aerobic lipophilic diphtheroids of cutaneous origin (Pollock, Wainwright & Manson, 1949; Smith, 1970). However, SSL do not stimulate the growth *in vitro* of skin bacteria and the observed stimulatory effect of oleic acid on the growth of certain strains of skin bacteria may be balanced *in vivo* by the inhibitory effects of other fatty acids in the SSL. The relative inhibitory/stimulatory activities for individual species on skin of SSL taken from various sites of skin have not been studied in any detail. If the ratio of oleic acid to other fatty acids varies between skin sites (and again, this has not been systematically studied) then this may determine to a certain extent the microbial species found on a given area of skin.

Somerville (1969 *a, b,*) suggested that the more varied microflora of infants and children as compared with those of adults might be due to the relatively low concentrations of free fatty acids on the skin of children. Thus, streptococci are very sensitive to the inhibitory effects of free fatty acids and this genus is rarely found on adult skin but it occurs not infrequently on infants' skin.

(vi) *Bacteriocins and other inhibitors*

A little-investigated but possibly very important factor in the ecology of the skin bacteria is the role of bacteriocins. These are a class of antibacterial substances, named by Jacob *et al.* (1953) and readily distinguishable from antibiotics. Bacteriocins are produced by many genera of Gram negative and Gram positive bacteria. Those produced by Gram negative bacteria act only against strains of the same or closely related species whilst those from Gram positive bacteria often have a much wider spectrum of activity, even against species of other genera although activity against Gram negative organisms has yet to be demonstrated.

Bacteriocins form a chemically heterogeneous group of substances ranging from polypeptides or simple proteins to proteins complexed with carbohydrates and lipids and phage-like particles of M.W. $>2 \times 10^5$. Their mode of action is unlike that of antibiotics in that they are irreversibly adsorbed on the outer walls of the cells of sensitive bacteria and are bactericidal but not bacteriostatic. Much of our knowledge of bacteriocins has been gained from studies on those of the Gram negative bacteria, especially the colicins of the Enterobacteriaceae. There have been several excellent reviews of bacteriocins such as those of Ivánovics (1962), Reeves (1965), Bradley (1967), Korzybski, Kowszyk-Gindifer & Kurylowicz (1967), Nomura (1967) and Reeves (1972).

The observation that staphylococci could also produce substances inhibitory to other strains and species of bacteria was first made in the late 19th century. The phenomenon of throat infections by certain strains of staphylococci providing resistance to infection by the more virulent *Coryn. diphtheriae* has also been known for a long time. Modern interest in the nature and properties of the antibacterial substances produced by staphylococci stems from the work of Fredericq (1946) and Jennings & Sharp (1947). Fredericq observed 5 antibacterial substances each with its own spectrum of activity against other strains of staphylococci. Jennings & Sharp found 26 strains of staphylococci out of 205 studied (199 of human origin) which were inhibitory to *Coryn. diphtheriae, Coryn. xerosis,* and *Staph. aureus.* Seven of these 26 active strains were coagulase negative. Other antibacterial products such as lysostaphin (Schindler & Schuhardt, 1964) and lysozyme (Jeljaszewicz, 1967) are secreted by staphylococci but not by strains normally found on healthy skin.

In recent years several reports of the formation of bacteriocins by various strains of Gram positive bacteria found on human skin have been published. Some of these bacteriocins, especially from staphylococci, have been highly purified and they often have antibacterial activity towards other genera e.g. diphtheroids. That strains of *Staph. aureus* of human origin produce bacteriocins which inhibit growth of other bacterial species such as diphtheroids has been known for a long time (Florey *et al.,* 1949; Parker & Simmons, 1959; Barrow, 1963; Dajani & Wannamaker, 1969; Gagliano & Hinsdill, 1970). However, *Staph. aureus* is not a major resident of healthy adult skin. Of greater interest and relevance to the ecology of skin are the bacteriocins and other inhibitors produced by those strains of Gram positive micrococci isolated from skin which are generally non-pathogenic residents, since such inhibitory substances may play an important role in determining the microbial ecology of skin.

In 1950 Murray & Loeb described inhibition of *Strep. pyogenes* and other Gram positive bacteria by substances produced by 2 strains of *Micrococcus epidermidis.* One of these inhibitory sustances was subsequently partially purified and shown to be stable to autoclaving at 121° for 20 min at pH 2, 7, or 12. It was slowly dialysable through cellophane (Loeb, Mayer & Murray, 1950). The antibacterial activity of this inhibitor was destroyed by trypsin but not by pepsin. The authors suggested that the substance was a polypeptide of low M.W. Evans *et al.* (1950) noted inhibition *in vitro* of *Coryn. acnes* by 2 strains of facultative anaerobic micrococci isolated from surface scrapings of healthy human skin. Pillsbury & Rebell (1952), using a crude screening test, were unable to find antibacterial activity against a strain of *Staph. aureus* when they tested 100 strains of normal skin micrococci. Ten strains of skin diphtheroids were also tested against a strain of β-haemolytic streptococcus, and again no inhibition by any strain was shown. Possibly, unsuitable test organisms were used because some strains of *Staph. aureus* and β-haemolytic streptococci are residents on the skin and these would be resistant to the bacteriocins and other inhibitors of the normal residents.

Hamon & Péron (1963) observed that 20 strains of *Micrococcus pyogenes* produced 'staphylococcins' with wide spectra of activities against *Listeria, Bacillus,* and *Coryn. diphtheriae.* They were proteins or polypeptides and acted by destruction of the cell cytoplasmic membranes. Lachowicz & Walczak (1966) also described the production of an inhibitor by non-pathogenic strains of staphylococci which they called 'staphylococcin'. It could be separated by gel-permeation chromatography into 2 proteins of M.W. 10^4 and 3×10^4, each protein having bactericidal activity against pathogenic strains of staphylococci and some other Gram positive bacteria. Again, a staphylococcin was obtained from a strain of *Staph. epidermidis* by Jetten, Vogels & de Windt (1972). This showed inhibitory activity towards *Staph. aureus* Oxford 209P, other strains of staphylococci and other Gram positive bacteria whose natures were not disclosed in the report. It was purified to homogeneity as shown by polyacrylamide gel electrophoresis and found to consist of protein (41.8%), carbohydrate (34%), and lipid (21.9%) with a M.W. between 1.5×10^5 and 4×10^5. In the presence of sodium dodecyl sulphate (0.1% aqueous) the staphylococcin was dissociated into sub-units of M.W. between 10^4 and 2.5×10^4. This staphylococcin was present in both the cell pellet and supernatant fraction of liquid cultures. The activities of all supernatant fractions were greatly increased by dialysis, indicating the presence of a low M.W. substance inhibitory to the staphylococcin, either produced by the organism or in the media as one of its constituents. Crude preparations were stable to heating at 120° for 15 min, but the purified fraction was fully destroyed under these conditions (Jetten & Vogels, 1972a). Bovine serum albumin (0.5% aqueous solution) gave full protection against this heat-lability. Proteolytic enzymes such as trypsin, chymotrypsin and pronase (bacterial), and the lipid solvents hexane and phenol, destroyed its activity. The mode of action was typical of a bacteriocin, viz. adsorption on the cell walls and membranes followed by death but not lysis of the cell (Jetten & Vogels, 1972b). A similar inhibitor of M.W. 2×10^5 and composed of protein, lipid and carbohydrate was obtained from disrupted, heat-killed cells of a strain of *Staph. aureus* obtained from turkey skin. It was inactivated only slowly by heat or proteases (Gagliano & Hinsdill, 1970).

Hsu & Wiseman (1967) found that 4.9% of 1065 strains of coagulase positive, and 8.5% of 387 strains of coagulase negative, staphylococci (all 1452 strains being of human origin) produced substances inhibitory to *Staph. aureus* Oxford 209P and other strains of Gram positive bacteria, including diphtheroids and mycobacteria. These inhibitors appeared to act by reducing the rate of cell division and were not bacteriolytic. The generic name epidermidins was proposed for them and the epidermidins from 2 strains of *Staph. epidermidis* (coagulase negative) were later separated and purified from the cell-free culture broth (Hsu & Wiseman, 1971). Each of the 2 staphylococcal strains studied secreted 2 epidermidins, one soluble in 0.1 M aqueous zinc chloride, the other insoluble. When obtained 'pure' i.e. at constant, maximum, specific antibacterial activity, the epidermidins were colourless, crystalline solids. They were readily dialysable, resistant to heat and extremes of pH and were only slightly inactivated by incubation with trypsin solution at 37° for 6 h. Chemical analysis indicated that they contained no lipid, carbohydrate, nucleic acid or halogen and their negative reactions with Lowry and ninhydrin agents indicated absence of protein. Neither C-terminal nor amino acid N-terminal groups could be detected.

Selwyn & Ellis (1972) reconsidered certain aspects of human skin microbiology. Among their findings was the isolation from the skin surface of 20 out of 98 normal adults of 32 strains of Micrococcaceae which secreted substances inhibitory to a wide range of bacterial species and genera including coagulase negative and coagulase positive staphylococci, streptococci, pneumococci, diphtheroids, bacilli, clostridia and neisseriae but not enterobacteria. The inhibitory substances appeared to be polypeptides. They were without bacteriolytic activity. These substances were regarded by Selwyn & Ellis (1972) as playing a key role in the ecology of skin microflora.

The coagulase negative staphylococci and micrococci may dominate the skin surface by virtue of the bacteriocins and other inhibitors they produce. The growth of *Staph. aureus,* a bacterial species which one would expect to be as readily adaptable to the dry environment on skin as the normal residents, is strongly inhibited by these inhibitory substances. As yet, there is no direct evidence to indicate that the bacteriocins and other inhibitors of the Gram positive bacteria on skin play a significant role in the prevention of the colonization of skin by Gram negative species. Nevertheless, Shehadeh & Kligman (1963) observed that when the growth of the resident Gram positive cocci and diphtheroids of human axillary skin was completely suppressed by repeated topical applications of both neomycin and aluminium chlorhydroxide, the total number of Gram negative rods increased within a few weeks to a steady level almost equivalent to the original total number of both Gram positive and Gram negative species. After cessation of treatment the original mixed population rapidly re-established itself. These observations are indirect evidence that the Gram positive bacteria on axillary skin inhibit the growth of Gram negative species and therefore that antibacterial substances active against Gram negative bacteria are produced by the Gram positive bacteria normally resident on the skin.

The phenomenon of bacterial interference was the basis of a method proposed by Eichenwald *et al.* (1965) to control cross-infection of the skin and nasal mucosa of newborn children by staphylococci. It was observed that an avirulent, penicillin-sensitive strain (No. 502A) of *Staph. aureus,* obtained from a nurse and infants in the hospital nurseries who had been remarkably free from staphylococcal disease, was capable of colonizing the nasal mucosa and the umbilicus at the skin junction following inoculation, and that this organism subsequently reduced drastically the incidence of infection of these body sites by virulent strains of *Staph. aureus.*

This procedure for preventing cross-infection in infants has been confirmed (Anon., 1968) but it is not without some hazards (Blair & Tull, 1969). However, no antibacterial substance seems to be produced by this strain (Anthony & Wannamaker, 1967) and therefore its mode of interference does not appear to be by production of a bacteriocin or other inhibitor.

(c) *Population-density control of resident micro-organisms*

Under normal conditions members of the bacterial flora on skin are in constant equilibrium. We have already discussed some of the factors known to contribute to this situation. Additional substances whose natures are unknown may be produced by some strains of skin bacteria which inhibit and/or stimulate the growth of other strains on skin. In a preliminary study of inhibitory substances produced by skin bacteria, Woodroffe & Lindy (unpublished) examined the interactions of 6 strains of human skin staphylococci *in vitro*. Four of these strains were equally inhibitory to one another and capable of inhibiting the remaining 2 strains which, in turn, could hold a balance between one another. Thus it would be possible to maintain an equilibrium between these bacterial strains.

The opposite effect, satellitism, viz. growth enhancement of one organism by another, is a little-explored phenomenon of the microbial ecology of skin. Numerous examples of microbial satellitism have been observed (Rosebury, Gale & Taylor, 1954) but rarely specifically in organisms which are major residential species on skin. A notable exception was the observation by Selwyn & Ellis (1972) that the growth of normal skin coagulase negative staphylococci *in vitro* was enhanced by satellite organisms from skin of 12 of 98 human adult subjects studied. The satellites were species of *Staphylococcus* (2), *Micrococcus* (5), and *Corynebacterium* (6). Whether they exerted their effects by production of essential nutrients or destruction of toxic substances such as peroxides or bacteriocins was not studied. The role of satellitism in the ecology of skin has yet to be studied systematically. It could prove to be a significant factor of skin microbiology.

5. Conclusions

The study of the microbial ecology of the skin and hair must depend on efficient and reproducible recovery of flora from its substrate followed by provision of growth conditions *in vitro* closely resembling those *in vivo*. There are 3 major factors which determine the ecology of skin bacteria: first, the micro-climate and micro-environment of an area can favour the establishment of certain species; secondly, the prevention by resident bacteria of the establishment of undesirable invaders; thirdly, the maintenance of equilibrium between the residents. Demonstration of these activities *in vitro*, cannot establish the precise situation occurring *in vivo* although certain guide lines can be established. Thus, the evidence available to us leads to some generalized conclusions. Undesirable microbial invaders are discouraged from colonizing skin by one or more of 3 phenomena: (a) the presence of the free fatty acids of skin surface lipids, (b) low hydration of the *stratum corneum*, (c) the presence of bacteriocins or other inhibitors produced by resident bacteria, in particular by staphylococci and micrococci.

Population-density control of the resident bacteria is determined by bacterial metabolism of sebum triglycerides to inhibitory or stimulatory free fatty acids, by availability of water and by inhibitory substances which are active within species.

6. References

ANON (1965). Shower-bath: duty or pleasure. *Lancet* **1**, 474.
ALY, R., MAIBACH, H. I., SHINEFIELD, H. R. & STRAUSS, W. G. (1972). Survival of pathogenic organisms on human skin. *J. invest. Derm.* **58**, 205.
ANTHONY, B. F. & WANNAMAKER, L. W. (1967). Bacterial interference in experimental burns. *J. exp. Med.* **125**, 319.
BAIRD-PARKER, A. C. (1963). A classification of micrococci and staphylococci based on physiological and biochemical tests. *J. gen. Microbiol.* **30**, 409.
BAIRD-PARKER, A. C. (1965). The classification of staphylococci and micrococci from world-wide sources. *J. gen. Microbiol.* **38**, 363.
BARROW, G. I. (1963). Microbial antagonism by *Staphylococcus aureus*. *J. gen. Microbiol.* **31**, 471.
BEHRENDT, H. & GREEN, M. (1971). *Patterns of Skin pH from Birth through Adolescence*. Springfield, Illinois: C. C. Thomas.
BLAIR, E. B. & TULL, A. H. (1969). Multiple infections among newborn resulting from colonization with *Staphylococcus aureus* 502A. *Amer. J. Path.* **52**, 42.
BLANK, I. H. (1965). Survival of bacteria on the skin. In *Skin Bacteria and their Role in Infection*. Eds H. Maibach & G. Hildick-Smith. New York: McGraw-Hill.
BLANK, I. H. & DAWES, R. K. (1958). The water content of the stratum corneum. IV. The importance of water in promoting bacterial multiplication on cornified epithelium. *J. invest. Derm.* **31**, 141.
BRADLEY, D. E. (1967). Ultrastructure of bacteriophages and bacteriocins. *Bact. Rev.* **31**, 230.
ANON. (1968). Bacterial interference in the nursery. *British Medical Journal* **3**, 264.
BURKE, R. C., LEE, T. H. & BUETTNER-JANUSCH, V. (1966). Free amino acids and water-soluble peptides in stratum corneum and skin surface film in human beings. *Yale J. Biol. Med.* **38**, 355.
COLTMAN, C. A., ROWE, N. J. & ATWELL, R. J. (1966). The amino acid content of sweat in normal adults. *Am. J. clin. Nutr.* **18**, 373.
DAJANI, A. S. & WANNAMAKER, L. W. (1969). Demonstration of a bactericidal substance againstt β-hemolytic streptococci in supernatant fluids of staphylococcus cultures. *J. Bact.* **97**, 985.
DUNCAN, W. C., McBRIDE, M. E. & KNOX, J. M. (1969). Bacterial flora. The role of environmental factors. *J. invest. Derm.* **52**, 479.
EICHENWALD, H. F., SHINEFIELD, H. R., BORIS, M. & RIBBLE, J. C. (1965). 'Bacterial interference' and staphylococci colonization in infants and adult. *Ann. N. Y. Acad. Sci.* **128**, 365.
EVANS, C. A., SMITH, W. M., JOHNSTON, E. A. & GIBLETT, E. R. (1950). Bacterial flora of the normal human skin. *J. invest. Derm.* **15**, 305.
EVANS, H. E., AKPATA, S. O. & BAKI, A. (1970). Factors influencing the establishment of the neonatal bacterial flora. I. The role of host factor. *Archs environ. Hlth.* **21**, 514.
FLOREY, H. W., CHAIN, E., HEATLEY, N. G., JENNINGS, M. A., SANDERS, A. G., ABRAHAM, E. P. & FLOREY, M. E. (1949). The *Staphylococcus*. In *Antibiotics. 1*, London: Oxford University Press.
FREDERICQ, P. (1946). Sur la sensibilité et l'activité antibiotique des staphylocoques. *C. r. Séanc. Soc. Biol.* **140**, 1167.

FREINKEL, R. K. & SHEN, Y. (1969). The origin of free fatty acids in sebum. II. Assay of the lipases of the cutaneous bacteria and effects of pH. *J. invest. Derm.* **53**, 422.
GAGLIANO, V. J. & HINSDILL, R. D. (1970). Characterization of a *Staphylococcus aureus* bacteriocin. *J. Bact.* **104**, 117.
GALBRAITH, H., MILLER, T. B., PATON, A. M. & THOMPSON, J. K. (1971). Antibacterial activity of long chain fatty acids and the reversal with calcium, magnesium, ergocalciferol and cholesterol. *J. appl. Bact.* **34**, 803.
HAMON, Y. & PÉRON, Y. (1963). Quelques remarques sur les bactériocines produites par les microbes. *C. r. Acad. Sci. Ser. D.* **257**, 1191.
HASSINEN, J. B., DURBIN, G. T. & BERNHART, F. W. (1951). The bacteriostatic effects of saturated fatty acids. *Archs Biochem. Biophys.* **31**, 183.
HECZKO, P. B., KASPROWICZ, A. & KUCHARCZYK, J. (1971). Biochemical and toxic properties of *Staphylococcus epidermidis* strains of human origin. *Exp. Med. Microbiol.* **23**, 185.
HSU, C. -Y. & WISEMAN, G. M. (1967). Antibiotic substances from staphylococci. *Can. J. Microbiol.* **13**, 947.
HSU, C. -Y. & WISEMAN, G. M. (1971). Purification of epidermidins, new antibiotics from staphylococci. *Can. J. Microbiol.* **17**, 1223.
HURLEY, H. J. & SHELLEY, W. B. (1960). *The Human Apocrine Sweat Gland in Health and Disease.* Springfield, Illinois: C. C. Thomas.
HURST, A., STUTTARD, L. W. & WOODROFFE, R. C. S. (1960). Disinfectants for use in bar-soaps. *J. Hyg., Camb.* **58**, 159.
IVÁNOVICS, G. (1962). Bacteriocins and bacteriocin-like substances. *Bact. Rev.* **26**, 108.
JACOB, F., LWOFF, A., SIMONOVITCH, A. & WOLLMAN, E. L. (1953). Définition de quelques termes relatifs à la lysogénie. *Annls Inst. Pasteur, Paris* **84**, 222.
JELJASZEWICZ, J. (1967). Biological attributes of staphylococcal products. In *Proceedings of the 5th International Congress of Chemotherapy,* Vol. 6, Suppl. II. Eds K. H. Spitzy & H. Haschek. Vienna: Verlag der Wiener Medizinischen Academie.
JENNINGS, M. A. & SHARP, A. E. (1947). Antibacterial activity of the *Staphylococcus. Nature, Lond.* **159**, 133.
JETTEN, A. M. & VOGELS, G. D. (1972a). Nature and properties of a *Staphylococcus epidermidis* bacteriocin. *J. Bact.* **112**, 243.
JETTEN, A. M. & VOGELS, G. D. (1972b). Mode of action of a *Staphylococcus epidermidis* bacteriocin. *Antimicrobial Agents Chemother.* **2**, 456.
JETTEN, A. M., VOGELS, G. D. & de WINDT, F. (1972). Production and purification of a *Staphylococcus epidermidis* bacteriocin. *J. Bact.* **112**, 235.
JIRKA, M. & KOTAS, J. (1957). The occurrence of mucoproteins in human sweat. *Clin. Chim. Acta* **2**, 292.
JOIRIS, E. (1957). Activité antibactérienne des lipides du sébum humaine *vis-à-vis* des staphylocoques. *C. r. Séanc. Soc. Biol.* **151**, 1049.
KARABINOS, J. V. & FERLIN, H. J. (1954). Bactericidal activity of certain fatty acids. *J. Am. Oil Chem. Soc.* **31**, 228.
KELLUM, R. E. (1967). Human sebaceous gland lipids. Analysis by thin-layer chromatography. *Archs Derm.* **95**, 218.
KLEŇHA, J. & KRS, V. (1967). Lysozyme in mouse and human skin. *J. invest. Derm.* **49**, 396.
KLIGMAN, A. M. (1965). The bacteriology of normal skin. In *Skin Bacteria and their Role in Infection.* Eds H. I. Maibach & G. Hildick-Smith. New York: McGraw-Hill.
KORZYBSKI, T., KOWSZYK-GINDIFER, Z. & KURYLOWICZ, W. (1967). *Antibiotics. Origin, nature and properties,* Vol. 1. Oxford: Pergamon.
LACHOWICZ, T. & WALCZAK Z. (1966). Preliminary report on purification of staphylococcin. *Postępy Mikrobiol.* **5**, 213.
LADEN, K. (1965). A comparative chemical study of dandruff flakes, skin scrapings and callus. *J. Soc. cosm. Chem.* **16**, 491.
LADEN, & SPITZER, R. (1967). Identification of a natural moisturizing agent in skin. *J. Soc. cosm. Chem.* **18**, 351.

LEWIS, C. A. & HAYWARD, B. J. (1971). Human skin surface lipids. In *Modern Trends in Dermatology-4,* Ed. P. Borrie. London: Butterworth.

LOEB, L. J., MAYER, A. & MURRAY, R. G. E. (1950). An antibiotic produced by *Micrococcus epidermidis. Can. J. Res.* **28,** 212.

LOVELL, L. D. (1945). Skin bacteria: their location with reference to skin sterilization. *Surg. Gynec. Obstet.* **80,** 174.

LOWBURY, E. J. L. (1969). Gram negative bacilli on the skin. *Br. J. Derm.* **81,** *Suppl. 1,* 55.

MAIBACH, H. I. & HILDICK-SMITH, G. (1965). *Skin Bacteria and their Role In Infection.* New York: McGraw-Hill.

MARPLES, M. J. (1965). *The Ecology of the Human Skin.* Springfield, Illinois: C. C. Thomas.

MARPLES, M. J. (1969). The normal flora of the human skin. *Br. J. Derm.* **81,** *Suppl. 1,* 2.

MARPLES, M. J. (1973). The normal microbial flora of the skin. In *The Normal Microbial Flora of Man. Soc. appl. Bact. Symp. Ser. No. 3.* London: Academic Press. (This volume).

MARPLES, R. R. (1965). The effect of hydration on the bacterial flora of the skin. In *Skin Bacteria and their Role in Infection.* Eds H. I. Maibach & G. Hildick-Smith. New York: McGraw-Hill.

MARPLES, R. R., DOWNING, D. T. & KLIGMAN, A. M. (1971). Control of free fatty acids in human surface lipids by *Corynebacterium acnes. J. invest. Derm.* **56,** 127.

MARPLES, R. R., DOWNING, D. T. & KLIGMAN, A. M. (1972). Influences of *Pityrosporum* species in the generation of free fatty acids in human surface lipids. *J. invest. Derm.* **58,** 155.

MONTES, L. F. & WILBORN, W. H. (1969). Location of bacterial skin flora. *Br. J. Derm.* **81,** *Suppl. 1,* 23.

MURRAY, R. G. E. & LOEB, L. J. (1950). Antibiotics produced by micrococci and streptococci that show selective inhibition within the genus *Streptococcus. Can. J. Res.* **28,** 177.

NICOLAIDES, N. (1965). Skin lipids. II. Lipid class composition of samples from various species and anatomical sites. *J. Am. Oil Chem. Soc.* **42,** 691.

NOBLE, W. C. (1966). *Staphylococcus aureus* on the hair. *J. clin. Path.* **19,** 570.

NOBLE, W. C. (1968). Observations on the surface flora of the skin and on the skin pH. *Br. J. Derm.* **80,** 279.

NOBLE, W. C. (1969). Skin carriage of the Micrococcaceae. *J. clin. Path.* **22,** 249.

NOMURA, M. (1967). Colicins and related bacteriocins. *Ann. Rev. Microbiol.* **21,** 257.

OGAWA, H., MIYAZAKI, H. & KIMURA, M. (1971). Isolation and characterization of human skin lysozyme. *J. invest. Derm.* **57,** 111.

ORFANOS, C. E., SCHLOESSER, E. & MAHRLE, G. (1971). Hair destroying growth of *Corynebacterium tenuis* in the so-called trichomycosis axillaris. *Archs Derm.* **103,** 632.

PARKER, M. T. & SIMMONS, L. E. (1959). The inhibition of *Corynebacterium diphtheriae* and other Gram positive organisms by *Staphylococcus aureus. J. gen. Microbiol.* **21,** 457.

PILLSBURY, D. M. & KLIGMAN, A. M. (1954). Some current problems in cutaneous bacteriology. In *Modern Trends in Dermatology-2,* Ed. R. M. B. McKenna. London: Butterworth.

PILLSBURY, D. M. & REBELL, G. (1952). The bacterial flora of the skin. Factors influencing the growth of resident and transient organisms. *J. invest. Derm.* **18,** 173.

POLLOCK, M. R., WAINWRIGHT, S. D. & MANSON, E. E. D. (1949). The presence of oleic acid-requiring diphtheroids on human skin. *J. Path. Bact.* **61,** 274.

PRICE, P. B. (1938). The bacteriology of normal skin: a new quantitative test applied to a study of the bacterial flora and the disinfectant action of mechanical cleansing. *J. infect. Dis.* **63,** 301.

PUHVEL, S. M. & REISNER, R. M. (1970). Effect of fatty acids on the growth of *Corynebacterium acnes in vitro. J. invest. Derm.* **54,** 48.

REEVES, P. (1965). The bacteriocins. *Bact. Rev.* **29**, 24.
REEVES, P. (1972).*The bacteriocins.* London: Chapman and Hall. New York, Berlin and Heidelberg: Springer-Verlag.
RICKETTS, C. R., SQUIRE, J. R. & TOPLEY, E. (1951). Human skin lipids with particular reference to the self-sterilizing power of the skin. *Clin. Sci.* **10**, 89.
ROBERTS, S. O. B. (1969). The mycology of the normal scalp. *Br. J. Derm.* **81**, 626.
ROIA, F. C. & VANDERWYK, R. W. (1969). Resident microbial flora of the human scalp and its relationship to dandruff. *J. Soc. cosm. Chem.* **20**, 113.
ROSEBURY, T. (1962). *Microorganisms Indigenous to Man.* New York: McGraw-Hill.
ROSEBURY, T., GALE, D. & TAYLOR, D. F. (1954). An approach to the study of interactive phenomena among microorganisms indigenous to man. *J. Bact.* **67**, 135.
ROTHMAN, S., SMILJANIČ, A., SHAPIRO, A. L. & WEITKAMP, A. W. (1947). The spontaneous cure of tinea capitis in puberty. *J. invest. Derm.* **8**, 81.
SARKANY, I. & GAYLARDE, C. C. (1967). Skin flora of the newborn. *Lancet* **1**, 589.
SARKANY, I. & GAYLARDE, C. C. (1968). Bacterial colonisation of the skin in the newborn. *J. Path. Bact.* **95**, 115.
SCHEIMANN. L. G., KNOX, G., SHER, D. & ROTHMAN, S. (1960). The role of bacteria in the formation of free fatty acids on the human skin surface. *J. invest. Derm.* **34**, 171.
SCHEUPLEIN, R. J. & BLANK, I. H. (1971). Permeability of the skin. *Physiol. Rev.* **51**, 702.
SCHINDLER, C. A. & SCHUHARDT, V. T. (1964). Lysostaphin: a new bacteriolytic agent for the staphylococcus. *Proc. natn. Acad. Sci. U.S.A.* **51**, 414.
SELWYN, S. & ELLIS, H. (1972). Skin bacteria and skin disinfection reconsidered. *Br. Med. J.* **1**, 136.
SEUTTER, E. & SUTORIUS, A. H. M. (1972). The quantitative analysis of some constituents of crude sweat. II. Zinc, copper, iron, sialic acid content and oxidative activity. *Dermatologica* **145**, 203.
SEUTTER, E., TRIJBELS, J. M. F., SUTORIUS, A. M. M. & URSELMANN, E. J. M. (1970). The sweat gland as a mucous gland. *Dermatologica* **141**, 397.
SHEHADEH, N. H. & KLIGMAN, A. M. (1963). The effect of topical antibacterial agents on the bacterial flora of the axilla. *J. invest. Derm.* **40**, 61.
SMITH, R. F. (1970). Comparative enumeration of lipophilic and non-lipophilic cutaneous diphtheroids and cocci. *Appl. Microbiol.* **19**, 254.
SMYLIE, H. G. & WEBSTER, C. U. (1960). 'Phisohex'. *Br. Med. J.* **1**, 201.
SMYLIE, H. G., WEBSTER, C. U. & BRUCE, M. L. (1959). 'Phisohex' and safer surgery. *Br. Med. J.* **2**, 606.
SOMERVILLE, D. A. (1969a). The normal flora of the skin in different age groups. *Br. J. Derm.* **81**, 248.
SOMERVILLE, D. A. (1969b). The effect of age on the normal bacterial flora of the skin. *Br. J. Derm.* **81**, *Suppl. 1*, 14.
SOMERVILLE, D. A. (1972). The microbiology of the cutaneous diphtheroids. *Br. J. Derm.* **86**, *Suppl. 8*, 16.
STEIGLEDER, G. K. (1964). Das Vorkommen von Enzymen an der Hautoberfläche. *Fette Seifen Anstrichmitt.* **66**, 691.
STRAUSS, J. S. & EBLING, F. J. (1970). Control and function of skin glands in mammals. *Mem. Soc. Endocr.* **18**, 341.
STUTTARD, L. W. (1961). Release of bacteria from surgeons' hands. *Br. Med. J.* **1**, 591.
SUMMERS, M. M., LYNCH, P. F. & BLACK, T. (1965). Hair as a reservoir of staphylococci. *J. clin. Path.* **18**, 13.
ULRICH, J. A. (1965). Dynamics of bacterial skin populations. In *Skin Bacteria and their Role in Infection.* Eds H. I. Maibach & G. Hildick-Smith. New York: McGraw-Hill.
VANFRAECHEM, J. & POORTMANS, J. R. (1968). Protéines plasmatiques de la sueur humaine normale. *Revues fr. Etud. clin. biol.* **13**, 383.
WEARY, P. E. (1968). *Pityrosporum ovale.* Observations on some aspects of host-parasite interrelationship. *Archs Derm.* **98**, 408.

WEINER, J. S. & HELLMAN, K. (1960). The sweat glands. *Biol. Rev.* **35**, 141.
WILLIAMSON, P. (1965). Quantitative estimation of cutaneous bacteria. In *Skin Bacteria and their Role in Infection.* Eds H. I. Maibach & G. Hildick-Smith. New York: McGraw-Hill.

Effects of Soaps, Germicides and Disinfectants on the Skin Flora

R. R. MARPLES

*Central Public Health Laboratory, Colindale,
London NW9 5HT, England*

CONTENTS

1. Introduction 35
2. Effects of systemic antibiotics 36
   (a) Oral demethylchlortetracycline and the flora of the axilla 36
   (b) Other systemic antibiotics 36
3. Effects of simple washing 38
4. Effects of antibacterial agents in soaps 39
5. Occlusion effects 40
   (a) Testing topical antibacterial agents 41
   (b) Prolonged occlusion and pyoderma 43
6. Experimental infections and antimicrobial agents 43
7. Clinical trials 44
8. Discussion 45
9. References 45

1. Introduction

THE MICROBIAL FLORA of human skin undergoes major disturbances as a result of the activities both of the subject and of his doctor. The skin is washed from time to time with soap and water, and various cosmetic and toilet preparations are applied to it that may contain antimicrobial agents; the doctor may treat the patient, knowingly or unknowingly, with an even greater variety of substances poisonous to the micro-organisms. Active agents may be deposited directly on an area of skin from soap or other vehicle, may be transferred to it from a distant skin site, or may reach it via the bloodstream after oral or parenteral treatment. The study of the resulting alterations of the flora is a valuable method of investigating the ecological factors controlling the numbers and kinds of organisms found on the skin and of assessing the obstacles that must be overcome if a potentially harmful colonist is to succeed in entering the flora. Greatest interest lies in the effects of antibacterial agents because their use has clearly been of value in the prevention of neonatal and post-operative infections but, because of selection, at the cost of some infections that are less treatable than those caused by *Staphylococcus aureus*. An antibacterial agent reaching the skin may produce 3 sorts of effect. First, the total number of organisms may be reduced; this is usually the intended effect. Secondly, the proportion of resistant organisms within the normal flora may increase with or without a detectable change in the total microbial population density; and thirdly, organisms not found in the skin flora before treatment may colonize the treated site. These effects may be separated in time and may be seen most clearly when the delivery of the drug to the skin surface is slow, as when antibiotics are given systemically.

2. Effects of Systemic Antibiotics

(a) *Oral demethylchlortetracycline and the flora of the axilla*

The axilla normally supports a flora of $c.$ 10^6 aerobic bacteria/cm^2. The flora includes micrococci, diphtheroids and often a few enterobacteria. When we studied the flora of the axilla in healthy volunteers who received 600 μg of demethylchlortetracycline daily by mouth for 3 weeks, we found the clearest picture in those persons in whose flora diphtheroids predominated (Marples & Williamson, 1969). In these individuals the geometric mean aerobic count was depressed only during the first week of treatment and recovered to the pre-treatment level early in the second week. Thereafter, and in the 3-week recovery period, the total aerobic count remained steady. During the first week the resistant cocci, which were present initially, multiplied and dominated the flora from the fourth day of treatment for as long as treatment continued. When treatment was stopped, lipophilic diphtheroids slowly replaced the cocci but resistant cocci remained as a high proportion of the total cocci. It appeared that the lipophilic diphtheroids were able to suppress growth of the cocci. In general, the flora was simplified by demethylchlortetracycline, with a lower frequency of enterobacteria and miscellaneous diphtheroids at the end of treatment than was found initially. In 2 subjects, *Staph. aureus* resistant to tetracycline spread to both axillae in the first week of treatment; the source was thought to be the nose in one of these individuals while the other yielded a small number of resistant *Staph. aureus* from one axilla before treatment. In this individual, *Staph. aureus* multiplied to 2.4 x 10^6 by day 7 and a rash developed in the axilla. Despite continued treatment the density of *Staph. aureus* fell as resistant coagulase negative cocci multiplied, and the rash resolved. We concluded that the coagulase negative cocci were more adapted to life on the skin than *Staph. aureus*.

(b) *Other systemic antibiotics*

In a comparative study of several antibiotics, we found that 3 other tetracyclines had very similar effects to those of demethylchlortetracycline (Marples & Kligman, 1971). Ampicillin and penicillin G, on the other hand, produced very little change. We interpreted these findings as indicating that systemic administration of penicillin was not accompanied by excretion on to the skin surface.

Clindamycin produced different effects on the composition of the flora. Lipophilic diphtheroids maintained their dominance of the flora through the selection of resistant strains and there was some increase in the number of cocci but, even after 3 weeks of treatment, these amounted to only 40% of the total aerobic flora (Fig. 1). Enterobacteria are constitutionally resistant to this drug and the proportion of these organisms in the total flora increased progressively with time.

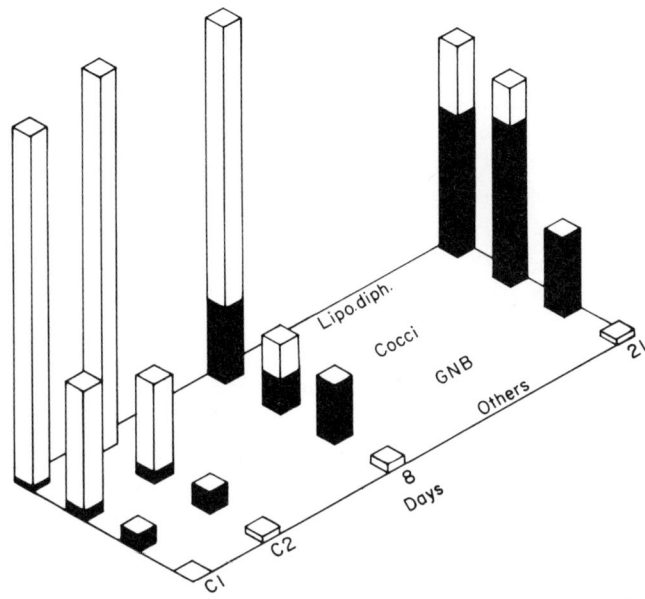

Fig. 1. Effects of clindamycin (200 mgm daily) on the flora of the axilla. Black indicates proportion of the group resistant to 10 μgm/ml of clindamycin.

In a study of the flora of the nose in acne patients we found a carriage rate of enterobacteria of 70% in patients receiving antibiotics and of 40% in control patients (Marples *et al.*, 1969). Even this rate in untreated patients was much higher than we had expected. Antibiotics have been shown to increase carriage rates of enterobacteria in the pharynx in children (Sprunt & Redman, 1968) and these may persist after treatment. Previous exposure to antibiotics may have been the cause of the high initial carriage rates.

These and other findings (Rebell *et al.*, 1950; Marples, 1965) suggest that the number of organisms on an area of skin is, to a great extent, determined by the physicochemical environment, particularly the amount of available water, but that the composition of this flora is determined mainly by bacterial interactions.

When the numbers are reduced by antibiotics, colonization by *Staph. aureus* or enterobacteria becomes a possibility. Once established this may be difficult to eradicate, as we have found with 'Gram negative' folliculitis (Leyden & Marples, 1973*b*).

3. Effects of Simple Washing

The dangers of not washing for the growing child have been widely proclaimed by mothers and teachers but have little experimental confirmation. Besides the cleansing effects, washing with soap may impair the power of the skin surface to repel colonists and affects the pH value of the skin (Bettley, 1960). Evans *et al.* (1950) found that abstention from washing for some days did not lead to an increase in bacterial count in the scapular region. In unpublished work, we found no increase in density of the organisms of the forearm after 2 weeks without washing. There is, of course, an immediate effect but recovery of the flora is usually rapid. In the axilla, where a dense flora exists, simple washing reduced the flora to a tenth of the initial density but recovery was virtually complete within 8 h (Marples & Williamson, unpublished data).

In these sites, washing by itself does not to any great extent disturb the structure of the micro-environment in which the organisms are living. The epithelial cells washed off with their organisms are those that would soon be lost by normal desquamation. This does not appear to be true of the scalp, where the effects of a simple detergent shampoo persist for several days (Kligman, Marples & McGinley, in preparation). We studied 2 sites on the scalp in 7 volunteers with moderate dandruff before, immediately after, and 1, 2, 4 and 7 days after, a single shampoo with a simple detergent formulation. The number of epithelial cells removed by the standard quantitative bacterial sampling technique, the corneocyte count, is a measure of the living space of the micro-organisms (McGinley *et al.*, 1969). As would be expected, the corneocyte count showed a small but significant fall after washing (from 1.27×10^6 to 7.18×10^5 cells/cm^2) but had fallen still further (to 2.8×10^5) one day later. From the 2nd to the 4th day the corneocyte count increased rapidly to 7.1×10^5 and by the 7th day was not significantly different from the pre-treatment value. The 3 main groups of micro-organisms did not respond in the same way. For simplicity these are shown in Fig. 2 as changes from the initial count set equal to 100%. The aerobic bacteria, essentially all cocci, were washed off by the shampoo leaving only 13% of the initial count but the number of anaerobic diphtheroids and of lipid-dependent yeasts was not significantly reduced. This suggests a more superficial location for the aerobic bacteria. By the next day the count of anaerobic diphtheroids and yeasts had also fallen. Aerobic bacteria and anaerobic diphtheroids recovered their numbers quickly and both exceeded the initial density by the fourth day. *Pityrosporum* was distinctly slower to recover.

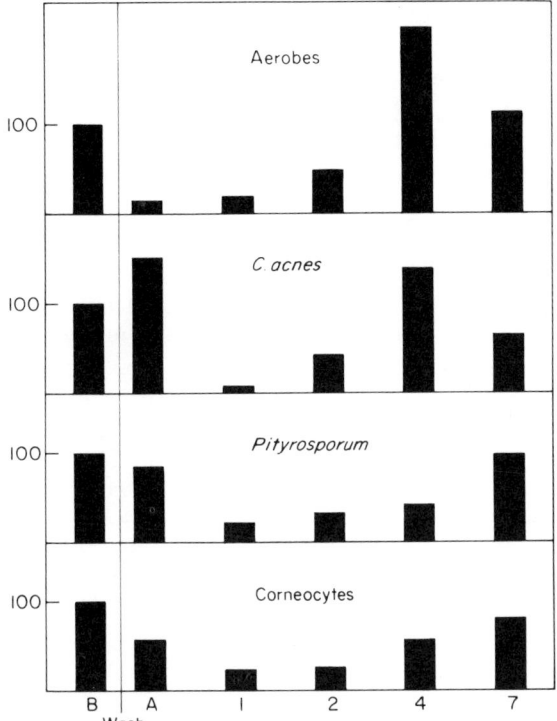

Fig. 2. Recovery of micro-organisms and corneocyte count before and at intervals after a single shampoo of the scalp. Aerobic bacteria and corneocytes were significantly reduced immediately after the wash while *C. acnes* and *Pityrosporum* showed no change (A). All were low 1 day after. Aerobes and *C. acnes* recovered in 4 days, *Pityrosporum* in 7 while corneocytes still were slightly reduced at the time.

The fall in the counts during the first day from the levels found immediately after the shampoo suggests that drying out of the scalp after removal of part of the lipid promotes further desquamation and loss of particles. The more rapid recovery of the bacterial counts than of the corneocyte count indicates that living space is not the only factor controlling the population.

4. Effects of Antibacterial Agents in Soaps

The efficacy of antibacterial soaps in reducing the numbers of organisms can be tested in the axilla. If different soaps are used on each axilla in the same volunteer, the transfer of the antibacterial agent from one site to the other may cause difficulties in interpretation (Marples & Kligman, 1973). We tested an experimental soap containing 2% of hexachlorophene in one axilla against 2 non-medicated soaps in the other axilla, by application twice daily for 10 days, with samples taken before and the day after the last wash. We observed a fall in density of 72% in the axilla treated with hexachlorophene. This differed significantly from the initial count and from the count in the axilla treated with the non-medicated soap. When, in another study, the same experimental soap was used in one axilla while the other was treated with one of 2 soaps containing mixtures of the more active salicylanilide and carbanilide bacteriostats, the count in both axillae fell by $> 95\%$ of the pre-treatment value but the 2 treatments could not be distinguished (Table 1). We felt that transfer of the more active anilide bacteriostats had affected the results by increasing the apparent activity of the hexachlorophene soap.

Table 1

Counts of bacteria in axillae treated with medicated or non-medicated soaps

Counts ($\times 10^6$) before and after 10 days treatment with a 2% hexachlorophene soap (Soap A), non-medicated soaps (Soap B) or strongly medicated soaps (Soap C)

	Soap A	v	Soap B	Soap A	v	Soap C
Before treatment	1.5	—NS—	1.8	1.4	—NS—	1.4
	*		NS	*		*
After 10 days	0.4	—*—	1.5	0.06	—NS—	0.04

*, Significant differences; NS, not significant differences.

Selection of resistant species by antibacterial soaps has been reported. Taplin demonstrated overgrowth first of enterobacteria, then of *Pseudomonas* spp. and finally of *Candida albicans* in his toewebs during excessive use of medicated soaps (Ehrenkranz et al., 1966). When a strong hexachlorophene preparation was used in the axilla, overgrowth of enterobacteria persisted for several days after last treatment (Marples & Kligman, 1969b). With the soap application described above, although enterobacteria increased in density during treatment in carriers, the number of subjects yielding these organisms did not increase.

5. Occlusion Effects

Occlusion of the forearm with plastic film is the basis of a number of methods of testing antimicrobial agents as pure compounds in simple solvents or in formulations including soaps, Occlusion prevents loss of moisture and permits an increase within 2 days in the bacterial count from 10^3 to 10^5, or even 10^7 when large areas are occluded (Marples, 1965). The stimulus to growth provided by an occlusive dressing is so great that the amount of antibacterial agent left after the repeated use of medicated soaps is insufficient to prevent growth. A reduction in population density relative to a control area can, however, be demonstrated (Marples & Kligman, 1969a). When a suspension of *Pseudomonas* sp. was applied to the skin and covered occlusively for 48 h on the forearms of subjects who had used a bland soap on one and a soap containing medication against Gram positive organisms on the other, twice daily for a week, the organism multiplied in 18 of 20 trials on the side treated with medicated soap but only in 10 of 23 trials with the bland soap (Marples, 1971). Sufficient agent was left to affect the survival of the inoculum.

(a) *Testing topical antibacterial agents*

Inhibition of growth stimulated by occlusion may be employed as a test of the efficacy of an antibacterial agent in the presence of human skin and its products. We termed this test the *simple occlusion test* and have shown its application to pure substances and to complete formulations (Marples & Kligman 1969a, 1973b). By occluding the skin for 2 days before applying the test agent, a dense flora can be induced making the test one of bactericidal rather than bacteriostatic action. This modification we termed the *dense-flora occlusion test*.

We applied 0.1 ml of 1%, 0.1% and 0.01% concentrations of a variety of topical agents to 5 cm squares on the forearms of healthy volunteers in both the simple occlusion test and the dense-flora modification. Bacitracin and chloramphenicol were effective at 0.01%, a dose of only 0.4 μgm/cm^2 in both tests (Table 2). A dose of 40 μgm/cm^2 of hexachlorophene was needed for

Table 2

Minimum effective dose* (μg/cm^2) in occlusion tests

Agent	Simple occlusion test	Dense-flora occlusion test
Bacitracin	0.4	0.4
Chloramphenicol	0.4	0.4
Hexachlorophene	40	40
Methylbenzethonium	4	>40
Neomycin	0.4	4
Providone iodine	40	>40
Pyrithione-Na	4	40
Tetracycline	4	40

*Significant reduction in count by comparison with control site effectiveness. The quaternary ammonium compound and providone iodine were effective in the simple test but not effective even at 40 μgm/cm^2 in the dense-flora occlusion test. The other compounds required higher doses in the second test than in the simple test to produce significant effects.

Paints can also be tested though the dose cannot be easily adjusted. The means of logarithms of the bacterial count (Table 3) again show the increased stringency of the dense-flora modification. The dye-compound formulation (Castellani's Paint) was more effective than basic fuchsin alone. Tincture of iodine was more effective than providone iodine in the simple occlusion test, in which the latter compound was effective in only some subjects.

Table 3

Log bacterial counts after occlusion tests of treatment with skin paints

Agent	Simple occlusion test	Dense-flora occlusion test
Alcohol	6.20	6.26
Basic fuchsin	3.98	5.46
Castellani's Paint	2.32	3.24
Crystal violet	1.98	3.13
Iodine tincture	4.30	4.53
Neocastaderm	Not done	4.11
Providone iodine	5.43	Not done
Thimerosal	4.88	5.94
Water	6.35	6.17

One reason for apparent failure of compounds in the dense-flora occlusion test was overgrowth of enterobacteria or of yeasts. Enterobacteria were numerous when hexachlorophene, methylbenzethonium chloride, tetracycline and basic fuchsin was applied but neomycin, as we have previously stated, sometimes selected yeasts.

Another modification, designed to test the ability of the agent to adhere to the skin, introduces a 3 day delay between treatment of the skin with the agent and the application of the occlusive dressing; *the persistence test*. The agent was applied 3 times a day for 3 days and the occlusive period was only 24 h. Chloramphenicol, neomycin, penicillin, pyrithione and gentamicin showed evidence of persistence while tetracycline, sulfamylon, hexachlorophene, erythromycin and several other chemotherapeutic agents did not. Again, failure often was due to overgrowth of enterobacteria, in particular at sites treated with erythromycin, hexachlorophene and the quaternary ammonium compound.

(b) *Prolonged occlusion and pyoderma*

When the skin of the forearm is occluded for periods longer than 48 h, the population density of micro-organisms does not increase further but cocci are replaced by diphtheroids (Marples, 1965). If an antibacterial agent has been applied, selection of resistant strains, present initially in very small numbers or contaminating the skin during dressing changes, is more likely. When we applied neomycin and wrapped the forearm completely for 1 week, 6 of 10 subjects developed a pyoderma due to resistant *Staph. aureus* (Marples & Kligman, 1969*b*). In a similar study with topical tetracycline, 1 of 4 subjects developed a severe rash due to *Candida albicans*. By inoculating large numbers of *Staph. aureus* on to occluded normal skin we were able to produce a pyoderma in *c.* 3 days, longer if a smaller inoculum was used (Singh *et al.*, 1971). Pre-treatment with 70% alcohol reduced the size of the inoculum needed to produce a reaction.

We compared the effect of applying 0.02 ml of a 5% solution of a medicated soap and a similar amount of a non-medicated soap to 5 cm square areas of skin that had been inoculated with *Staph. aureus* and occluded 3 days earlier. With the medicated soap there was no further increase in the lesion count but with the non-medicated soap new lesions continued to appear (Marples, 1971).

6. Experimental Infections and Antimicrobial Agents

A better model for *Staph. aureus* infections was discovered later. Removal of the horny layer of the skin by repeated application and removal of cellophane tape produces a mild and reproducible injury and leaves a site free of bacteria and with a slight exudation of serous fluid. Immediate inoculation and occlusion leads to a spreading infection and must be avoided. One day after injury such a stripped site can be safely inoculated with the organism and occluded for 24 h. Samples of the bacterial growth and of the exudate are taken, and the experiment is concluded by topical therapy and exposure (Marples & Kligman, 1972). Test treatments can be applied at the time of inoculation or, more rationally, 6 h later.

Occlusion alone is sufficient to permit an inoculum of *Candida albicans* to multiply on forearm skin. One day of occlusion is required with 10^5 cells although the dermatitis is not seen until 48 h after inoculation (Rebora *et al.*, 1973). Prevention of growth of the yeast and the reduction on the severity of the dermatitis can be estimated.

Both of these infection models were employed to investigate the activity of antimicrobial agents in a steroid-antibiotic formulation (Marples *et al.*, 1973). The results are summarised in Table 4. Each organism, as expected was reduced in number whenever the combination contained an antibiotic to which it was susceptible. In these cases the clinical severity relative to the untreated site was,

Table 4

Effect of a steroid antibiotic mixture and of its components in experimental infections with Staph. aureus *or* C. albicans

Compound or mixture tested	Organism reduced		Clinical severity reduced	
	Staph. aureus	C. albicans	Staph. aureus	C. albicans
Triamcinolone neomycin, nystatin, and gramicidin	Yes	Yes	Yes	Yes
Triamcinolone neomycin, and gramicidin	Yes	ND	Yes	ND
Nystatin	ND*	Yes	ND	Yes
Neomycin gramicidin	Yes	ND	Yes	ND
Triamcinolone	No	No	Yes	Yes
Cream Base	No	No	No	Slight
Untreated	No	No	No	No

* ND = Not done

of course, reduced. The sites treated with the steroid alone were of interest; with both organisms the steroid had no effect on the microbial count but significantly reduced the severity of the dermatitis. Because other substances have anti-inflammatory properties, microbial as well as clinical appraisal of topical formulations is necessary.

7. Clinical Trials

The final method of testing antimicrobial agents is the clinical trial. Because of the difficulties in selecting equivalent patients and in reducing extraneous variables, clinical trials are not suitable for comparing 2 active treatments but can separate active from inactive treatments. Even when the question to be answered is simple, a clinical trial is expensive and time-consuming and should be contemplated only when formulation has been decided upon. A popular design, which takes into account the variation between individuals, is to treat comparable sites with the active drug and the inactive base at the same time and to compare the 2 sites at the end of treatment, leaving any change from the initial state as a separate variable. With antibacterial agents this design may fail to distinguish active from inactive because one site may be contaminated from the other. We studied a neomycin cream and the cream base in this way, and also successively, in dermatitis lesions colonized by *Staph. aureus* (Marples & Kligman, 1973). When the placebo cream was used to treat one site and the neomycin cream another site at the same time on the same patient, the count of

Table 5

Staphylococcus aureus *count before and after parallel or successive treatments of skin lesion*

Sampling time	No. of bacteria/cm^2 in tests by			
	Parallel treatment		Successive treatment	
	Placebo	Neomycin	First Placebo	Second Neomycin
Before treatment	83,700	34,000	970,000	1,080,000
After treatment	101	22	1,080,000	43

Staph. aureus fell significantly in both sites while the placebo first caused no change in count. Neomycin following, reduced the count to very low levels (Table 5).

8. Discussion

The main effect of applying antimicrobial substances to human skin is to reduce the numbers of organisms that are susceptible to the drug. For the purpose of prophylaxis against colonization with *Staph. aureus*. and for deodorancy, application in a soap is adequate and seems to lead to few side effects. When greater amounts of active agent are supplied, temporary reductions of the flora to very low counts may occur. The resistant flora which, given time, inevitably will develop, is often innocuous but occasionally a more harmful colonist may become established. There is a real danger that such an incident may not be recognized as a side effect of treatment and that the same treatment will be continued or increased to overcome the 'exacerbation'; considerable morbidity may then follow (Leyden & Marples, 1973*a*). The frequency of superinfections of this sort will depend on the type of patient and disease as well as the antimicrobial agents used. Reactions are unlikely in normal individuals; in patients who have extensive exudative dermatitis, however, colonization by potential or actual pathogens is likely. The occurrence of clinical disease as a response to antimicrobial therapy is a clear indication of the protective value to the host of his normal microbial flora.

9. References

BETTLEY, F. R. (1960). Some effects of soap on the skin. *Br. Med. J.* **1,** 1675.
EHRENKRANZ, N. J., TAPLIN, D. & BUTT, P. (1967). Antibiotic resistant bacteria in the nose and skin: colonization and cross infection. In *Antimicrobial Agents and Chemotherapy.* Ed. G. L. Hobby. Amer. Soc. Microbiol.

EVANS, C. A., SMITH, W. M., JOHNSTONE, E. A. & GIBLETT, E. R. (1950). Bacterial flora of the normal human skin. *J. invest. Derm.* **15**, 305.

LEYDEN, J. J. & MARPLES, R. R. (1973*a*). Ecological principles and antibiotic therapy in chronic dermatoses. *Archs Derm.* **107**, 208.

LEYDEN, J. J. & MARPLES, R. R. (1973*b*). Gram negative folliculitis. A complication of antibiotic therapy in *acne vulgaris*. *Br. J. Derm.* **88**, 533.

McGINLEY, K. J., MARPLES, R. R. & PLEWIG, G. (1969). A method of visualizing and quantitating the desquamating portion of the human stratum corneum. *J. invest. Derm.* **53**, 107.

MARPLES, R. R. (1965). The effect of hydration on the bacterial flora of the skin. In *Skin Bacteria and their Role in Infection.* Eds H. I. Maibach & G. Hildick-Smith. New York: McGraw-Hill.

MARPLES, R. R. (1971). Antibacterial cosmetics and the microflora of human skin. *Devs ind. Microbiol.* **12**, 178.

MARPLES, R. R., FULTON, J. E., LEYDEN, J. J. & McGINLEY, K. J. (1969). Effect of antibiotics on the nasal flora in acne patients. *Archs Derm.* **99**, 647.

MARPLES, R. R. & KLIGMAN, A. M. (1969*a*). *In vivo* methods for appraising antibacterial agents. *TGA Cosmetics J.* **1**, 26.

MARPLES, R. R. & KLIGMAN, A. M. (1969*b*). Pyoderma due to resistant *Staphylococcus aureus* following topical application of neomycin. *J. invest. Derm.* **53**, 11.

MARPLES, R. R. & KLIGMAN, A. M. (1971). Ecological effects of oral antibiotics on the microflora of human skin. *Archs Derm.* **103**, 148.

MARPLES, R. R. & KLIGMAN, A. M. (1972). Bacterial infection of superficial wounds: a human model for *Staphylococcus aureus.* In *Epidermal Wound Healing.* Eds H. I. Maibach & D. T. Rovee. Chicago Year Book Medical Publishers.

MARPLES, R. R. & KLIGMAN, A. M. (1973*a*). Limitations of paired comparisons of topical drugs. *Br. J. Derm.* **88**, 61.

MARPLES, R. R. & KLIGMAN, A. M. (1973*b*). Evaluation of topical antibacterial agents on human skin. (Under review).

MARPLES, R. R., REBORA, A. E. & KLIGMAN, A. M. (1973). Topical steroid-antibiotic combinations. Assay of use in experimentally induced human infections. *Archs Derm.* **108**, 237.

MARPLES, R. R. & WILLIAMSON, P. (1969). Effects of systemic demethylchlortetracycline on human cutaneous microflora. *Appl. Microbiol.* **18**, 228.

REBELL, G., PILLSBURY, D. M., DE ST. PHALLE, M. & GINSBURG, D. (1950). Factors affecting the rapid disappearance of bacteria placed on the normal skin. *J. invest. Derm.* **14**, 247.

REBORA, A. E., MARPLES, R. R. & KLIGMAN, A. M. (1973). Experimental infection with *Candida albicans. Archs Derm.* **108**, 69.

SINGH, G., MARPLES, R. R. & KLIGMAN, A. M. (1971). Experimental *Staphylococcus aureus* infections in humans. *J. invest. Derm.* **57**, 149.

SPRUNT, K. & REDMAN, W. (1968). Evidence suggesting importance of role of interbacterial inhibition in maintaining balance of normal flora. *Ann. intern. Med.* **68**, 579.

The Normal Microbial Flora of the Mouth

J. M. HARDIE AND G. H. BOWDEN

*Dental Bacteriology Laboratory and MRC Dental Epidemiology Unit,
The London Hospital Medical College,
Turner Street, London E1 2AD, England*

CONTENTS

1. Introduction 47
2. The oral environment 48
3. The development of the oral microflora 49
4. The microbial flora of different parts of the mouth 50
    (a) Tongue 51
    (b) Saliva 52
    (c) Dental plaque 53
    (d) Gingival crevice 59
5. The oral flora of primitive man 61
6. Some ecological determinants 62
    (a) Anaerobiosis 62
    (b) Adherence and aggregation 64
    (c) Microbial inter-relationships 64
7. The principal groups of micro-organisms found in the mouth 67
    (a) Gram positive cocci 67
    (b) Gram negative cocci 70
    (c) Gram positive rods and filaments 71
    (d) Gram negative rods and filaments 72
    (e) Other micro-organisms 73
8. References 74

1. Introduction

THE SUBJECT of this review is large and the relevant published work is scattered widely amongst the medical, dental and microbiological literature. Frequently, the reported studies have been directed towards gaining some insight into disease processes in the mouth, rather than developing an understanding of the normal microbial flora. The situation is complicated by the fact that the two major dental diseases, dental caries and chronic periodontal disease, are ubiquitous and both appear to be caused by the normal commensal flora rather than some extrinsic infectious agent. As far as possible, the disease aspects of the oral flora will be ignored here, and an attempt made to consider the mouth as an ecological environment which supports the growth of a variety of micro-organisms.

Inevitably, some aspects of the subject will be given more prominence than others in this review. For wider coverage of the oral flora in general, and especially for a more detailed account of the older literature, the books by Bissett & Davis (1960), Burnet & Scherp (1968), Nolte (1973) and Rosebury (1962) are recommended. A very comprehensive survey of the bacteriology of the oral cavity was conducted by Morris (1953*a, b;* 1954*a, b, c, d*) and this series of papers contains much useful information on the composition and properties of the oral flora. Recent review articles have been contributed by Socransky (1970), Socransky & Manganiello (1971) and Davies (1972).

2. The Oral Environment

The mouth cannot be regarded as a single, uniform environment. Rather, it presents a series of different ecological situations, each of which may be suitable for colonization by different micro-organisms. In addition, the nature of the various micro-environments within the mouth may change from time to time during the life of an individual.

At birth the mouth consists of a cavity enclosed by the soft tissues of the lips, cheeks, tongue and palate, all of which are kept moist by the secretions of the various salivary glands. After a few months, the primary teeth begin to erupt and this event produces a significant change in the environment which is reflected by some observable changes in the oral flora. The primary dentition is usually complete by the age of 3 and the dental situation is relatively stable until about 6 years, when the permanent teeth begin to erupt. The changeover from primary to permanent dentition is usually complete by the age of 12. During the mixed dentition stage, the local conditions in the mouth are bound to change as teeth are shed and new ones erupt.

Apart from the natural eruption and shedding of teeth, the ecological situation within the mouth may be altered by numerous other environmental factors such as diet and nutrition (Carlsson & Sundstrom, 1968; Bowen, 1970; Carlsson & Egelberg, 1965; Littleton *et al.,* 1967), antibiotic therapy (Handelman & Hawes, 1964, 1965; Lobene *et al.,* 1969; Bibby, 1970), radiation of salivary glands (Llory *et al.,* 1972) and various types of dental treatment (Shklair *et al.,* 1956; Kesel *et al.,* 1958; Shklair & Mazzarella, 1961; Sakamaki & Bahn, 1968). Other factors which may affect the composition of the oral flora, particularly salivary constituents, are reviewed in the chapter by Jenkins & Geddes in this symposium.

The following areas within the mouth of a dentate person may be expected to provide more or less different environmental conditions:

(i) The lips, cheek and palate
(ii) The tongue
(iii) The tooth surface
    (a) Smooth surfaces (buccal and lingual)

(b) Approximal surfaces (mesial and distal)
  (c) Pits and fissures (occlusal)
  (d) Gingival areas
 (iv) Saliva

When studying the oral flora, it is important to select a suitable method of sampling the area of interest (Bowden & Hardie, 1971). Collecting saliva or samples from soft tissues is relatively simple, but problems arise when attempting to collect material from inaccessible parts of the tooth surface. Approximal areas may be sampled using dental floss (Thomson, 1971; Loesche & Syed, 1973) or with a specially shaped strip of abrasive tape (Hardie et al., 1972). Occlusal fissures are difficult to sample satisfactorily, but Loe et al. (1973) have recently described an elegant method for investigating artificial fissures *in vivo*.

Any reports on the microbial flora of the mouth should state the type of sample and the collection method employed so that the results can be compared with other similar studies.

3. The Development of the Oral Microflora

Few investigators have studied in detail the development of the whole oral flora from birth, although several have reported the occurrence of individual bacterial species in the infant's mouth. A comprehensive review on this topic has been published recently by Socransky & Manganiello (1971), in which the authors emphasize the variability of the flora in different parts of the mouth.

At birth, the oral cavity is usually sterile, but from 6-10 h later the number of organisms detected appears to increase rapidly (Lewkowitz, 1901; Brailovsky-Lounkevitch, 1915; Kostecka, 1924). McCarthy et al. (1965) studied the development of the oral flora in 51 newborn and 44 one month-old infants, and collected further specimens from the latter group at 8 and 12 months. Species of *Streptococcus, Staphylococcus, Veillonella* and *Neisseria* were present in all subjects by 12 months of age, whilst *Actinomyces, Lactobacillus, Nocardia (Rothia)* and *Fusobacterium* species were cultured from more than half. *Bacteroides, Leptotrichia, Candida, Corynebacterium* species and coliforms were isolated from less than half of the infants at a year. Streptococci were found to be numerically dominant throughout the time period studied, although the percentage fell from 98 to 70% of the cultivable flora by the end of the first year of life. *Streptococcus salivarius* was especially common in the mouths of the newborn. Staphylococci occurred in very low numbers in about half of the newborn group and were invariably observed in older children. The presence of teeth did not appear to be essential for the establishment of filamentous bacteria, since examples of *Fusobacterium, Leptotrichia, Actinomyces* and *Nocardia (Rothia)* species were all isolated on occasion from pre-dentate infants.

However, the frequency of isolation of *Fusobacterium* and *Actinomyes* species did increase after the teeth had started to erupt.

The importance of the presence of erupted teeth for the establishment of *Strep. sanguis* in the mouths of infants has been demonstrated by Carlsson *et al.* (1970*a*). The mouths of 27 infants were studied during their first year of life, and in most cases *Strep. sanguis* became established only after the eruption of teeth. An interesting corollary to this was the observation that both *Strep. sanguis* and *Strep. mutans* tended to disappear from the mouths of edentulous subjects when they stopped wearing their dentures for a few days, but both species could be recovered again shortly after the dentures were replaced (Carlsson *et al.*, 1969).

In contrast to the relatively late appearance of *Strep. sanguis* in the infants mouth, *Strep. salivarius* appears to be a regular inhabitant from the second day of life (McCarthy *et al.*, 1965; Carlsson *et al.*, 1970*b*). Carlsson's group were unable to isolate *Strep. mutans* from any of the 27 infants studied during their first year of life.

Only fragmentary information exists on the composition of the oral flora during childhood and adolescence. Several workers have studied the prevalence of *Bacteroides melaninogenicus* at different ages, and it appears that this species can be more commonly isolated with increasing age (Bailit *et al.*, 1964, Kelstrup 1966). By the early teens, *B. melaninogenicus* can be isolated from most mouths, especially from the gingival crevice region. A similar increase of numbers with age also occurs with oral spirochaetes (Socransky & Manganiello, 1971).

4. The Microbial Flora of Different Parts of the Mouth

The majority of investigations on the flora of the mouth have been concerned either with saliva or dental plaque; relatively few studies have been reported on samples taken from soft tissue areas, although the fact that epithelial cells from the mouth are usually heavily colonized with micro-organisms is frequently observed. It was established many years ago that the oral flora varies from one site to another (Bibby, 1938), and it appears that certain bacteria have a specific 'primary ecologic niche' within the oral cavity (Socransky & Manganiello, 1971). In this section, some information on the microbial composition of the tongue, saliva, dental plaque and gingival crevice material is reviewed.

It should be borne in mind that accurate quantitation of the mixed flora of the mouth is extremely difficult to achieve for a number of technical reasons, not least of which is the problem of dispersing the material sufficiently well for viable counts to be performed. In most reported studies on viable counts, a variety of selective media have been used and the identification procedures employed have often been rudimentary. Results are usually expressed either as counts per unit amount of material, or, more commonly, as percentages of total viable counts. Few workers would claim that their differential counts give more than an indication of the relative proportions of micro-organisms within a given sample.

(a) *Tongue*

A study on the cultivable flora of the tongue was reported by Gordon & Gibbons (1966), and the results of their survey are summarized in Table 1.

Table 1
The predominant cultivable flora of the tongue

Organisms	Mean %	Range
Gram + facultative cocci	44.8	32.1-64.9
Gram + anaerobic cocci	4.2	0-8.9
Gram + facultative rods	13.0	10.5-17.9
Gram + anaerobic rods	7.4	0-14.6
Gram − anaerobic cocci	16.0	10.4-21.0
Gram − facultative cocci	3.4	0-7.0
Gram − anaerobic rods	8.2	2.1-17.3
Gram − facultative rods	3.2	0-8.6
No. of samples from which above micro-organisms were isolated	6	

From Gordon & Gibbons (1966)

Streptococcus salivarius comprised 53.5% of the total facultative streptococci growing aerobically, but only 30% of those cultivated anaerobically. Overall, this species comprised 8.2% of the total cultivable flora. Notably lacking in the flora of the 6 tongue specimens examined were spirochaetes, which were not seen in dark-field preparations, and *Bacteroides melaninogenicus* which was isolated from only one sample.

An unusual group of Gram positive, catalase variable cocci, which resembled staphylococci, was isolated from 5 of 6 tongues, and these organisms averaged 3.6% of the cultivable flora. In a subsequent paper, Gordon (1967) reported the properties of 24 strains of this coccus, which was referred to as *Staphylococcus salivarius*. Bowden (1969) examined 4 strains in detail, and described the composition of their cell walls and extracellular polysaccharide slime. This organism has been shown to have a DNA base ratio of 55.4-58.3%, which is more characteristic of *Micrococcus* than *Staphylococcus* species, and it has recently been referred to as *Micrococcus mucilagenosus* (Kocur et al., 1971; Bergan et al., 1970). The presence of this species can almost invariably be demonstrated in tongue smears by immunofluorescence (Bowden & Hardie, unpublished data), but it is seen only in low numbers, if at all, in dental plaque.

(b) *Saliva*

The overall bacterial composition of whole saliva as reported in three independent studies (Richardson & Jones, 1958; Slack & Bowden, 1965; Gordon & Jong, 1968) is shown in Table 2. In all these investigations streptococci

Table 2
The predominant cultivable flora of human saliva

Organisms	Gordon & Jong (1968)	Slack & Bowden (1965)	Richardson & Jones (1958)
Streptococci	41.0	56.3	16.3
Streptococcus salivarius	4.6	19.7	10.0
Anaerobic Gram + cocci	13.0	ND	ND
Neisseria sp.	1.2	29.2	1.8
Veillonella sp.	15.9	2.9	15.4
Gram + rods and filaments	16.6	8.28	–
Gram – anaerobic rods	4.8	3.63	–
Gram – facultative rods	2.3	–	0.01
Number of samples from which above micro-organisms were isolated	6	12	140

Selected organisms expressed as percentage of total viable count. ND = not detected; – = not reported.

predominate numerically and *Strep. salivarius* appears to comprise a significant proportion of the total count in each case. Krasse (1953, 1954) has also shown that *Strep. salivarius* is present in high numbers in saliva, as well as on the tongue, but is far less numerous in dental plaque. Gibbons *et al.* (1964) found that this species averaged 47.4% of the facultative streptococci in saliva, 55.3% of those isolated from the tongue, and 10.7% from the cheek. These authors agreed with Krasse's earlier suggestion that the tongue is the principal source of salivary bacteria, whereas dental plaque and gingival crevice debris apparently contribute little to the salivary flora. It may be worth commenting that *Strep. salivarius* is usually recognized by the production of large, mucoid colonies on sucrose-containing selective media. Although the identification of this species on the basis of colony morphology may be correct in the majority of cases, other streptococci can occasionally produce similar mucoid variants (Edwardsson, 1970; Thomson, 1971).

Hadi & Russell (1968) have shown that salivary levels of fusobacteria vary from a mean level of 2.72×10^5/ml for normal subjects to 1.79×10^6/ml in patients suffering from acute ulcerative gingivitis (AUG, Vincent's gingivitis). These authors compared their results with those of several other surveys, which showed a range in salivary *Fusobacterium* level from 6.0×10^4/ml (Slanetz & Reynolds, 1952) to 1.8×10^5/ml (Baird-Parker, 1959) or 8.4×10^5/ml (Omata & Disraely, 1956). It was also observed (Hadi & Russell, 1969) that significantly higher numbers of fusobacteria could be isolated from gingival crevice material of subjects with AUG or advanced chronic periodontal disease than from subjects with normal, healthy gingivae.

Since it has been recognized that the salivary flora does not necessarily represent very closely the microbial composition of different areas in the mouth, the use of saliva samples has become far less common than it used to be. For studies on the aetiology of dental disease it is now more usual to collect samples of dental plaque.

(c) *Dental plaque*

The term 'gelatinous microbial plaques' was first used by Black in 1898 to describe the aggregations of bacteria which collect on the teeth. Dental plaque forms rapidly on tooth surfaces and, if not removed by tooth brushing within a few hours of formation, it builds up into a thick, adherent film of material which cannot be washed off with a stream of water. This plaque consists essentially of bacteria, embedded in an organic matrix which is derived partly from salivary glycoprotein and partly from microbial extracellular polymers. Dental plaque can be demonstrated very clearly in the mouth by staining with a disclosing solution such as erythrosin. Contrary to popular belief, food debris does not appear to contribute significantly to the mass of dental plaque. In some situations the plaque becomes calcified, and is then referred to as calculus or tartar.

Dental plaque has received a considerable amount of attention in the literature, especially in the last 10 years or so, and there is strong evidence linking the presence of plaque with caries and periodontal disease. Much of the literature and many aspects of current thinking on the development, microbiology, structure and biochemistry of dental plaque are reviewed in the book edited by McHugh (1970).

One aspect which has been investigated by several workers is the development of dental plaque on a clean surface. Such studies have been of 2 basic types. In the first type, natural teeth are cleaned and polished until no visible surface deposits or integuments remain, and are then allowed to develop plaque for varying periods without being disturbed by tooth brushing. Some of these studies have been based largely on microscopic observations, whilst others have employed cultural techniques. The second type of investigation depends upon the use of various artificial devices which can be placed in the mouth and removed for study after suitable time intervals. The advantage of this approach is that the same sample may be divided and used for both microscopic and cultural examination.

Table 3
Development of dental plaque on clean tooth surfaces

	R 1*	H 2	R 3	H 4	R 5	H 7	R 7	R 9	H 14	H 21	H 28	H 90
Streptococci	46.0	50.4	69.0	38.7	52.5	57.0	48.0	36.0	42.8	39.0	50.2	16.5
Gram + rods and filaments	9.3	11.6	2.4	27.2	12.6	10.2	25.3	31.2	22.2	37.9	30.3	51.0
Actinomyces spp.	0.18	1.3	0.93	1.0	6.05	0	18	23	1.7	14.1	13.1	36.0
Neisseria spp.	9.1	12.4	7.7	16.6	6.4	22.1	3.7	1.7	5.2	11.0	4.0	1.7
Veillonella spp.	1.46	0.1	3.83	0.1	15.5	0.1	13.8	12.5	0.1	0.1	0.1	0.1
Fusobacterium spp.	0.03	1.4	0.11	1.0	0.3	1.1	0.85	0.9	3.3	2.4	2.7	1.5

Pooled data from studies by Ritz (1967) (R) and Howell *et al.* (1965) (H)
Selected organisms expressed as percentage of total viable count
* Plaque age (days)

The results of 2 cultural studies of plaque development on natural tooth surfaces have been combined and are shown in Table 3 (Howell et al., 1965; Ritz, 1967), and some comparable results from experiments with an artificial device are illustrated in Table 4 (Slack & Bowden, 1965).

Table 4
Experimental plaque development (from Slack & Bowden, 1965)

Organism	Age of plaque in days					Mean plaque age (All samples)
	1	3	5	7	14	
Streptococcus spp.	96.9	89.1	75.4	82.1	78.9	84.4
Lactobacillus spp.	0.19	5.7	11.6	3.5	5.0	5.5
Neisseria spp.	2.4	1.5	0.7	0.1	0.4	1.1
Veillonella spp.	0.2	3.4	7.8	2.2	1.5	2.9
Fusobacterium spp.	<0.01	<0.01	0.08	0.48	0.05	0.1
Bacteroides spp.	0	0.6	1.9	0.9	0	0.8
Leptotrichia spp.	0	<0.01	<0.01	<0.01	2.3	0.2
ANO$_2$ filaments	0	<0.01	0.1	2.6	11.4	1.6
Nocardia (Rothia) spp.	0.04	0.13	2.7	0.008	0.43	0.66
No. of samples yielding above micro-organisms	7	8	8	6	3	32

Selected organisms expressed as percentage of total viable count

In most studies the same general trends have been observed. Early plaque consists predominantly of streptococci, neisseriae and a few Gram positive rods and filaments. As the plaque develops, especially after *c*. 7 days, the number of anaerobes increases, the aerobic species tend to decrease, and there is a reduction in the overall proportion of streptococci. The actual proportion of streptococci appears to be higher (75-97%) in the study using an artificial device (Table 4) than in those based on natural tooth surfaces. This may be due to sampling difficulties in the latter case, or possibly because the artificial surface is completely free of micro-organisms at the start of the experiment, whereas the cleaned and polished natural tooth may still harbour pockets of bacterial plaque material in enamel surface defects.

Plaque which has been allowed to develop for 14 days or more is usually more filamentous in appearance than early plaque, and may yield high counts of vibrios and spirochaetes in addition to other anaerobes.

Electron microscopic observations on plaque development (Saxton, 1973) indicate that bacteria colonize the clean tooth surface within a few hours of exposure to the oral environment. However, little information is currently available on the identity or source of those bacteria which initiate plaque formation or on what selective factors operate in determining the pattern of colonization. There is some indication that *Strep. sanguis* may be one of the important organisms in early colonization of the tooth (Carlsson, 1965), and this species has been shown to have the ability to adhere to tooth enamel (van Houte et al., 1970).

Several reports in recent years have given an overall picture of the microbial composition of established or 'mature' dental plaque, and the results of 5 such studies are summarized in Table 5. It can be seen that there appears to be an

Table 5
The predominant cultivable flora of human dental plaque

Organism	Results from surveys*				
	1	2	3	4	5
Streptococci	17.8†	27	37.8	22.9	16.5
Gram + rods and filaments	22.5	41	35.3	42.1	51.8
Actinomyces spp.	27.1	–	14.2	35.5	36.0
Lactobacilli	4.2	–	–	–	–
Neisseria spp.	ND	0.4	ND	1.48	1.7
Veillonella spp.	28.1	6	6	13.07	<0.1
Gram – anaerobic rods	ND	10	16.9	7.79	10.3
Fusobacteria	ND	4	6.8	0.39	1.5
No. of samples yielding above micro-organisms	6	5	11	59	6

* *Key to different surveys*
(1) Loesche & Syed (1973); (2) Gibbons, Socransky et al. (1964); (3) Loesche, Hocket & Syed (1972); (4) Bowden & Hardie (1973); (5) Howell et al. (1965).
† Figures expressed as percentage of total viable count. ND, none detected; –, not reported

overall similarity between the figures obtained in each investigation, although some differences are apparent. The 2 most numerous groups of bacteria in all the quoted studies are streptococci and Gram positive rods and filaments, most of which are probably *Actinomyces* species. *Veillonella* spp. and Gram negative anaerobic rods (presumptive *Bacteroides* species) also represent a significant proportion of flora in most cases.

Figures such as those in Table 5 are obtained by averaging the results of several observations, often made upon samples from different tooth surface areas, and whilst they give a general view of the composition of dental plaque, individual variations are obscured. In Table 6 the results of a recent unpublished study carried out in this laboratory are summarized, and it can be seen that a wide range of percentage counts for each group of bacteria was observed. Streptococci and *Actinomyces* spp. formed the largest groups and were isolated from every sample, but none of the other types of bacteria was recovered in every case. This particular study was concerned exclusively with plaque samples obtained from the distal surfaces of the upper first premolar teeth of ten 14-year old children, using a special sampling device (Hardie et al., 1972). The rather high proportions of *Actinomyces* spp. recovered from these specimens may be due to a peculiarity of the site examined, or, alternatively, may be a reflection of the counting methods employed.

Table 6
The microbial composition of approximal dental plaque

Organism	Mean % of viable count	Range
Genera		
Streptococci	22.9	0.4-70.0
Gram + rods (mainly Actinomyces)	42.07	4.0-81.0
Gram − rods (mainly Bacteroides)	7.79	0-66.0
Neisseria	1.48	0-44.0
Veillonella	13.07	0-59.0
Fusobacterium	0.39	0- 5.4
Individual Species		
Streptococcus mutans	2.17	0-23.0
Strep. sanguis	5.9	0-64.0
Strep. salivarius	0.67	0-33.0
Strep. milleri	0.51	0- 7.0
Actinomyces israeli	16.5	0-78.0
A. viscosus/naeslundi	19.05	0-74.0

Selected organisms expressed as percentage of total viable count. Samples obtained from distal surfaces of upper first premolar teeth. Results of 59 samples from ten 14-year old children.

Site variations in the plaque flora from different tooth surfaces are known to exist (Handleman & Hess, 1969), although up to now these have not been investigated in great detail. In a small pilot study we have examined the composition of plaque collected from 3 adjacent sites, c. 2 mm apart, on freshly extracted teeth. These sites were; (A) the contact area, (B) the gingival crevice area (2 mm below the contact area) and (C) the buccal surface. The results obtained from 3 teeth studied in this way are shown in Table 7.

It is quite clear from this table that wide variations exist in the relative proportions of some bacteria isolated from closely adjacent sites on the same tooth. Such findings substantiate the opinion, based on microscopic examination of sections, that dental plaque is made up of a heterogeneous collection of bacteria which are often arranged in discrete micro-colonies. The variations in microbial composition of plaque in different sites may explain why certain areas

Table 7

Variations of cultivable plaque flora on 3 sites of the same tooth

Subject	Site	Percentage of total viable count of				
		Streptococci	Actinomyces	Bacteroides	Fusobacterium	Veillonella
1	A	11.4	0.04	8.5	0.85	11.4
	B	65.5	5.7	9.0	11.4	0
	C	0.4	0.56	0.08	0.26	73.0
	Mean	25.7	2.1	5.9	4.2	28.1
2	A	8.6	85.3	0.65	0	1.7
	B	71.5	18.7	8.0	0.93	0
	C	14.3	56.6	0	0	16.6
	Mean	31.2	53.5	2.9	0.3	6.1
3	A	30.0	14.0	2.0	0	23.0
	B	65.0	7.0	25.0	3.0	14.0
	C	64.0	25.0	0	0.5	10.0
	Mean	53.0	15.3	9.0	1.2	15.6
Mean for all sites		36.6	23.6	5.9	1.9	16.6

Site A, contact area; Site B, gingival crevice below contact; Site C, buccal surface

of enamel develop carious lesions, whilst other parts of the tooth are unaffected.

Many of the reported bacteriological studies on dental plaque have been based upon the examination of fairly large, often pooled, samples. It would seem from the preceding discussion that such studies, whilst giving a general impression of the plaque microflora, completely obscure small, and possibly important, site variations. It is hoped that more precise sampling techniques will be employed in future investigations, in order to obtain specimens from reasonably well-defined areas.

Cultural studies give no information about the organization and spatial relationships of micro-organisms in dental plaque. Several investigators have reported the structure of plaque as it appears in the light, or electron, microscope, but such pictures give little information about the identity of the structures seen. The fluorescent antibody technique appears at present to offer the best chance of identifying bacteria *in situ* within plaque, although few investigations using this approach have been published (Ritz, 1969). The majority of studies have been made on smears of plaque material (Snyder *et al.,* 1967; Slack *et al.,* 1971; Collins *et al.,* 1973). Technical difficulties still exist both with regard to suitable preparation of plaque material for fluorescent microscopy and the development of a wide enough range of specific antisera, but it is hoped that, in the future, immunofluorescence will provide much useful information on the distribution of micro-organisms within plaque and their relationship to tooth enamel and periodontal tissues.

Electron microscopic examination of plaque has revealed the presence of strange 'corn cob' arrangements of bacteria (Jones, 1972), which seem to correspond morphologically to the organism known as *Leptothrix racemosa* (Nolte, 1973). A recent study by Listgarten *et al.* (1973) indicates quite clearly that these 'corn cobs' consist of 2 distinct morphological types of bacteria, a central filament surrounded by cocci, although the identity of these species remains to be established.

The type of dental plaque which accumulates in occlusal fissures has not received the same attention as the other tooth surfaces, mainly because of the technical difficulties involved in sampling such inaccessible areas. Löe *et al.* (1973) have recently described an elegant method of implanting artificial Mylar film fissures in the mouth, and Theilade *et al.* (1973) have reported the microbial composition of material accumulating in such fissures. The bacteria isolated consisted mainly of streptococci (77-89%), with few lactobacilli. Spirochaetes were absent, and few filaments and fusiforms were observed, indicating that the flora was different from that found on other tooth surfaces.

Few investigators have conducted longitudinal studies on dental plaque over a prolonged period of time although such studies would seem to be highly relevant to establishing the relationship between the microflora and dental disease. Early studies of this type on the aetiology of caries were reported by Hemmens *et al.* (1946) and Harrison (1948), and similar, but short-term, experiments have been made on gingivitis (Theilade *et al.*, 1966). Longitudinal studies on the flora of plaque and its association with the onset of dental caries are currently in progress, but it will take several years of work before conclusions can be drawn.

(d) *Gingival crevice*

The bacterial population of the gingival crevice region, which is essentially a particular area of dental plaque, and its relationship to periodontal disease has been reviewed by Socransky (1970) and Carlsson (1971a). Total microscopic counts of gingival debris have been shown to average 1.7×10^{11} micro-organisms/g wet weight, whilst total viable counts yielded 1.6×10^{10}/g aerobically and 4×10^{10}/g anaerobically (Socransky *et al.*, 1963). The approximate bacterial composition of gingival plaque (Gibbons *et al.*, 1963; Socransky, 1970) and material from periodontal pockets (Dwyer & Socransky, 1968) are shown in Table 8.

The data reported by Gibbons and Socransky's group was obtained by examination of material from adult subjects, 20 with normal gingivae and 20 with periodontal disease. De Aranjo & MacDonald (1964) made a similar investigation on the gingival crevice flora of 5 children, aged 3-7 years. Forty-six per cent of the isolates examined were Gram positive, facultative or anaerobic

Table 8
Bacteria isolated from the human gingival crevice region

Type of bacteria	Percentage of total cultivable flora	
	Socransky, 1970	Dwyer & Socransky, 1968
Gram positive facultative cocci	28.8	36.0
Gram positive anaerobic cocci	7.4	15.1
Gram positive anaerobic rods	20.2	4.4
Gram positive facultative rods	15.3	12.2
Gram negative anaerobic rods	16.1	19.5
Gram negative facultative rods	1.2	0.9
Gram negative anaerobic cocci	10.7	9.0
Gram negative facultative cocci	0.4	2.9
Spiral forms	1-3	not recorded

(After Socransky, 1970 and Dwyer & Socransky, 1968)

rods, 16% streptococci, 16% *Veillonella,* and the remainder were described as *Peptostreptococcus, Neisseria* spp. and Gram negative facultative rods. Although the overall picture appeared to resemble that described for adults, *Bacteroides melaninogenicus* and spirochaetes were observed less regularly in the samples from children.

Bailit *et al.* (1964) studied the prevalence of *B. melaninogenicus* in the gingival crevice of 320 children, ranging from 5-15 years old. Few children yielded positive cultures before the age of 6, at which age 17 of 40 subjects harboured the organism. There was general increase in prevalence with age, especially noticeable between 7-9 years, and by the age of 13 *c.* 100% of cultures were positive for *B. melaninogenicus.* Kelstrup (1966) also noted an increased prevalence of *B. melaninogenicus* with age, although he found lower levels of this organism in the younger age groups than that reported by Bailit *et al.* (1964).

From the relatively few detailed studies on the gingival crevice microflora, it appears that anaerobic bacteria constitute a large and important part. Some of the nutritional factors and bacterial inter-relationships which are thought to be of relevance to the ecology of this particular area have been discussed by Loesche (1968). *B. melaninogenicus* requires both haemin (Evans, 1951) and Vitamin K (Lev, 1958, Gibbons & MacDonald, 1960) for growth, and these are provided in the gingival region partly by blood and inflammatory exudate from the host tissue, and partly by the synthesis of Vitamin K by other micro-organisms present. More recently it has been shown that succinate may act as a growth factor for some strains of *B. melaninogenicus* (Lev *et al.*, 1971).

Amino acids derived from host tissues or dietary constituents may be utilized for energy production by some bacteria in the gingival crevice, including *B.*

melaninogenicus (Wahren & Gibbons, 1970) and *Fusobacterium nucleatum* (Jackins & Barker, 1951; Loesche & Gibbons, 1968).

Spirochaetes are especially numerous in gingival crevice of periodontal pocket and are particularly fastidious in their growth requirements. In addition to a suitably low E_h, of the order of -185 mV or lower, several growth factors have been shown to be required. *Treponema microdentium* requires spermine, spermidine or putrescine, which may be produced by some 'diphtheroids', and isobutyrate which is produced by fusobacteria (Socransky et al., 1964). *Treponema denticola*, which is also found in the gingival crevice, has been shown to require α-2 globulin (Socransky & Hubersak, 1967) and this is provided by gingival fluid.

Vibrio sputorum requires hydrogen or formate as an energy source (Loesche, 1968), which may be provided by *Fusobacterium* or *Bacteroides* spp. (Loesche & Gibbons, 1965).

As pointed out by Carlsson (1971a), our understanding of the microbiology of the gingival crevice region is far from complete. Many of the micro-organisms which can be isolated by routine anaerobic techniques are difficult to characterize or identify with any degree of certainty at present, and it is likely that improving technology will allow the recovery of even more types which will confuse the issue still further.

5. The Oral Flora of Primitive Man

Most reported studies on the oral microflora have originated either from European countries or the U.S.A. and there is little information available concerning geographical or racial variations. A few studies have been reported on the oral flora of primitive peoples, but some of these, especially the older ones, have been based largely on the examination of stained smears. Clement (1957) reviewed much of the older literature on this subject and concluded that, in general terms, the oral flora of 'caries-immune' primitive mouths appeared to resemble that of the 'caries-susceptible' European mouth. Much interest has been attracted to the various observations that primitive tribes appear to be almost immune to dental caries, the probable explanation being the nature of their diets which usually lack refined carbohydrates.

The early study by Pickerill & Champtaloup (1913) on Maori children and Guatemalan Mayan Indians (Morhart et al., 1970) indicated a preponderance of Gram negative bacteria, and Clement (1957) regarded the presence of aciduric Gram negative rods as particularly noteworthy. Other primitive populations investigated included Eskimos (Waugh & Waugh, 1936). Aborigines (Featherstone, 1958, 1959, 1960) and Kalahari Desert Bushmen (Clement & Rae, 1953; Clement et al., 1956). The relevance of these studies in relation to the effect of foods on oral bacterial populations have been discussed by Bowen (1970).

We have recently commenced a small study on the dental plaque microflora of primitive peoples in Papua, New Guinea, in association with Dr. R. G. Schamschula and Dr. D. E. Barmes.*

The dental caries experience of these populations have been studied previously by Barmes (1969), and some aspects of their oral flora (Schamschula & Barmes 1970a, b) have also been reported. Preliminary investigations indicate that there may be much higher numbers of facultative, Gram negative rods in these samples than are usually isolated from the mouths of Western people. However, much further work is required before detailed results can be published.

6. Some Ecological Determinants

(a) *Anaerobiosis*

The numerous reports on the microflora of dental plaque and gingival crevice material indicate that a high proportion of the micro-organisms present are anaerobic. In the studies of Socransky *et al.* (1963), using conventional anaerobic techniques, only 20-25% of the bacteria which were counted microscopically could be cultivated. More rigorous anaerobic techniques have been described, such as the Hungate roll-tube method (Hungate, 1950; 1966) and the anaerobic chamber (Drasar, 1967; Leach *et al.*, 1971), and the use of such methods greatly increases the recovery of viable bacteria from oral samples (Aranki *et al.*, 1969), Gordon *et al.*, 1971). By using the Hungate technique, 70.4% of the total microscopic count could be cultivated from 8 samples of gingival material (Gordon *et al.*, 1971). Aranki *et al.* (1969) investigated the relative recovery of anaerobes from replicate samples of gingival crevice material, using conventional methods and a modified anaerobic glove box technique. The latter technique increased the total counts obtained between 2.8 to 4.2 times compared to conventional methods, and the percentage of anaerobes recovered was increased 3.4- to 19.6-fold. The authors suggest that the critical stage in the technique is the exposure of samples to oxygen during manipulative stages such as dispersion, dilution and plating, rather than the actual incubation period.

Very few studies on the flora of the mouth have been conducted under ideal conditions, with minimal exposure of the samples to oxygen. One of the difficulties encountered in experimental work is preservation of fastidious anaerobes during the period between collection of oral samples and arrival in the laboratory, in order to prevent exposure to oxygen. Several transport media have been described (Møller, 1966; Gastrin *et al.*, 1968) and Syed & Loesche (1972) have recently recommended a Reduced Transport Fluid (for composition, see Table 9). The preservation of oral streptococci in this transport medium has been evaluated by Rundell *et al.* (1973).

* This work is receiving financial support from the World Health Organization.

Table 9
Composition of Reduced Transport Fluid (RTF)

Constituent	Volume required (ml)
Stock mineral solution 1	75
Stock mineral solution 2	75
0.1 M EDTA	10
8% Na_2CO_3	5
1% dithiothreitol (freshly prepared)	20
0.1% rezazurin (optional)	1
Distilled water	814

Stock mineral solution 1 comprises K_2HPO_4, 0.6%. Stock mineral solution 2 consists of (%): NaCl, 1.2; $(NH_4)_2SO_4$, 1.2; $KHPO_4$, 0.6; $MgSO_4$, 0.25. (Syed & Loesche, 1972)

In our own studies, a transport medium containing 2% of bovine serum and 0.1% of cysteine hydrochloride has been found suitable for the recovery of anaerobic bacteria (Bowden & Hardie, 1971) but because no growth inhibiting agents are added this may not be ideal for prolonged storage of samples.

The oxygen sensitivity of various anaerobic bacteria was investigated by Loesche (1969), who classified them into 2 groups. Those strains designated strict anaerobes were incapable of surface growth on agar at pO_2 levels $>0.5\%$. Such strains included *Treponema macrodentium, T. denticola, T. oralis, Selenomonas ruminatium, Butyrivibrio fibrisolvens, Succinivibrio dextrinosolvens* and *Lachnospira multiparus*. Moderate anaerobes were those which could grow in the presence of 2-8% of oxygen and survived exposure to atmospheric oxygen for periods of 60-90 min. *Bacteroides fragilis, B. melaninogenicus, B. oralis, Fusobacterium nucleatum, Clostridium novyi* type A and *Peptostreptococcus elsdenii* were included in the moderately anaerobic group. Strains of *Vibrio sputorum* and *V. fetus* were also tested, but since these grew better in the presence of low levels of oxygen they were regarded as microaerophilic.

The observed variations in the distribution of bacterial species within the mouth may be due partly to differences in oxygen tension or E_h in different locations. Eskow & Loesche (1971) have demonstrated that the oxygen tension on the anterior surface of the tongue is 16.4%, the posterior surface of the tongue 12.4%, the maxillary buccal fold 0.4% and the mandibular buccal fold 0.3%. The oxidation-reduction (O/R) potentials of developing plaque, periodontal pockets and healthy gingival sulci have been investigated by Kenney & Ash (1969). The mean O/R potential for 10 periodontal pockets was −47.6 mV (range +12 to −57), whereas healthy gingival sulci from the same subjects gave a mean reading of +72.6 mV (range +10 to +113). Healthy sulci from a control group of subjects gave a comparable figure of 74.4 mV. The O/R potentials of the periodontal pockets were significantly different from those of the healthy gingival sulci and appeared to provide conditions suitable for the growth of anaerobes. In the same study, the E_h of developing plaque in 2 subjects showed a steady decrease from an initial level of over +200 mV to a level of −112 or −141 mV after 4 or 7 days development. This fall in E_h was accompanied by the typical increase in the complexity of the plaque flora that has been reported by many other workers.

(b) *Adherence and aggregation*

Recently there has been considerable emphasis on relative ability of oral micro-organisms to adhere to different surfaces, and the research group at the Forsyth Dental Centre in Boston have published several papers on this interesting aspect of oral ecology. It has been established by Carlsson (1968b) that *Strep. sanguis* is one of the organisms which preferentially colonizes the tooth surface, and this finding seems to be substantiated by the observation that this species adheres particularly well to teeth (van Houte *et al.*, 1970; 1971). *Strep. salivarius,* on the other hand, appears to have a marked affinity for epithelial surfaces but shows little adherence to the tooth surface (Gibbons & van Houte, 1971). There is evidence to suggest that the phenomenon of adherence may be due in part to the possession of an outer, trypsin-sensitive 'fuzzy coat' which can be observed in several species of streptococci with the electron microscope (Ellen & Gibbons, 1972; Gibbons *et al.*, 1972; Liljemark *et al.*, 1972).

The ability of other bacteria to adhere to oral surfaces, including *Veillonella* and *Neisseria* spp. (Liljemark *et al.*, 1971) and lactobacilli (van Houte *et al.*, 1972) has also been studied. Much of the information from these various studies has been summarized by Gibbons (1972) and some of the salient points are shown in Table 10.

Another phenomenon which may influence the colonization of oral surfaces is that of salivary-induced aggregation of bacteria (Gibbons & Spinell, 1970). Work is in progress in several laboratories to elucidate the chemical nature of the active fraction in saliva which produces this effect. Inhibition of adherence of *Strep. mutans* by antibody has been investigated by Olsen, Bleiweis & Small (1972).

(c) *Microbial inter-relationships*

Few studies have been reported concerning the interaction of micro-organisms in the mouth. Carlsson (1970a, b; 1971b; 1972) has studied the individual growth requirements of *Strep. salivarius, Strep. sanguis* and *Strep. mutans,* and subsequently carried out experiments on the growth of the latter 2 species in mixed culture (Carlsson, 1971c). Under anaerobic conditions a strain of *Strep. mutans* requires *p*-aminobenzoic acid for growth, whereas a strain of *Strep. sanguis* did not require it. In a mixed culture system, using media free from

Table 10
Ability of bacteria to adhere to oral surfaces as related to their proportions found indigenously

	Proportions found indigenously			Experimentally observed adherence		
	Tooth	Tongue	Cheek	Tooth	Tongue	Cheek
Streptococcus mutans	low to high	low	low	low to high	low	low
Strep. sanguis	high	moderate	moderate	high	moderate	moderate
Strep. salivarius	low	high	moderate	low	high	moderate
Neisseria spp.	low	low	low	low	low	low
Veillonella spp.	low	high	low	low	high	low

(After Gibbons, 1972)

p-aminobenzoic acid, the presence of *Strep. sanguis* supported the growth of the *Strep. mutans* strain, indicating that the missing growth factor was being provided. Carlsson suggests that this relationship may be regarded as a type of parasitism of one strain upon the other. Further work upon the metabolism of these streptococci has been reported more recently by Yamada & Carlsson (1973).

The production of extracellular polysaccharides of the glucan and fructan type by many of the streptococci found in the mouth has been well documented (Gibbons & Nygaard, 1968), and no attempt will be made here to review all the literature on this subject. It is thought that these polymers, which are mainly synthesized from sucrose, contribute significantly to the matrix of dental plaque and may be important in the initial colonization of the tooth surface. Parker & Creamer (1971) have studied the utilization of some extracellular polymers by other bacteria. A streptococcal levan and an amylopectin produced by a *Neisseria* sp. were utilized more readily by some strains of streptococci and lactobacilli than was sucrose. Levans in particular seem to be labile in dental plaque and may be rapidly hydrolysed by other bacteria (de Costa & Gibbons, 1968; van Houte & Jansen 1968; Leach *et al.*, 1972). Little information is available concerning the breakdown and utilization of extracellular glucose polymers in plaque. Interactions between extracellular polysaccharides from *Strep. mutans* have been investigated by Kelstrup & Funder-Nielsen (1972).

The utilization of lactate by *Veillonella* spp. (Rogosa *et al.*, 1965) has attracted some attention, since this may theoretically reduce the damaging effect of lactic acid produced by bacteria adjacent to the tooth. Some experimental evidence in support of this has been provided recently by Mikx *et al.* (1972). In experimental caries studies on gnotobiotic animals, oral inoculation with mixtures of *Strep. mutans* or *Strep. sanguis* together with veillonella produced less caries than mono infection with either of the streptococcal species alone.

It is likely that there are many other examples of utilization of metabolic products by other bacteria in the mouth, although few have been documented. The importance of nutrition and microbial inter-dependence in the gingival crevice area has been reviewed by Loesche (1968), and some of the relevant factors have been discussed in the section relating to the flora of the gingival region.

In addition to the utilization of metabolic products, there is also the possibility that some of these products may inhibit the growth of other micro-organisms. It has long been considered that the normal commensal flora of the body in some way helps to prevent establishment and colonization by extrinsic organisms. Some evidence also exists for the inhibition of normal residents by other members of indigenous oral flora. For instance, Parker (1970) has shown in paired culture experiments that some combinations of oral bacteria have a stimulatory effect upon growth, whilst other result in growth inhibition.

Production of bacteriocins by oral streptococci has been demonstrated by several workers, including Green & Dodd (1956), Kelstrup & Gibbons (1969), Kelstrup *et al.* (1970), Donoghue (1972), Rogers (1972) and Schlegel & Slade (1972). Holmberg & Hallander (1972) showed interference *in vitro* between Gram positive bacteria isolated from dental plaque. *Strep. sanguis* in particular produced a broad spectrum of inhibitory action on other streptococci, lactobacilli, corynebacteria and actinomycetes. In subsequent investigation (Holmberg & Hallander, 1937a), the bactericidal action of *Strep. sanguis* was found to be caused by the production of hydrogen peroxide. This was produced under aerobic growth conditions and was also responsible for the observed α-haemolysis of this species.

The demonstration *in vitro* of any microbial interaction, whether it be stimulatory or inhibitory, is interesting but it is difficult to extrapolate the results of such experiments to the situation *in vivo*. At the present time it is probably fair to say that we understand very little about the effect of microbial products on the regulation of the microbial flora of different areas in the mouth.

7. The Principal Groups of Micro-organisms found in the Mouth

It is clear from the large amount of literature on various aspects of the microbial ecology of the human mouth that the flora is extremely complex, and in many instances detailed studies have been hampered by a relative lack of information on the taxonomy and nomenclature of bacteria isolated. Lack of precision in identifying isolates, which unfortunately is common, often makes comparisons between different reports difficult.

It is not intended in this section to give a comprehensive review of the isolation, characterization and classification of oral micro-organisms. However, in order to provide a starting point for those unfamiliar with this aspect of microbiology, a brief summary of some of the main groups of bacteria commonly isolated from the mouth will be given, together with a few references.

(a) *Gram positive cocci*

(i) *Streptococci*

The streptococci have probably received more attention than any other group of micro-organisms found in the mouth. β-Haemolytic species, such as *Strep. pyogenes,* are relatively uncommon in healthy individuals although they can be isolated from the saliva of people suffering from streptococcal sore throats (Ross, 1971). Enterococci are occasionally isolated from the normal mouth, but usually do not constitute a large proportion of the streptococcal flora.

As discussed previously, streptococci represent a large part of the normal oral flora, the actual percentage varying from one site to another. Most species found are either of the 'viridans' type or are non-haemolytic. The classification of greening streptococci is notoriously difficult, although recent taxonomic studies by Colman (1968), Colman & Williams (1965, 1972), Carlsson (1968a),

Guggenheim (1968), Rogers (1969) and Drucker & Melville (1971) have helped to clarify the situation. In our laboratory we employ a system for identification which is based largely on the work of Colman & Williams (1972). By using a simple series of 7 physiological tests (Table 11), the following species can be recognized.

Colman & Williams (1972)	Carlsson's Groups (1968a)
Strep. mutans	II
Strep. sanguis	I : B
Strep. mitior (or Strep. sanguis/2)	I : A
Strep. salivarius	III
Strep. milleri	?

In recent investigation (Hardie & Bowden, unpublished data), c. 90% of over 400 freshly isolated streptococci from dental plaque could be identified as one of these 5 species. A few of the remaining 10% of strains were enterococci, and the remainder are, as yet, unidentified.

Streptococcus mutans (Clark, 1924) has been extensively studied in recent years, because it is highly effective in producing dental caries in experimental animals (Fitzgerald, 1968) and may be an important aetiological agent in human caries (Krasse *et al.*, 1968; de Stoppelaar *et al.*, 1969; Littleton *et al.*, 1970; Ikeda *et al.*, 1973). The physiological and colonial characteristics of this species have been described by Edwardsson (1968, 1970), and Bratthall (1970) has shown that there are at least 5 serological types, designated *a-e*. The subdivision of *Strep. mutans* into different types is also supported by DNA base ratio and hybridization studies (Coykendall, 1970; 1971) and by cell wall analysis (Bleiweis *et al.*, 1971; Hardie & Bowden, in press). Recent work on the antigenic structure of *Strep. mutans* has been reported by Mukasa & Slade (1973) and van der Rijn & Bleiweis (1973).

Streptococcus mutans produces both soluble and insoluble extracellular glucose polymers from sucrose, and these have been examined in detail by several workers (Gibbons & Banghart, 1967; Guggenheim & Schroeder, 1967; Gibbons & Nygaard, 1968; Guggenheim, 1970 *a, b*). It is thought that such polymers produced by oral streptococci constitute a significant part of the matrix of dental plaque (Critchley *et al.*, 1967; Wood 1967, 1969; Fitzgerald & Jordan, 1968).

Streptococcus sanguis (White & Niven, 1946) and some strains of *Strep. mitior* (Schottmüller, 1903) (or *Strep. sanguis* type 2) also characteristically produce extracellular polyglucans from sucrose. These streptococci are common in dental plaque, and are sometimes found to be the causative organisms in cases of bacterial endocarditis. The serology and antigenic structure of *Strep. sanguis* is not entirely clear, especially the relationship to Lancefield Group H (Washburn *et al.*, 1946; Dodd, 1949; Porterfield, 1950; Farmer, 1954), and this rather complex situation has been reviewed recently by Colman (1970).

Table 11
Simple scheme for identifying the most numerous types of oral streptococci

	Strep. mutans	Strep. sanguis (type 1)	Strep. mitior (or Strep. sanguis type 2)	Strep. salivarius	Strep. milleri
Fermentation of:					
mannitol	+	−	−	−	−
sorbitol	+ or − (1)	−	−	−	−
Production of:					
ammonia from arginine	− (2)	+	−	−	+
dextran from sucrose	+	+	+ (4)	− (5)	−
hydrogen peroxide	− (3)	+	+	−	−
acetoin from glucose	+	−	−	v	+
Hydrolysis of aesculin	+	+	−	v	+

+, usually positive; −, usually negative; v, variable reactions
(1) Sorbitol negative strains usually belong to serotype *a* or *d*; (2) serotype *b* strains are arginine positive; (3) some strains are positive, usually serotype *a* or *d*; (4) some strains negative; (5) *Strep. salivarius* strains produce levan from sucrose.

Streptococcus salivarius strains produce extracellular polyfructan (levan) from sucrose, and this results in the appearance of large, characteristically mucoid colonies when they are grown on sucrose-containing media. If the typical colony morphology is not shown, identification of this species can be difficult because negative or variable results are obtained in many of the physiological tests which are useful for distinguishing other oral streptococci. Williams (1956) has shown that some strains carry the Lancefield Group K antigen. The levans produced by *Strep. salivarius* have been shown to be highly labile and may be rapidly utilized by other bacteria (De Costa & Gibbons, 1968; Manly & Richardson, 1968; Leach et al., 1972).

Streptococcus milleri (Guthof, 1956) is the name preferred by Colman & Williams (1972) for a group of serologically heterogeneous streptococci which includes strains containing the Lancefield A, C, F and G antigens. Isolates which appear to belong to this species are present in dental plaque, although no strains of this type were reported in Carlsson's study (1968a). The validity of the species remains to be established by further studies.

In many of the studies reported in the dental literature, identification of oral streptococci has been based largely on colony morphology, often on sucrose-containing media such as *mitis-salivarius* agar. Since each of the common species can produce several colony types (Thomson, 1971), this approach can be very misleading and diagnosis should always be supported by supplementary tests.

As more details become available about the antigenic structure of various oral streptococci, rapid diagnosis should be possible by immunofluorescence. This technique has already been used for the detection of different serotypes of *Strep. mutans* in oral specimens (Zinner & Jablon, 1968; Bratthall, 1972a; Grenier et al., 1973), and Bratthall (1972b) has used it to investigate the distribution of these types in different parts of the world.

(ii) *Staphylococci and micrococci*

Catalase positive, Gram positive cocci are quite frequently isolated from the mouth, although they are not usually present in high numbers. The properties of a number of oral isolates have been reported by Morris (1954b) and Pike et al., (1962) and the literature reviewed by Burnett & Scherp (1968). As mentioned previously, *Micrococcus mucilagenosus (Staph. salivarius)* appears to be a characteristic inhabitant of the dorsum of the tongue.

(b) *Gram negative cocci*

Small, anaerobic Gram negative cocci belonging to the genus *Veillonella* are common in the mouth, often in relatively high numbers. These organisms have been fully described (Rogosa, 1964, 1965; Rogosa & Bishop 1964a, b,) and provide few difficulties for isolation or identification. The 2 species recognized are *V. parvula* and *V. alkalescens*.

Aerobic or facultative, oxidase positive, Gram negative cocci are also common in the mouth and may constitute a significant proportion of the organisms which initially colonize clean enamel surfaces (Ritz, 1970). Colonies of many of these presumptive *Neisseria* isolates are sticky probably because of the presence of extracellular slime or capsular material. The properties of a number of *Neisseria* isolates have been reported by Morris (1954b), and according to Pike et al. (1962), most strains from the mouth can be classified either as *N. pharyngis* or *N. catarrhalis*. However, it is often difficult to identify oral neisseriae with any degree of confidence and further taxonomic studies in this area would be extremely useful.

(c) *Gram positive rods and filaments*

Numerous investigators have shown that the bacterial flora of the mouth includes large numbers of pleomorphic Gram positive rods and filaments. Often strains from this heterogeneous group of organisms are described simply as 'diphtheroids' and these may include aerobic, facultative and anaerobic species. Identification and taxonomy of these isolates is often difficult, although in many cases it is possible to arrive at a diagnosis if sufficient care is taken (Rasmussen et al., 1966).

Many of the Gram positive rods isolated from the mouth are *Actinomyces species*, including *A. israeli, A. naeslundi, A. viscosus* and *A. odontolyticus*. The extensive literature on this genus has been reviewed elsewhere and will not be repeated again here (Pine, 1970; Bowden & Hardie 1971, 1973). Mycobacteria may also be isolated from oral specimens (Mills 1972), but few workers have reported their presence.

Aerobic or facultative filaments include *Rothia dentocariosa* (known formerly as *Nocardia salivae*) and *Bacterionema matruchotii* (previously *Leptotrichia dentium*). Both species are commonly isolated from oral samples although they usually represent a low proportion of the total viable count. A numerical taxonomic study on these organisms has recently been published (Holmberg & Hallander, 1973b) and their classification is also discussed by Cross & Goodfellow (1973).

Little useful information can be given concerning the occurrence of true Corynebacteria in the mouth. From time to time presumptive *Corynebacterium* species are isolated but the classification of these strains is far from clear. Spore-bearing rods do not appear to be a major component of the human oral flora, although van Reenan & Coogan (1970) have reported the isolation of clostridia from the mouth. *Bacillus* spp. are occasionally found, but are usually considered to be transient inhabitants.

Lactobacilli in the mouth have attracted much attention in the past, although their numbers are often low. It appears that the presence of large, unrestored carious cavities increases the lactobacillus count considerably. Both homofermentative and heterofermentative species may be isolated, and their identification can often be achieved by standard physiological tests (Sharpe et al., 1966; Hansen, 1968). However, other biochemical techniques are of value in many cases (Rogosa, 1970; Williams, 1971; Williams, in press).

Other anaerobic Gram positive rods and filaments isolated include *Leptotrichia buccalis, Propionibacterium, Arachnia, Bifidobacterium,* and *Eubacterium* spp. and some of the literature in these genera has been described previously (Bowden & Hardie, 1971). By using anaerobic jars containing 95% of $H_2^- + 5\%$ of CO_2, together with pre-reduced media, no great difficulty should be experienced in isolating these bacteria. Occasionally, rather unusual anaerobes can be isolated which are difficult to identify with certainty and it is possible that some of these are *Eubacterium* spp. The Laboratory Manual produced by the Anaerobic Laboratory, at Virginia Polytechnic Institute (Holdeman & Moore, 1972) is extremely useful when attempting to characterize and identify anaerobic isolates.

(d) *Gram negative rods and filaments*

The presence of aerobic and facultative rods has been reported in several studies on the oral flora, although coliforms do not appear to be regular or predominant inhabitants of the mouth in those Western communities that have been investigated thoroughly. Most genera have been isolated at one time or another, but in the majority of instances these are probably transient inhabitants. Sutter *et al.* (1966) have reported the presence of *Pseudomonas* spp. (mainly *Ps. aeruginosa*) in *c.* 8% of saliva samples from 350 individuals.

One genus which has been overlooked by many workers is *Haemophilus* although occasional isolations were reported by Morris (1954*d*). Sims (1970) has shown that haemophili are frequently present in saliva and on mucosal surfaces, and they may also be isolated from dental plaque. In our own studies on dental plaque a few haemophili have occasionally been isolated although no particular effort has been made to look for them. It is likely that the use of a suitable selective medium, such as chocolate agar containing bacitracin (10 units/ml) and cloxacillin (5 µg/ml), would increase the isolation rate. Sims (1970) found the mean salivary level, based on 100 specimens, to be 31.8×10^2 haemophili/ml. Most of the isolates (92.8%) were V-factor dependent, whilst the remaining strains required V and X factors.

The isolation and properties of anaerobic Gram negative rods and filaments has been reviewed previously (Bowden & Hardie, 1971). *Fusobacterium* and *Bacteroides* spp. are isolated regularly from dental plaque, especially from the gingival crevice region. The taxonomy of *Bacteroides* is in a particularly confused state at present although several taxonomic studies and identification schemes have been published (MacDonald, 1953; Sawyer *et al.*, 1962; Loesche *et al.*, 1964; Loesche & Gibbons, 1965; Suzuki *et al.*, 1966; Courant & Gibbons, 1967; Barnes & Goldberg, 1968; Speirs, 1971; Werner *et al.*, 1971 *a, b*). Unfortunately, even those *Bacteroides* strains which produce dark-brown or black pigmented colonies, usually referred to as *B. melaninogenicus,* are apparently heterogeneous in their properties and may not belong to a single species. Taxonomic studies on black pigment-producing isolates from the mouth are currently in progress in this laboratory.

(e) *Other micro-organisms*

This review has been confined almost exclusively to a consideration of the bacterial flora of the mouth. However, for the sake of completeness it should be mentioned that several other types of micro-organism occur regularly in the oral cavity.

Yeasts, especially *Candida albicans,* are commonly present in healthy individuals (Schmitt, 1971) and may give rise to several types of local inflammatory condition under certain circumstances (Lehner, 1966). The literature on this aspect has been reviewed by Burnett & Scherp (1968).

Mycoplasmas can be isolated from the human mouth and reports have been published by Morton *et al.* (1951), Dienes & Madoff (1953), Shklair *et al.* (1961), Razin *et al.* (1964), Taylor-Robinson *et al.* (1964), del Guidice & Carski (1968), Fox *et al.* (1969), Klein *et al.* (1969), Engel & Kenny (1970) and Lyon & Nemes (1971).

Protozoa, such as *Entamoeba gingivalis, Trichomonas* and *Selenomonas* spp., are often seen during microscopic examination of oral specimens and may be cultivated by appropriate techniques. Further information may be obtained from the work of Wantland *et al.* (1963).

Spirochaetes are common in the gingival crevice region and several types may be cultivated in the laboratory (Blake, 1968). Socransky *et al.* (1969) have described some morphological and biochemical criteria by which small oral spirochaetes may be differentiated into 3 distinct species, *Treponema denticola, T. macrodentium* and *T. oralis.* Few laboratories attempt to grow or identify oral spirochaetes routinely, diagnosis being based in most cases upon microscopy alone. It is possible that in future diagnosis may be facilitated by serology or by gas chromatographic techniques (Farshy *et al.*, 1970).

With the exception of *Herpesvirus hominis* (*Herpes simplex* virus) which is carried by many people, most viruses are regarded as transient when isolated from the mouth (Burnett & Scherp,1968). Elsewhere in this symposium, those viruses which are associated with healthy individuals, are discussed by Read & Tyrrell.

8. References

ARANKI, A., SYED, S. A., KENNEY, E. B. & FRETER, R. (1969). Isolation of anaerobic bacteria from human gingiva and mouse cecum by means of a simplified glove box procedure. *Appl. Microbiol.* **17,** 568.

BAILIT, H. L., BALDWIN, D. C. & HUNT, E. E., JR. (1964). Increasing prevalence of gingival *Bacteroides melaninogenicus* with age in children. *Arch. oral Biol.* **9,** 435.

BAIRD-PARKER, A. C. (1959). The classification of oral *Leptotrichia* and *Fusobacterium* species. Ph. D. Thesis, University of Birmingham.

BARMES, D. E. (1969). Caries etiology in Sepik villages: trace element micronutrient and macronutrient content of soil and food. *Caries Res.* **3,** 44.

BARNES, E. M. & GOLDBERG, H. S. (1968). The relationships of bacteria within the family Bacteroidaceae as shown by numerical taxonomy. *J. gen. Microbiol.* **51,** 313.

BERGAN, T., BØVRE, K. & HOVIG, B. (1970). Reisolation of *Micrococcus mucilagenosus* Migula 1900. *Acta Path. et Microbiol. Scand.* **78B,** 85.

BIBBY, B. G. (1938). The bacterial flora in different parts of the mouth. *J. dent. Res.* **17,** 471.

BIBBY, B. G. (1970). Antibiotics and dental caries. In 'Dietary Chemicals vs. Dental Caries'. Advances in Chemistry Series 94. American Chemical Society: Washington, D.C.

BISSET, K. A. & DAVIS, G. H. G. (1960). *The Microbial Flora of the Mouth.* London, Heywood & Co. Ltd.

BLACK, G. V. (1898). Dr. Black's conclusions reviewed again. *Dent. Cosmos* **40,** 440.

BLAKE, G. C. (1968). The microbiology of acute ulcerative gingivitis with reference to the culture of oral trichomonads and spirochaetes. *Proc. R. Soc. Med.* **61,** 131.

BLEIWEIS, A. S., CRAIG, R. A., ZINNER, D. D. & JABLON, J. M. (1971). Chemical composition of purified cell walls of cariogenic streptococci. *Infect. and Immun.* **3,** 189.

BOWDEN, G. H. (1969). The components of the cell walls and extracellular slime of four strains of *Staphylococcus salivarius* isolated from human dental plaque. *Arch. oral Biol.* **14,** 685.

BOWDEN, G. H. & HARDIE, J. M. (1971). Anaerobic organisms from the human mouth. In *Isolation of Anaerobes.* Soc. Appl. Bact. Tech. Ser. No. 5. Eds. D. A. Shapton & R. G. Board. London: Academic Press.

BOWDEN, G. H. & HARDIE, J. M. (1973). Commensal and pathogenic *Actinomyces* species in man. In *Actinomycetales: Characteristics and Practical Importance.* Soc. Appl. Bact. Symp. Ser. No. 2. Eds. G. Sykes & F. A. Skinner. London: Academic Press.

BOWEN, W. H. (1970). Effects of foods on oral bacterial populations in man and animals. *J. dent. Res.* **49,** 1276.

BRAILOVSKY-LOUNKEVITCH, Z. A. (1915). Contribution à l'etude de la flore microbienne habituelle de la bouche normal (nouveau-nés, enfants, adultes). *Ann. Inst. Pasteur* **29,** 379.

BRATTHALL, D. (1970). Demonstration of five serological groups of streptococcal strains resembling *Streptococcus mutans. Odont. Revy* **21,** 143.

BRATTHALL, D. (1972*a*). Immunofluorescent identification of *Streptococcus mutans. Odont. Revy* **23,** 181.

BRATTHALL, D. (1972*b*). Demonstration of *Streptococcus mutans* strains in some selected areas of the world. *Odont. Revy* **23,** 401.

BURNETT, G. W. & SCHERP, H. W. (1968). Oral Microbiology and Infectious Disease, 3rd ed. Baltimore: Williams & Wilkins.

CARLSSON, J. (1965). Zooglea-forming streptococci, resembling *Streptococcus sanguis,* isolated from dental plaque in man. *Odont. Revy* **16,** 348.

CARLSSON, J. (1968*a*). A numerical taxonomic study of human oral streptococci. *Odont. Revy* **19,** 137.

CARLSSON, J. (1968*b*). Plaque formation and streptococcal colonization on teeth. *Odont. Revy* **19,** Suppl. 14.

CARLSSON, J. (1970a) Chemically defined medium for growth of *Streptococcus sanguis*. *Caries Res.* **4,** 297.
CARLSSON, J. (1970b). Nutritional requirements of *Streptococcus mutans*. *Caries Res.* **4,** 305.
CARLSSON, J. (1971a). Bacterial populations associated with the periodontium. In *The prevention of Periodontal Disease*, Eds J. E. Eastoe, D. C. A. Picton & A. G. Alexander. London: Henry Kimpton.
CARLSSON, J. (1971b). Nutritional requirements of *Streptococcus salivarius*. *J. Gen. Microbiol.* **67,** 69.
CARLSSON, J. (1971c). Growth of *Streptococcus mutans* and *Streptococcus sanguis* in mixed culture. *Arch. oral Biol.* **16,** 963.
CARLSSON, J. (1972). Nutritional requirements of *Streptococcus sanguis*. *Arch. oral Biol.* **17,** 1327.
CARLSSON, J. & EGELBERG, J. (1965). Effect of diet on early plaque formation in man. *Odont. Revy* **16,** 112.
CARLSSON, J. & SUNDSTRÖM, B. (1968). Variations in composition of early dental plaque following ingestion of sucrose and glucose. *Odont. Revy* **19,** 161.
CARLSSON, J., SÖDERHOLM, G. & ALMFELDT, I. (1969). Prevalence of *Streptococcus sanguis* and *Streptococcus mutans* in the mouth of persons wearing full-dentures. *Arch. oral Biol.* **14,** 243.
CARLSSON, J., GRAHNEN, H., JONSSON, G. & WIKNER, S. (1970a). Establishment of *Streptococcus sanguis* in the mouths of infants. *Arch. oral Biol.* **15,** 1143.
CARLSSON, J., GRAHNEN, H., JONSSON, G. & WIKNER, S. (1970b). Early establishment of *Streptococcus salivarius* in the mouths of infants. *J. dent. Res.* **49,** 415.
CLARKE, J. K. (1924). On the bacterial factor in the aetiology of dental caries. *Brit. J. exp. Path.* **5,** 141.
CLEMENT, A. J. (1957). The bacteriology of the primitive mouth. *J. Dent. Assn. S. Afr.* **12,** 281.
CLEMENT, A. J. & RAE, W. (1953). Field studies in the Southern Kalahari. III. A bacteriological investigation of the saliva of various Bakalagadi school children and bushmen. *J. Dent. Assn. S. Afr.* **8,** 64.
CLEMENT, A. J., PLOTKIN, R. & FOSDICK, L. S. (1956). Oral conditions of primitive bushmen in the Western Kalahari Desert. *J. dent. Res.* **35,** 780.
COLLINS, P. A., GERENCSER, M. A. & SLACK, J. M. (1973). Enumeration and identification of *Actinomycetaceae* in human dental calculus using the fluorescent antibody technique. *Arch. oral Biol.* **18,** 145.
COLMAN, G. (1968). The application of computers to the classification of streptococci. *J. gen. Microbiol.* **50,** 149.
COLMAN, G. (1970). The classification of streptococcal strains. *Ph.D. Thesis,* University of London.
COLMAN, G. & WILLIAMS, R. E. O. (1965). The cell walls of streptococci. *J. gen. Microbiol.* **41,** 375.
COLMAN, G. & WILLIAMS, R. E. O. (1972). Taxonomy of some human viridans streptococci. In *Streptococci and Streptococcal Diseases*. Ed. L. W. Wannamaker & J. M. Matsen. New York and London: Academic Press.
COURANT, P.R. & GIBBONS, R. J. (1967). Biochemical and immunological heterogeneity of *Bacteroides melaninogenicus*. *Arch. oral Biol.* **12,** 1605.
COYKENDALL, A. L. (1970). Base composition of deoxyribonucleic acid isolated from cariogenic streptococci. *Arch. oral Biol.* **15,** 365.
COYKENDALL, A. L. (1971). Genetic heterogeneity in *Streptococcus mutans*. *J. Bact.* **106,** 192.
CRITCHLEY, P., WOOD, J. M., SAXTON, C. A. & LEACH, S. A. (1967). The polymerization of dietary sugars by dental plaque. *Caries Res.* **1,** 112.
CROSS, T. & GOODFELLOW, M. (1973). Taxonomy and Classification of the Actinomycetes. In *Actinomycetales: Characteristics and Practical Importance*. Soc. Appl. Bact. Symp. Ser. No. 2. Eds. G. Sykes & F. A. Skinner. London: Academic Press.

DAVIS, R. M. (1972). General ecology of the commensal microflora of the mouth. In *Host resistance to commensal bacteria. The response to dental plaque.* Ed. T. MacPhee. Edinburgh and London: Churchill Livingstone.

DEL GUIDICE, R. A. & CARSKI, T. R. (1968). Characterization of a new Mycoplasma species of human origin. *Bact. Proc.* p. 67.

DIENES, L. & MADOFF, S. (1953). Differences between oral and genital strains of human pleuropneumonia-like organisms. *Proc. Soc. Exp. Biol. Med.* **82,** 36.

DE ARANJO, W. C. & MACDONALD, J. B. (1964). Gingival crevice microbiota of preschool children. *Arch. oral Biol.* **9,** 227.

DE COSTA, T. & GIBBONS, R. J. (1968). Hydrolysis of levan by human plaque streptococci. *Arch. oral Biol.* **13,** 609.

DE STOPPELAAR, J. D., VAN HOUTE, J. & BACKER DIRKS, O. (1969). The relationship between extracellular polysaccharide producing streptococci and smooth surface caries in 13 year old children. *Caries Res.* **3,** 190.

DODD, R. L. (1949). Serologic relationship between streptococcus Group H and *Streptococcus sanguis. Proc. Soc. Exp. Biol. Med.* **70,** 598.

DONOGHUE, H. D. (1972). Antagonisms among bacteria isolated from dental plaque. (Abstract). *J. Dent. Res.* **51,** 1238.

DRASAR, B. S. (1967). Cultivation of anaerobic intestinal bacteria. *J. Path. Bact.* **94,** 417.

DRUCKER, D. B. & MELVILLE, T. H. (1971). The classification of some oral streptococci of human or rat origin. *Arch. oral Biol.* **16,** 845.

DWYER, D. M. & SOCRANSKY, S. S. (1968). Predominant cultivable micro-organisms inhabiting periodontal pockets. *Brit. Dent. J.* **124,** 560.

EDWARDSSON, S. (1968). Characteristics of caries-inducing human streptococci resembling *Streptococcus mutans. Arch. oral Biol.* **13,** 637.

EDWARDSSON, S. (1970). The caries-inducing property of variants of *Streptococcus mutans. Odont. Revy* **21,** 153.

ELLEN, R. P. & GIBBONS, R. J. (1972). M Protein-associated adherence of *Streptococcus pyogenes* to epithelial surfaces: a prerequisite for virulence. *Infection and Immunity* **5,** 826.

ENGEL, L. D. & KENNY, G. E. (1970). *Myoplasma salivarium* in human gingival sulci. *J. Periodont. Res.* **5,** 163.

ESKOW, R. N. & LOESCHE, W. J. (1971). Oxygen tensions in the human oral cavity. *Arch. oral Biol.* **16,** 1127.

EVANS, R. J. (1951). Haematin as a growth factor for a strict anaerobe *Fusiformis melaninogenicus. Proc. Soc. Gen. Microbiol.* **5,** XIX.

FARMER, E. D. (1954). Serological subdivisions among the Lancefield group H streptococci. *J. Gen. Microbiol.* **11,** 131.

FARSHY, D. C., THOMAS, M. L. & MOSS, C. W. (1970). Characterization of treponemes by gas chromatography. *Brit. J. Venereal Diseases* **46,** 441.

FEATHERSTONE, J. L. (1958). The oral lactobacilli of Central Australian Aborigines. *Austral. Dent. J.* **3,** 179.

FEATHERSTONE, J. L. (1959). The oral lactobacilli of Central Australian Aborigines. II. Differentiation of species. *Austral. Dent. J.* **4,** 39.

FEATHERSTONE, J. L. (1960). The oral lactobacilli of Central Australian Aborigines. III. Species isolated from Nomads in Arnhem Land. *Austral. Dent. J.* **5,** 204.

FITZGERALD, R. J. (1968). Dental caries research in gnotobiotic animals. *Caries Res.* **2,** 139.

FITZGERALD, R. J. & JORDAN, H. V. (1968). Polysaccharide producing bacteria and caries. In *Art and Science of Dental Caries Research*. Ed. Harris, pp. 79-86. Academic Press.

FOX, H., PURCELL, R. H. & CHANOCK, R. M. (1969). Characterization of a newly identified Mycoplasma (Mycoplasma oral type 3) from the human oropharynx. *J. Bact.* **98,** 36.

GASTRIN, B., KALLINGS, L. O. & MARCETIC, A. (1968). The survival time for different bacteria in various transport media. *Acta path. microbiol. Scand.* **74,** 371.

GIBBONS, R. J. (1972). Ecology and cariogenic potential of oral streptococci. In *Streptococci and Streptococcal Diseases.* Eds. L. W. Wannamaker & J. M. Matsen. New York and London: Academic Press.
GIBBONS, R. J. & BANGHART, S. B. (1967). Synthesis of extracellular dextran by cariogenic bacteria and its presence in human dental plaque. *Arch. oral Biol.* **12**, 11.
GIBBONS, R. J. & MACDONALD, J. B. (1960). Hemin and Vitamin K compounds as required factors for cultivation of certain strains of *Bacteroides melaninogenicus. J. Bact.* **80**, 164.
GIBBONS, R. J. & NYGAARD, M. (1968). Synthesis of insoluble dextran and its significance in the formation of gelatinous deposits by plaque-forming streptococci. *Arch. oral Biol.* **13**, 1249.
GIBBONS, R. J. & SPINELL, D. M. (1970). Salivary-induced aggregation of plaque bacteria. In *Dental Plaque.* Ed. W. D. McHugh, Livingstone: Edinburgh.
GIBBONS, R. J. & VAN HOUTE, J. (1971). Selective bacterial adherence to oral epithelial surfaces and its role as an ecological determinant. *Infection and Immunity* **3**, 567.
GIBBONS, R. J., SOCRANSKY, S. S., SAWYER, S., KAPSIMALIS, B. & MACDONALD, J. B. (1963). The microbiota of the gingival crevice area of man. II. The predominant cultivable organisms. *Arch. oral Biol.* **8**, 281.
GIBBONS, R. J., KAPSIMALIS, B. & SOCRANSKY, S. S. (1964). The source of salivary bacteria. *Arch. oral Biol.* **9**, 101.
GIBBONS, R. J., SOCRANSKY, S. S., DE ARANJO, W. C. & VAN HOUTE, J. (1964). Studies on the predominant cultivable microflora of dental plaque. *Arch. oral Biol.* **9**, 365.
GIBBONS, R. J., VAN HOUTE, J., & LILJEMARK, W. F. (1972). Parameters that affect the adherence of *Streptococcus salivarius* to oral epithelial surfaces. *J. Dent. Res.* **51**, 424.
GORDON, D. F. JR. (1967). Reisolation of *Staphylococcus salivarius* from the human oral cavity. *J. Bact.* **94**, 1281.
GORDON, D. F. JR. & GIBBONS, R. J. (1966). Studies of the predominant cultivable micro-organisms from the human tongue. *Arch. oral Biol.* **11**, 627.
GORDON, D. F. JR. & JONG, B. B. (1968). Indigenous flora from human saliva. *Appl. Microbiol.* **16**, 428.
GORDON, D. F., STUTMAN, M. & LOESCHE, W. J. (1971). Improved isolation of anaerobic bacteria from the gingival crevice area of man. *Appl. Microbiol.* **21**, 1046.
GREEN, G. E. & DODD, M. C. (1956). Inhibition of oral lactobacilli by streptococci. *J. Bact.* **72**, 690.
GRENIER, E. M., EVELAND, W. C. & LOESCHE, W. J. (1973). Identification of *Streptococcus mutans* serotypes in dental plaque by fluorescent antibody techniques. *Arch. oral Biol.* **18**, 707.
GUGGENHEIM, B. (1968). Streptococci of dental plaques. *Caries Res.* **2**, 147.
GUGGENHEIM, B. (1970*a*). Enzymatic hydrolysis and structure of water-insoluble glucan produced by glucosyltransferases from a strain of *Streptococcus mutans. Helv. odont. Acta* **14**, suppl. V, 89.
GUGGENHEIM, B. (1970*b*). Extracellular polysaccharides and microbiol plaque. *Int. dent. J.* **20**, 657.
GUGGENHEIM, B. & SCHROEDER, H. E. (1967). Biochemical and morphological aspects of extracellular polysaccharides produced by cariogenic streptococci. *Helv. odont. Acta* **11**, 131.
GUTHOF, O. (1956). Pathogenic strains of *Streptococcus viridans*; streptococci found in dental abscesses and infiltrates in the region of the oral cavity. *Zentrabl. Bakteriol. Parasitenk. Abt 1.* **166**, 553.
HADI, A. W. & RUSSELL, C. (1968). Quantitative estimations of fusiforms in saliva from normal individuals and cases of acute ulcerative gingivitis. *Arch. oral Biol.* **13**, 1371.
HADI, A. W. & RUSSELL, C. (1969). Fusiforms in gingival material. *Brit. dent. J.* **126**, 82.
HANDELMAN, S. L. & HAWES, R. R. (1964). The effect of long-term systematic antibiotic therapy on the antibiotic resistance of the salivary flora. *J. oral Ther.* **1**, 23.

HANDELMAN, S. L. & HAWES, R. R. (1965). The effect of long-term antibiotic administration on the numbers of salivary organisms. *Arch. oral Biol.* **10**, 353.
HANDELMAN, S. L. & HESS, C. (1969). Bacterial populations of selected tooth surface sites. *J. Dent. Res.* **48**, 67.
HANSEN, P. A. (1968). Type strains of Lactobacillus species. A report by the Taxonomic Subcommittee on Lactobacilli and closely related organisms. Rockville, Maryland: American Type Culture Collection.
HARDIE, J. M., BOWDEN, G. H., BURT, B. A., WALLER, D. F. & SLACK, G. L. (1972). A method of collecting dental plaque. Pilot study. (Abstract). *Caries Res.* **6**, 270.
HARRISON, R. W. (1948). Lactobacilli versus streptococci in the etiology of dental caries. *J. Amer. Dent. Ass.* **37**, 391.
HEMMENS, E. A., BLAYNEY, J. R., BRADEL, S. F. & HARRISON, R. W. (1946). The microbial flora of dental plaque in relation to the beginning of caries. *J. dent. Res.* **25**, 195.
HOLDEMAN, L. V. & MOORE, W. E. C. (1972). Anaerobe Laboratory Manual. Blacksburg, Virginia: The Virginia Polytechnic Institute and State University Anaerobe Laboratory.
HOLMBERG, K. & HALLANDER, H. O. (1972). Interference between Gram positive micro-organisms in dental plaque. *J. dent. Res.* **51**, 588.
HOLMBERG, K. & HALLANDER, H. O. (1973a). Production of bactericidal concentrations of hydrogen peroxide by *Streptococcus sanguis*. *Arch. oral Biol.* **18**, 423.
HOLMBERG, K. & HALLANDER, H. O. (1973b). Numerical taxonomy and laboratory identification of *Bacterionema matruchotii*, *Rothia dentocariosa*, *Actinomyces naeslundii*, *Actinomyces viscosus* and some related bacteria. *J. gen. Microbiol.* **76**, 43.
HOWELL, A., RIZZO, A. & PAUL, F. (1965). Cultivable bacteria in developing and mature human dental calculus. *Arch. oral Biol.* **10**, 307.
HUNGATE, R. E. (1950). The anaerobic mesophilic celluloytic bacteria. *Bact. Rev.* **14**, 1.
HUNGATE, R. E. (1966). *The Rumen and its Microbes*. New York: Academic Press.
IKEDA, T., SANDHAM, H. J. & BRADLEY, E. L. JR. (1973). Changes in *Streptococcus mutans* and lactobacilli in plaque in relation to the initiation of dental caries in negro children. *Arch. oral Biol.* **18**, 555.
JACKINS, H. C. & BARKER, H. A. (1951). Fermentative processes of the fusiform bacteria. *J. Bact.* **61**, 101.
JONES, S. J. (1972). A special relationship between spherical and filamentous microorganisms in mature human dental plaque. *Arch. oral Biol.* **17**, 613.
KELSTRUP, J. (1966). The incidence of *Bacteroides melaninogenicus* in human gingival sulci, and its prevalence in the oral cavity at different ages. *Periodontics* **4**, 14.
KELSTRUP, J. & GIBBONS, R. J. (1969). Bacteriocins from human and rodent streptococci. *Arch. oral Biol.* **14**, 251.
KELSTRUP, J. & FUNDER-NIELSEN, T. D. (1972). Molecular interactions between the extracellular polysaccharides of *Streptococcus mutans*. *Arch. oral Biol.* **17**, 1659.
KELSTRUP, J., RICHMOND, S., WEST, C. & GIBBONS, R. J. (1970). Finger printing human oral streptococci by bacteriocin production and sensitivity. *Arch. oral Biol.* **15**, 1109.
KENNEY, E. B. & ASH, M. M. (1969). Oxidation-reduction potential of developing plaque, periodontal pockets and gingival sulci. *J. Periodont.* **40**, 630.
KESEL, R. G., SHKLAIR, I. L., GREEN, G. H. & ENGLANDER, H. R. (1958). Further studies on lactobacilli counts after elimination of carious lesions. *J. Dent. Res.* **37**, 50.
KLEIN, J. O., BUCKLAND, D. & FINLAND, M. (1969). Colonization of newborn infants by Mycoplasmas. *New Eng. J. Med.* **280**, 1025.
KOCUR, M., BERGAN, T. & MORTENSEN, N. (1971). DNA base composition of Gram positive cocci. *J. gen. Microbiol.* **69**, 167.
KOSTECKA, F. (1924). Relation of the teeth to the normal development of the microbial flora in the oral cavity. *Dental Cosmos* **66**, 927.
KRASSE, B. (1953). The proportional distribution of different types of streptococci in saliva and plaque material. *Odont. Revy* **4**, 304.
KRASSE, B. (1954). The proportional distribution of *Streptococcus salivarius* and other streptococci in various parts of the mouth. *Odont. Revy* **5**, 203.

KRASSE, B., JORDAN, H. V., EDWARDSSON, S., SVENSSON, I. & TRELL, L. (1968). The occurrence of certain 'caries-inducing' streptococci in human dental plaque material. *Arch. oral Biol.* **13**, 911.
LEACH, P. A., BULLEN, J. J., & GRANT, I. D. (1971). Anaerobic CO_2 cabinet for the cultivation of strict anaerobes. *Appl. Microbiol.* **22**, 824.
LEACH, S. A., APPLETON, J., DADA, O. & HAYES, M. L. (1972). Some factors affecting the metabolism of fructan by human oral flora. *Arch. oral Biol.* **17**, 137.
LEHNER, T. (1966). Classification and clinico-pathological features of candida infections in the mouth. In *Symposium on Candida Infection.* Eds H. I. Winner and Rosalinde Hurley. London: E and S Livingstone.
LEV, M. (1958). Apparent requirement for vitamin K of rumen strain of *Fusiformis nigrescens. Nature Lond.* **87**, 203.
LEV, M., KENDALL, K. C. & MILFORD, A. F. (1971). Succinate as a growth factor for *Bacteroides melaninogenicus. J. Bact.* **108**, 175.
LEWKOWICZ, X. (1901). Recherches sur la flore microbienne de la bouche des nourrissons. *Arch. Med. Exp.* **13**, 633.
LILJEMARK, W. F. & GIBBONS, R. J. (1971). Ability of *Veillonella* and *Neisseria* species to attach to oral surfaces and their proportions present indigenously. *Infect. and Immunity* **4**, 264.
LILJEMARK, W. F. & GIBBONS, R. J. (1972). Proportional distribution and relative adherence of *Streptococcus miteor* (mitis) on various surfaces in the oral cavity. *Infect. and Immunity* **6**, 852.
LISTGARTEN, M. A., MAYO, H. & AMSTERDAM, M. (1973). Ultrastructure of the attachment device between coccal and filamentous micro-organisms in 'corn cob' formations of dental plaque. *Arch. oral Biol.* **18**, 651.
LITTLETON, N. W., McCABE, R. M. & CARTER, C. H. (1967). Studies of oral health on persons nourished by stomach tube. II. Acidogenic properties and selected bacterial components of plaque material. *Arch. oral Biol.* **12**, 601.
LITTLETON, N. W., KAKEHASHI, S. & FITZGERALD, R. J. (1970). Recovery of specific 'caries-inducing' streptococci from carious lesions in the teeth of children. *Arch. oral Biol.* **15**, 461.
LLORY, H., DAMMRON, A., GIOANNI, M. & FRANK, R. M. (1972). Some population changes in oral anaerobic micro-organisms, *Streptococcus mutans* and yeasts following irradiation of the salivary glands. *Caries Res.* **6**, 298.
LOBENE, R. L., BRION, M. & SOCRANSKY S. S. (1969). Effect of erythromycin on dental plaque and plaque-forming micro-organisms of man. *J. Periodont.* **40**, 287.
LÖE, H., KARRING, TH., & THEILADE, ELSE (1973). An *in vivo* method for the study of the microbiology of occlusal fissures. *Caries Res.* **7**, 120.
LOESCHE, W. J. (1968). Importance of nutrition in gingival crevice microbial ecology. *Periodontics* **6**, 245.
LOESCHE, W. J. (1969). Oxygen sensitivity of various anaerobic bacteria. *Appl. Microbiol.* **18**, 723.
LOESCHE, W. J. & GIBBONS, R. J. (1965). A practical scheme for identification of the most numerous oral Gram negative anaerobic rods. *Arch. oral Biol.* **10**. 723.
LOESCHE, W. J. & GIBBONS, R. J. (1968). Amino acid fermentation by *Fusobacterium nucleatum. Arch. oral Biol.* **13**, 191.
LOESCHE, W. J., HOCKETT, R. N. & SYED, S. A. (1972). The predominant cultivable flora of tooth surface plaque removed from institutionalized subjects. *Arch. oral Biol.* **17**, 1311.
LOESCHE, W. J., SOCRANSKY, S. S. & GIBBONS, R. J. (1964). *Bacteroides oralis*: proposed new species isolated from the oral cavity of man. *J. Bact.* **88**, 1329.
LOESCHE, W. J. & SYED, S. A. (1973). The predominant cultivable flora of carious plaque and carious dentine. *Caries Res.* **7**, 201.
LYON, T. C. JR., & NEMES, J. L. (1971). Antibiotic sensitivities of oral Mycoplasma. *J. dent. Res.* **50**, 1678.
MACDONALD, J. B. (1953). The motile non-sporulating anaerobic rods of the oral cavity. Thesis. Faculty of Dentistry, University of Toronto.

MANLY, R. S. & RICHARDSON, D. T. (1968). Metabolism of levan by oral samples. *J. dent. Res.* **47**, 1080.

McCARTHY, C., SNYDER, M. L. & PARKER, P. B. (1965). The indigenous oral flora of man–1. The newborn to the 1 year old infant. *Arch. oral Biol.* **10**, 61.

McHUGH, W. D. (1970). Dental plaque. Edinburgh and London: E and S Livingstone.

MIKX, F. H. M., HOEVEN, J. S. VAN DER, KONIG, K. G., PLASSCHAERT, A. J. M. & GUGGENHEIM, B. (1972). Establishment of defined microbial ecosystems in germ free rats. 1. The effect of the interaction of *Streptococcus mutans* or *Streptococcus sanguis* with *Veillonella alkalescens* on plaque formation and caries activity. *Caries Res.* **6**, 211.

MILLS, C. C. (1972). Occurrence of *Mycobacterium* other than *Mycobacterium tuberculosis* in the oral cavity and in sputum. *Appl. Microbiol.* **24**, 307.

MØLLER, J. R. A. (1966). Microbiological examination of root canals and periapical tissues of human teeth. Akademiforlaget-Goteborg, Goteborg, Sweden.

MORHART, R. E., MATA, L. J., SINSKEY, A. J. & HARRIS, R. S. (1970). A microbiological and biochemical study of gingival crevice debris obtained from Guatemalan Mayan Indians. *J. Periodont.* **41** 644.

MORRIS, E. O. (1953a). The bacteriology of the oral cavity. I. The distribution of bacteria in and on the dental enamel. *Brit. Dent. J.* **XCV**, 77.

MORRIS, E. O. (1953b). The bacteriology of the oral cavity. IIA. Methods used in the study of the oral flora. IIB. Lactobacillus. *Brit. Dent. J.* **XCV**, 259.

MORRIS, E. O. (1954a). The bacteriology of the oral cavity. III. Streptococcus. *Brit. Dent. J.* **XCVI**, 95.

MORRIS, E. O. (1954b). The bacteriology of the oral cavity. IV (A) *Micrococcus*, (B) *Neisseria* and (C) *Veillonella*. *Brit. Dent. J.* **XCVI**, 259.

MORRIS, E. O. (1954c). The bacteriology of the oral cavity. V. Corynebacterium and Gram positive filamentous organisms. *Brit. Dent. J.* **XCVII**, 29.

MORRIS, E. O. (1954d). The bacteriology of the oral cavity. VI. Fusiforms, *Bacillus*, *Bacterium* and *Haemophilus* General Conclusions.

MORTON, H. F., SMITH, P. F. & WILLIAMS, N. B. (1951). Isolation of pleuropneumonia-like organisms from human saliva: A newly detected member of the oral flora. *J. dent. Res.* **30**, 415.

MUKASA, H. & SLADE, H. D. (1973). Structure and immunological specificity of the *Streptococcus mutans* Group b cell wall antigen. *Infect. and Immun.* **7**, 578.

NOLTE, W. A. (1973). *Oral Microbiology*. 2nd Edition. St Louis: The C. V. Mosby Company. London: Henry Kimpton.

OLSON, G. A., BLEIWEIS, A. S. & SMALL, P. A. JR. (1972). Adherence inhibition of *Streptococcus mutans*: an assay reflecting a possible role of antibody in dental caries prophylaxis. *Infect. and Immun.* **5**, 419.

OMATA, R. R. & DISRAELY, M. N. (1956). A selective medium for oral fusobacteria. *J. Bact.* **72**, 677.

PARKER, R. B. (1970). Paired culture interaction of the oral microbiota. *J. dent. Res.* **49**, 804.

PARKER, R. B. & CREAMER, H. R. (1971). Contribution of plaque polysaccharides to growth of cariogenic micro-organisms. *Arch. oral Biol.* **16**, 855.

PICKERILL, H. P. & CHAMPTALOUP, S. T. (1913). The bacteriology of the mouth in Maori children. *Brit. Med. J.* **II**, 1482.

PIKE, E. B., FREER, J. H., DAVIS, G. H. G. & BISSET, K. A. (1962). The taxonomy of Micrococci and Neisseriae of oral origin. *Arch. oral Biol.* **7**, 715.

PINE, L. (1970). Classification and phylogenetic relationship of microaerophilic actinomycetes. *Int. J. Syst. Bact.* **20**, 445.

PORTERFIELD, R. S. (1950), Classification of the streptococci of subacute bacterial endocarditis. *J. gen. Microbiol.* **4**, 92.

RASMUSSEN, E. G., GIBBONS, R. J. & SOCRANSKY, S. S. (1966). Taxonomic study of 50 Gram positive anaerobic diphtheroids isolated from the oral cavity. *Arch. oral Biol.* **11**, 573.

RAZIN, S., MICHMANN, J. & SHIMSHONI, Z. (1964). The occurrence of Mycoplasma (Pleuropneumonia-like organisms, PPLO) in the oral cavity of dentulous and edentulous subjects. *J. Dent. Res.* **43,** 402.
RICHARDSON, R. L. & JONES, M. (1958). A bacteriologic census of human saliva. *J. Dent. Res.* **37,** 697.
RITZ, H. L. (1967). Microbial population shifts in developing human dental plaque. *Arch. oral Biol.* **12,** 1561.
RITZ, H. L. (1969). Fluorescent antibody staining of *Neisseria, Streptococcus,* and *Veillonella* in frozen sections of dental plaque. *Arch. oral Biol.* **14,** 1073.
RITZ, H. L. (1970). The role of aerobic *Neisseriae* in the initial formation of dental plaque. In *Dental Plaque*, Ed. W. D. McHugh. Edinburgh: Livingstone.
ROGERS, A. H. (1969). The proportional distribution and characteristics of streptococci in human dental plaque. *Caries Res.* **3,** 238.
ROGERS, A. H. (1972). Effect of the medium on bacteriocin production among strains of *Streptococcus mutans. Appl. Microbiol.* **24,** 294.
ROGOSA, M. (1964). The genus *Veillonella,* 1. General cultural, ecological and biochemical considerations. *J. Bact.* **87,** 162.
ROGOSA, M. (1965). The genus *Veillonella.* 4. Serological grouping, and genus and species emendations. *J. Bact.* **90,** 704.
ROGOSA, M. (1970). Characters used in the classification of lactobacilli. *Internat. J. Syst. Bact.* **20,** 519.
ROGOSA, M. & BISHOP, F. S. (1964*a*). The genus *Veillonella.* 2. Nutritional studies. *J. Bact.* **87,** 574.
ROGOSA, M. & BISHOP, F. S. (1964*b*). The genus *Veillonella.* 3. Hydrogen sulphide production by growing cultures. *J. Bact.* **88,** 37.
ROGOSA, M., KRICHEVSKY, M. I. & BISHOP, F. S. (1965). Truncated glycolytic system in Veillonella. *J. Bact.* **90,** 164.
ROSEBURY, T. (1962). Micro-organisms indigenous to man. New York: McGraw-Hill.
ROSS, P. W. (1971). Beta-haemolytic streptococci in saliva. *J. Hygiene* **69,** 347.
RUNDELL, B. B., THOMSON, L. A., LOESCHE, W. J. & STILES, H. M. (1973). Evaluation of a new transport medium for the preservation of oral streptococci. *Arch. oral Biol.* **18,** 871.
SAKAMAKI, S. T. & BAHN, A. N. (1968). Effect of orthodontic banding on localized oral lactobacilli. *J. Dent. Res.* **47,** 275.
SAWYER, S. J., MACDONALD, J. B. & GIBBONS, R. J. (1962). Biochemical characteristics of *Bacteroides melaninogenicus. Arch. oral Biol.* **7,** 685.
SAXTON, C. A. (1973). Scanning electron microscope study of the formation of dental plaque. *Caries Res.* **7,** 102.
SCHAMSCHULA, R. G. & BARMES, D. E. (1970*a*). The lactobacillus flora of saliva and plaque in primitive peoples in Papua–New Guinea. *Austral. Dent. J.* **15,** 28.
SCHAMSCHULA, R. G. & BARMES, D. E. (1970*b*). A study of the streptococcal flora of plaque in caries free and caries active primitive peoples. *Austral. Dent. J.* **15,** 377.
SCHMITT, J. A. (1971). Epidemiological investigations of oral *Candida albicans. Mycopath. et Mycol. applicata* **43,** 65.
SCHOTTMULLER, H. (1903). Die Artunterscheidung der für den Menschen pathogenen Streptokokken durch Blutagar. *München. Med. Wochenschr.* **50,** 849.
SHARPE, M. E., FRYER, T. F. & SMITH, D. G. (1966). Identification of the Lactic Acid Bacteria. In *Identification Methods for Microbiologists.* Soc. Appl. Bact. Symp. Ser. No. 1 Eds B. M. Gibbs & F. A. Skinner. London: Academic Press.
SHKLAIR, I. L., ENGLANDER, H. R., STEIN, L. M. & KESEL, R. G. (1956). Preliminary report on the effect of complete mouth rehabilitation on oral lactobacilli counts. *J. Amer. Dent. Ass.* **53,** 155.
SHKLAIR, I. L. & MAZZARELLA, M. A. (1961). Effects of full-mouth extraction on oral microbiota. *D. Progress.* **1,** 275.
SHKLAIR, I. L., MAZZARELLA, M. A., GUTEKUNST, R. R. & KIGGINS, E. M. (1961). Isolation and incidence of Pleuropneumoniae-like organisms from the human oral cavity. *J. Bact.* **38,** 785.

SCHLEGEL, R. & SLADE, H. D. (1972). Bacteriocin production by transformable Group H streptococci. *J. Bact.* **112,** 824.
SIMS, W. (1970). Oral Haemophili. *J. Med. Microbiol.* **3,** 615.
SLACK, G. L. & BOWDEN, G. H. (1965). Preliminary studies of experimental dental plaque in vivo. *Adv. Fluor. Res. dent. Caries Prev.* **3,** 193.
SLACK, J. M., LANDFRIED, S. & GERENCSER, M. A. (1971). Identification of *Actinomyces* and related bacteria in dental calculus by the fluorescent antibody technique. *J. dent. Res.* **50,** 78.
SLANETZ, L. W. & REYNOLDS, H. (1952). The bactericidal action of certain antiseptics on the oral bacteria. *J. dent. Res.* **31,** 35.
SNYDER, M. L., BULLOCK, W. W. & PARKER, R. B. (1967). Morphology of Gram positive filamentous bacteria identified in dental plaque by fluorescent antibody technique. *Arch. oral Biol.* **12,** 1269.
SOCRANSKY, S. S. (1970). Relationship of bacteria to the etiology of periodontal disease. *J. dent. Res.* **49,** 203.
SOCRANSKY, S. S., GIBBONS, R. J., DALE, A. C., BORTNICK, L., ROSENTHAL, E. & MACDONALD, J. B. (1963). The microbiota of the gingival crevice area of man. I. Total microscopic and viable counts of specific organisms. *Arch. oral Biol.* **8,** 275.
SOCRANSKY, S. S. & HUBERSAK, C. (1967). Replacement of ascitic fluid or rabbit serum requirement of *Treponema dentium* by α globulin. *J. Bact.* **94,** 1795.
SOCRANSKY, S. S., LISTGARTEN, M., HUBERSAK, C., COTMORE, J. & CLARK, A. (1969). Morphological and biochemical differentiation of three types of small oral spirochaete. *J. Bact.* **98,** 878.
SOCRANSKY, S. S., LOESCHE, W. J., HUBERSAK, C. & MACDONALD, J. B. (1964). Dependency of *Treponema microdentium* on other oral organisms for isobutyrate, polyamines and a controlled oxidation-reduction potential. *J. Bact.* **88,** 200.
SOCRANSKY, S. S. & MANGANIELLO, S. D. (1971). The oral microbiota of man from birth to senility. *J. Periodont.* **42,** 485.
SPEIRS, M. (1971). Classification systems of the Bacteroides group. *Med. Lab. Technology* **28,** 360.
SUTTER, V. L., HURST, V. & LANDUCCI, A. O. J. (1966). Pseudomonads in human saliva. *J. Dent. Res.* **45,** 1800.
SUZUKI, S., USHIJIMA, T. & ICHINOSE, H. (1966). Differentiation of Bacteroides from Sphaerophorus and Fusobacterium. *Japan J. Microbiol.* **10,** 193.
SYED, S. A. & LOESCHE, W. J. (1972). Survival of human dental plaque flora in various transport media. *Appl. Microbiol.* **24,** 638.
TAYLOR-ROBINSON, D., CANCHOLA, J., FOX, H. & CHANOCK, R. M. (1964). Newly identified Mycoplasma (M. orale) and its relationship to other human Mycoplasmas. *Am. J. Hyg.* **80,** 135.
THEILADE, E., WRIGHT, W. H., JENSEN, S. B. & LÖE, H. (1966). Experimental gingivitis in man. II. A longitudinal clinical and bacteriological investigation. *J. Periodont. Res.* **1,** 1.
THEILADE, E., LARSON, R. H. & KARRING, TH. (1973). Microbiological studies of plaque in artificial fissures implanted in human teeth. *Caries Res.* **7,** 130.
THOMSON, L. A. (1971). The development and testing of epidemiologic methods for sampling human dental plaque. Ph. D. Thesis, University of Michigan.
VAN DE RIJN, I. & BLEIWEIS, A. S. (1973). Antigens of *Streptococcus mutans.* 1. Characterization of a Serotype-Specific determinant from *Streptococcus mutans. Infect. and Immun.* **7,** 795.
VAN HOUTE, J., GIBBONS, R. J. & BANGHART, S. B. (1970). Adherence as a determinant of the presence of *Streptococcus salivarius* and *Streptococcus sanguis* on the human tooth surface. *Arch. oral Biol.* **15,** 1025.
VAN HOUTE, J., GIBBONS, R. J. & PULKKINEN, A. J. (1971). Adherence as an ecological determinant for streptococci in the human mouth. *Arch. oral Biol.* **16,** 1131.
VAN HOUTE, J., GIBBONS, R. J. & PULKKINEN, A. J. (1972). Ecology of human oral lactobacilli. *Infect. and Immun.* **6,** 723.

VAN HOUTE, J. & JANSEN, H. M. (1968). Levan degradation by streptococci isolated from human dental plaque. *Arch. oral Biol.* **13**, 827.
VAN REENAN, J. F. & COOGAN, M. M. (1970). Clostridia isolated from human mouths. *Arch. oral Biol.* **15**, 845.
WAHREN, A. & GIBBONS, R. J. (1970). Amino acid fermentation by *Bacteroides melaninogenicus. Antonie van Leeuwenhoek* **36**, 149.
WANTLAND, W. W., WANTLAND, E. M. & WINQUIST, D. L. (1963). Collection, identification and cultivation of oral protozoa. *J. dent. Res.* **42**, 1234.
WASHBURN, M. R., WHITE, J. C. & NIVEN, C. F. JR. (1946). *Streptococcus S. B. E:* Immunological characteristics. *J. Bact.* **51**, 723.
WAUGH, L. M. & WAUGH, D. S. (1936). Dental observations among Eskimo. VI. Bacteriologic findings in most primitive and populous native district of Alaska. *J. dent. Res.* **15**, 317.
WERNER, H., PULVERER, G. & REICHERTZ, C. (1971). The biochemical properties and antibiotic susceptibility of *Bacteroides melaninogenicus. Med. Microbiol. Immunol.* **157**, 3.
WERNER, H., NEUHAUS, F. & HUSSELS, H. (1971). A biochemical study of fusiform anaerobes. *Med. Microbiol. Immunol.* **157**, 10.
WHITE, J. C. & NIVEN, C. F. JR. (1946). *Streptococcus S.B.E.* A streptococcus associated with subacute bacterial endocarditis. *J. Bact.* **51**, 717.
WILLIAMS, R. A. D. (1971). Cell wall composition and enzymology of lactobacilli. *J. dent. Res.* **50**, 1104.
WILLIAMS, R. E. O. (1956). *Streptococcus salivarius* (Vel hominis) and its relation to Lancefield's Group K. *J. Path. Bact.* **72**, 15.
WOOD, J. M. (1967). The amount, distribution and metabolism of soluble polysaccharides in dental plaque. *Arch. oral Biol.* **9**, 91.
WOOD, J. M. (1969). The state of hexose sugar in human dental plaque and its metabolism by the plaque bacteria. *Arch. oral Biol.* **14**, 161.
YAMADA, T. & CARLSSON, J. (1973). Phosphoenolpyruvate carboxylase and ammonium metabolism in oral streptococci. *Arch. oral Biol.* **18**, 799.
ZINNER, D. D. & JABLON, J. M. (1968). Human streptococcal strains in experimental caries. In *Art and Science of Dental Caries Research.* Ed. R. H. Harris. New York: Academic Press.

Intrinsic and Extrinsic Factors Influencing the Flora of the Mouth

D. A. M. GEDDES AND G. N. JENKINS

Department of Oral Physiology, University of Newcastle upon Tyne, Markham Laboratories, Upper Claremont Street, Newcastle upon Tyne, NE2 4AJ, England

CONTENTS

1. Introduction 85
2. Establishment of flora 85
3. Physical effects of saliva 86
4. Mechanical removal of plaque 87
5. Composition of saliva 89
6. Plaque acid production from exogenous sugar 91
7. Microbial interactions 92
8. Antibacterial systems of non-bacterial origin 92
9. Dietary factors 93
10. Factors inferred from presence of caries 94
11. Effect of fluoride 95
12. Conclusions 96
13. References 96

1. Introduction

OUR AIM in this paper has been to bring together a wide range of information about intrinsic and extrinsic factors which have been shown to influence the oral flora of healthy human subjects. Priority has been given to effects known to occur *in vivo* but reference is also made to *in vitro* work which seems relevant to the maintenance of the oral flora *in situ*. Because the taxonomic criteria for many oral species are poorly defined and the efficiency of cultural methods employed varies widely from study to study, information on many points is fragmentary. In some instances, the effects of certain factors are known only on plaque metabolic activity and caries; the effects on the flora have only been inferred. In general, we have included only the more recent references from which earlier literature can be traced.

2. Establishment of Flora

Normally the mouth of the neonatal infant is sterile and the resident flora begins to colonize the oral cavity within the first day, possibly coinciding with the first feeding. When the development of the flora was followed at intervals over the first year of life streptococci were the only organisms cultured consistently; the average ranged from 98% of the cultivatable bacteria on the first day or two of life to 70% at the end of 12 months, by which time the flora of the majority of infants resembled in complexity that of adults (McCarthy, Synder & Parker, 1965). With increasing age *Nocardia (Rothia), Neisseria, Veillonella* and *Bacteroides* spp. were also established and there was a relative reduction in the percentage of aerobic streptococci. The increase of anaerobes in the edentulous mouth of infants in the absence of obvious 'anaerobic niches', such as gingival crevices around the crevical margins of erupted teeth, has been attributed to symbotic growth with oral aerobes (Hurst, 1957). Isolation of strains of *Streptococcus salivarius,* with the same bacteriocin patterns from babies and their mothers suggests transmission from mother to baby (Kelstrup *et al.,* 1970). However, *Strep. sanguis,* another member of the adult flora, was not established in the mouth prior to the eruption of teeth and it has been suggested that the tooth surface is necessary for the establishment of this organism (Carlsson *et al.,* 1970). Because of the complexity of the oral flora difficulty has been encountered in cultivating a high percentage of organisms giving the total direct count. Among important factors which must be controlled to obtain high viable colony counts are efficient dispersion methods which will also maintain viability, also the use of appropriate gaseous environments and media because the flora contains both aerobes and strict anaerobes, and mono and di-saccharide-utilizers as well as obligate lactate-utilizers (Gilmour & Poole, 1970).

3. Physical Effects of Saliva

The oral cavity is kept moist by saliva which flows continuously during waking hours, in the absence of exogenous stimuli, at a rate varying over a wide range (0.5-111.0 ml/h) or an average of 19 ml/h, i.e. <0.3 ml/min (Becks & Wainwright, 1943). This continuous flow leads to swallowing. In the only quantitative study of swallowing known to us, the rate was *c.* 30 swallows/h between meals in 15 subjects (Flanagan, Clement & Moorrees, 1963). The number of bacteria in resting saliva has been reported as being of the order of 10^8/ml, the total aerobes being slightly fewer than the anaerobes cultivated (reviewed by Handelman & Mills, 1965), the numbers being somewhat higher on waking (i.e. after a period when bacteria accumulate because the frequency of swallowing is low as a result of the slow flow-rate of saliva, averaging only 50 swallows during the whole period of sleep). Mechanically stimulated saliva also contains more bacteria, presumably because the method of stimulation dislodges more bacteria from the surface deposits. The counts are reported to decrease after a meal, gradually increasing between meals, implying that removal by swallowing of resting saliva does not keep pace with growth of bacteria in the mouth.

It must be emphasized that the flora suspended in samples of saliva is not exactly representative, either quantitatively or qualitatively, of the bacterial aggregations on any of the surfaces within the mouth. For example, the percentage distribution of *Strep. salivarius* differs between saliva, mucosal and tooth surfaces and is most abundant in saliva and on the tongue (Krasse, 1963). Furthermore, a review of the mean percentage of the cultivatable organisms in the adult oral cavity shows that other organisms such as *Bacteroides melaninogenicus* and spirochaetes, found in gingival crevice areas, are not usually cultivated from saliva while *Strep. mutans* is c. 20 times more abundant on the tooth surface than in saliva (Socransky & Manganiello, 1971). It follows, therefore, that data obtained from samples in which saliva is used as the source of oral organisms cannot be assumed to apply also to dental plaque. Plaque is the soft tenacious organic deposit, incorporating metabolically active bacteria, which adheres to the tooth surface and which is not removed by rinsing with water (Dawes, Jenkins & Tonge, 1963).

4. Mechanical Removal of Plaque

Because the presence of dental plaque is an essential factor, both in dental caries and, it has now been shown, in periodontal disease, factors affecting plaque accumulation have been extensively investigated.

Although foods thought to have a cleaning action, especially apples, have been advocated for years as a means of plaque control (and, by implication, caries control) a simple test shows their ineffectiveness. If the extent of plaque accumulation after, say 24 h abstention from tooth cleaning is evaluated by staining it *in situ* with a disclosing solution and the subject eats raw celery, carrot or an apple, it is observed that although plaque may be removed from the smooth surfaces it remains almost unaffected at the potential sites of caries, the gingival margin, the contact points and fissures. Several clinical trials have shown that cleaning foods have little or no protective effect against caries although gingival conditions were slightly improved and calculus production reduced in some of the trials (suggesting that the flora had been reduced or modified to some extent) but mechanical effects of massaging the gingivae may also have contributed: in other trials there were no clinical effects (for references, see Longhurst & Berman, 1973).

Tooth brushing removes plaque from easily accessible smooth surfaces, but, unless very thorough, leaves much on the sheltered gingival areas and at the interdental areas below the contact points between adjacent teeth. Even after thorough tooth brushing, plaque forms again within hours, indeed, organisms have been recovered from tooth surfaces 5 min after cleaning (Socransky *et al.*, 1971). The total amount of plaque accumulated in 24 h (35.3 ± 13.4 mg wet wt) could be almost halved (18.1 ± 9.2 mg wet wt) when subjects improved the efficiency with which they cleaned their teeth (Geddes, unpublished).

After rigorous cleaning by a dental prophylaxis (this entails scaling the tooth surface and polishing by an engine-driven rubber cup with pumice and clearing of the interdental areas) recolonization was slower, the plaque that developed during the first day comprised mainly streptococcus, neisseria and nocardia (rothia). By the 9th day after prophylaxis, the proportional distribution had shifted with a decrease in aerobic organisms such as neisseria and nocardia (rothia) and an increase in the population of anaerobes such as veillonella and facultative microaerophiles such as corynebacterium and filamentous *Actinomyces* (Ritz, 1967). These changes suggest that the growth of the anaerobic species was dependent upon prior growth of aerobic facultative types with concomitant thickening of the plaque providing a micro-environment suitable for anaerobic growth. This hypothesis was strengthened by fluorescent antibody staining of neisseria and veillonella in sections of plaque grown *in vivo* on Mylar strips for 2 to 10 days (Ritz, 1969) showing a predominance of neisseria in the younger plaque and outer layers with veillonella in the deeper layers of the older plaque.

When the percentage distribution of the major morphological types of organisms was studied before, and at various times up to 60 days after, prophylaxis (during the 60 days subjects continued their usual oral hygiene habits) by 30 days the distribution had returned to pre-prophylactic levels in the maxillary premolar interstitial areas but not in the mandibular incisor region. By 60 days, that region too had returned to pre-prophylactic levels (Handleman & Hess, 1970). On the other hand, when 5 subjects were given prophylaxis and then allowed to accumulate plaque without any oral hygiene for 23 days, longitudinal study of the percentage distribution of bacteria in smears from buccal surfaces of maxillary molar teeth showed that the final proportion distribution was reached after 10–15 days (von der Fehr, Löe & Theilade, 1970). In a similar study, 5 subjects produced a mean plaque weight of 65 ± 26 mg wet wt (Geddes, unpublished).

Presumably the situation most prevalent in the human mouth, assuming that most people practise some measure of oral hygiene, will be areas of the tooth surface covered with young, relatively aerobic streptococcal plaque merging into areas of 'older', relatively anaerobic flora with filamentous forms. Furthermore, the inner layers of plaque include a number of bacterial cells exhibiting thick cell walls and polysaccharide storage (Saxton & Critchley, 1970; van Houte & Saxton, 1971).

Not only does the bacterial composition of plaque differ from deposits upon the oral mucosa but also it is dependent upon the position of the tooth in the dental arch (Handleman & Hess, 1970) and the specific site upon the individual tooth surface, for example plaque from the supragingival or crevice area (Gibbons, Kapsimalis & Socransky, 1964; Salkind, Oshrain & Mandel, 1971) where the micro-environment produces an E_h value low enough for spirochaetes. Davies (1972) has published an extensive review of the general ecology of the commensal microflora of the mouth.

5. Composition of Saliva

The saliva which produces the moist environment for the oral flora is secreted by 3 major (paired) glands, *via* ducts entering the oral cavity from the parotid gland in the cheek opposite the maxillary molar teeth, and from the submandibular and sublingual glands under the tongue behind the mandibular incisor teeth, and from numerous minor glands situated in the mucosa of lips, cheek and palate. Upon entering the mouth, these sterile secretions are modified by the incorporation of desquamated epithelial cells from the oral mucosa, moribund polymorpho-nuclear leucocytes and frequently small quantities of gingival fluid, more like serum than saliva in composition. The resultant mixture is termed 'whole saliva' to which the submandibular secretions contribute the greatest volume followed by parotid, while sublingual and the minor glands each supply *c.* 10%. That large numbers of many species of bacteria do survive and can metabolize in the environment of whole saliva is self-evident but its contribution to growth of the resident flora is not so certain. When employed *in vitro* the unaugmented secretions at best support feeble growth of some species but also inhibit the growth of others (Williams & Powlen, 1959).

Although saliva contains nitrogenous compounds and the total protein concentration averages *c.* 300 mg/100 ml, the proportion of this which is available as a readily assimilable source of nitrogen or as free amino acids is probably low (Critchley, 1969) and variability in amino acid growth requirements within species of oral streptococci has been noted (Inward, Upstone & van Houte, 1970). Urea is present in saliva and can be utilized as a source of nitrogen (Biswas & Kleinberg, 1971). The glucose concentration of saliva is $<$ 1 mg/100 ml but salivary mucoids, each consisting of a polypeptide helix with mucopolysaccharide side chains, are present: these can, presumably, be utilized.

Pantothenic and nicotinic acids have been shown to be essential or stimulatory for the growth of some oral streptococci (Rogers, 1973). However, although water-soluble vitamins are present in saliva (Jenkins, 1970) the concentrations are possibly too low to support optimal growth.

Carbon dioxide has been shown to affect rates of both growth and glycolysis (Pine, 1956; Rogosa, 1964; Gaffney, 1965) and in the mouth will be available from expired air and from bacterial metabolism.

Saliva is hypotonic relative to serum, the main electrolytes in order of decreasing equivalency are, for cations, sodium, potassium and calcium; for anions, chloride, bicarbonate and phosphate. The concentration of these electrolytes is different for each set of glands and alters with increasing flow-rate, calcium and phosphate decreasing, sodium and bicarbonate increasing while potassium and chloride remain relatively constant. The concentration of fluoride is *c.* 0.02 p/m or less (but is surprisingly high in plaque, 20 p/m being typical). It is now known that salivary flow-rate exhibits circadian rhythms of high amplitude and calcium and phosphate concentrations have opposite circadian variations (Dawes, 1972; Ferguson & Fort, 1973) thus providing another source of fluctuation in the ionic environment *in vivo*. Recent work on the flow-rate and composition of the secretions from the minor glands indicates that they differ from the major glands secretions in, for example, a lower inorganic phosphate concentration and very low bicarbonate (Dawes & Wood, 1973). Because salivary secretions differ in composition and in the relative contribution each type of gland makes to the resultant mixture, the whole saliva present in the mouth varies with conditions. Any attempted recipe for 'artificial saliva' can provide only a gross approximation to the oral environment. Moreover, even when other factors are controlled, subject-to-subject variations are very large. To give one typical example, in estimations of inorganic phosphate in parotid secretions, collected without exogenous stimulation from 1026 healthy young, adult, male fasting subjects at 7.30h, the coefficient of variation was 45% (mean, 32 ± 15 mg of P/100 ml) (Shannon, 1967).

The pH of saliva is extremely sensitive to flow-rate and contributes to large subject-to-subject variation, the average value of 6.75 for unstimulated whole saliva range of 5.6–7.6 (Brawley, 1935) the higher figure probably arising from loss of CO_2 and possible production of NH_3 from urea. The range of pH values of plaque is even wider, values as high as 8.0 have been reported after overnight fasting when the plaque pH value is frequently higher than the pH value of the saliva bathing it (Kleinberg & Jenkins, 1964) while the pH value in carious cavities, even without recent exogenous sugar, can be as low as 4.6 and, unlike plaque, may remain at these low values for long periods (Caldwell & Bibby, 1958). Salivary buffers consist of bicarbonate and phosphates (Lilienthal, 1955) therefore the buffering capacity varies at different pH values. The bicarbonate content, low in unstimulated saliva, increases with stimulation, so that when sugar is present in the mouth the saliva thus stimulated will have high buffering capacity. The pH optima for growth *in vitro* of several oral streptococci (Drucker, 1970) are within the range of unstimulated saliva. Saliva is thought to be important in buffering the pH value of plaque since the decrease in plaque pH, after a sugar rinse, is larger and more prolonged than usual when saliva is prevented from reaching the plaque (Englander, Shklair & Fosdick, 1959). However, plaque is buffered by proteins as well as phosphates and under normal physiological conditions plaque pH (Kleinberg & Jenkins, 1964) and calcium phosphate levels (Kleinberg *et al.,* 1971) are influenced by the location of the tooth within the dental arch; mandibular secretions seemed to be more important than parotid secretions in influencing regional plaque levels.

When plaque is allowed to accumulate indefinitely it may undergo calcification and become calculus. Filamentous bacteria such as *Actinomyces* and *Leptotrichia* spp. have been implicated in its formation (Bowen & Gilmour, 1961). Susceptibility to calculus formation varies among individuals. There is evidence that plaque itself does not differ among those who do or do not form calculus readily, the difference lies in the power of saliva to calcify it (discussed by Jenkins, 1970). Calculus is preferentially deposited on tooth surfaces adjacent to the orifices of the major salivary glands.

6. Plaque Acid Production from Exogenous Sugar

When sugar is consumed *in vivo* plaque pH value falls rapidly to pH minima approaching pH 5.0 within 5 to 10 min (Stephan, 1940) owing to the production of both L(+) and D(−) lactic acid (Geddes, 1972a). As the pH value rises to the resting values, approaching neutrality, lactate decreases but acetate and propionate increase. Studies in our laboratory have shown that this pattern of acid production occurs when either glucose or sucrose is the exogenous sugar at dietary levels of carbohydrate excess when, presumably, low levels of utilizable nitrogen would be limiting growth. After overnight starvation and in plaque sampled >2h after exogenous nutrients were supplied, acetate and propionate levels were relatively high, $c.\ 3 \times 10^{-5}$ mmoles/mg wet weight of plaque, and the total volatile acid to total lactate ratio was 6:1 (Geddes, unpublished). Thus the effect of acid production may influence metabolism by varying the pH value as in lactic fermentation (Gunsalus & Niven, 1942) and by producing substrate for the obligate lactate-utilizers but, since acetate and propionate radicals have a toxic effect independent of pH effect (Bergeim, 1940; Meynell, 1963; de Stoppelaar & Gibbons, 1965), it has been suggested that they may serve as another factor regulating the proportion of bacterial types in plaque (Geddes, 1972b).

Even the addition of very low concentrations of sugar (0.5%) has some effect in lowering plaque pH value and if the additions are at, say, 3 min intervals over a period the decrease in pH value continues over that period (Kleinberg, 1961). In this experiment, drops of sugar solution were put on to the plaque direct, contact with the tongue being avoided, so that no stimulation of saliva flow occurred. During normal eating, rapidly flowing saliva with a high pH value bathes the plaque and it is likely that the pH value may remain high until after eating has finished and the saliva flow has decreased. The study of plaque pH during eating presents technical difficulties and has not been adequately studied. With foods containing starch, which tend to remain in the plaque for some time after eating, the pH value may not fall as quickly as with sugar but remain at the minimum for a longer time, an observation difficult to reconcile with the finding that starch is much less cariogenic than sugars in animal experiments. However, the relative cariogenicity has never been tested in man.

Lanke (1957) has studied the rate of sugar clearance from the human mouth by measuring the sugar content of saliva after eating various foods. Clearance times differed greatly among different people but each individual treated different foods in a similar way, if clearance time was long for one food it was also long for all foods. Movements of the tongue and lips after food increased the clearance rate which was longer for starchy foods than for sweets or sugars.

7. Microbial Interactions

Although microbial antagonisms are thought to be of importance in the regulation of the oral flora specific examples are scarce. The complexity of the interactions between different pairs of 7 indigenous oral bacteria, including *Strep. salivarius, Bacterionema matruchotii, Neisseria perflava, Odontomyces viscosus, Fusobacterium* and *Lactobacillus casei,* in batch cultures is shown by the observation that no one species stimulated nor inhibited all others while some species could be both stimulatory and inhibitory (Parker, 1970). *Streptococcus sanguis* has been shown to support the growth of *Strep. mutans in vitro* (Carlsson, 1971). In germ-free rats, inoculated simultaneously with *Veillonella alcalescens* and *Strep. mutans* or *Strep. sanguis* of oral origins developed significantly less caries than control animals inoculated with either streptococcus alone (Mikx *et al.,* 1972). One explanation is that lactate utilization reduces the caries attack. However, increased generation times for *Strep. salivarius* in mixed culture with *V. alcalescens in vitro* have been observed (Parker & Snyder, 1961). The commensal flora of the crevicular crevice together with crevicular fluid would seem to provide the necessary micro-environment for oral spirochaetes which are normally found only in that location (see Davies, 1972).

Intra-species inhibition by bacteriocins has also been suggested (Kelstrup & Gibbons, 1969) and the inhibitory effects of peroxide produced by *Strep. sanguis* on other oral organisms has been demonstrated *in vitro* (Holmberg & Hallander, 1973).

8. Antibacterial Systems of Non-Bacterial Origin

Salivary gland secretions contain a number of antibacterial factors. Lysozyme concentrations in saliva are similar to those in serum but lower than in tears. However, the effectiveness of lysozyme in saliva is probably reduced by the presence of mucin which is known to inhibit its action (Simmons, 1952; Hoerman, Englander & Shklair, 1956). Lysozyme has been shown to have no effect *in vitro* upon a broad spectrum of the indigenous flora (Gibbons, de Stoppelaar & Harden, 1966) but presumably is important in inhibiting contamination by lysozyme-sensitive organisms. The main immunoglobulin of salivary secretions is IgA including secretory IgA, but the concentration is <1% of that of serum. The concentration of IgG and IgM is even lower and saliva does not contain complement. Crevicular fluid, on the other hand, contains IgG, IgA, IgM and complement in approximately the same ratio as serum but at about one third the concentration (Shillitoe, 1972). However, the total volume of crevicular fluid present, even in subjects with established gingivitis, will be very small compared to the volume of saliva so that the antibacterial effects of IgA, IgM and complement would be expected to act only in the immediate area of the gingival crevice. The effectiveness of salivary IgA *in vivo* is uncertain but adsorption of IgA on to oral bacteria (Brandtzaeg, Fjellenger & Gjeruldsen, 1968) has been demonstrated and IgA agglutinates certain organisms. The fact that detectable antibody is present does not indicate the degree of antibacterial activity it exerts and the action of IgA in saliva must at present remain an open question.

Two thiocyanate-dependent antibacterial systems have been detected in saliva. One of these consists of thiocyanate, a peroxidase and H_2O_2, is inhibited by the presence of catalase and functions aerobically. The active reaction product was identified as sulphur dicyanoxide by Oram & Reiter (1966) but Hogg & Jago (1970) gave reasons for believing that it was either cyanosulphurous acid or cyansulphuric acid (HO_2SCN and HO_3SCN, respectively).

The other system consists of thiocyanate and an unidentified salivary protein which functions aerobically as well as anaerobically and is not inhibited by catalase (Dogon & Amdur, 1970). It has been noted that the saliva of smokers was higher in thiocyanate and gave greater inhibitions of lactobacilli in *in vitro* tests, than did saliva of non-smokers (Courant, 1967). We are unaware of any evidence that the antibacterial factor is more active *in vivo* in smokers than in non-smokers. Furthermore, the presence of many oral organisms which are catalase-positive might be expected to militate against the effectiveness of the peroxidase system *in vivo*.

The importance *in vivo* of all these antibacterial systems is not known and *in vitro* tests have been confined mostly to a limited number of strains of lactobacilli and streptococci.

Leucocytes are present in whole saliva having entered the mouth via the gingival crevice (Wright, 1964) where they are present in the serum-like crevicular fluid; however, once in saliva they become swollen, presumably owing to the relatively low osmotic pressure, and disintegrate. The absence of active leucocytes in saliva makes it unlikely that they function as antibacterial agents *in vivo* except, perhaps, in the special microenvironment of the gingival crevice.

9. Dietary Factors

Dental plaque forms even in subjects who have been nourished exclusively by stomach tube (Littleton, McCabe & Carter, 1967) but the amounts and frequency of dietary sugar (sucrose) intake affect plaque accumulation. Many oral organisms (including those shown to produce caries in gnotobiotic animals) produce extracellular polyglucans and levans in the presence of sucrose (Gibbons

& Banghart, 1967). Since polyglucans are not rapidly degraded by oral bacteria and are relatively insoluble in saliva they form much of the matrix of plaque when subjects consume sucrose. Levans, and other polysaccharides formed from sucrose by oral species, were also utilized by oral organisms, in some instances in preference to sucrose (Parker & Creamer, 1972). Many oral species including streptococci, diphtheroids, fusobacteria and bacteroides synthesize intracellular polysaccharide from a variety of exogenous sugars and subsequently utilize these endogenous stores (Gibbons & Socransky, 1962). On the other hand, total restriction of dietary carbohydrate for several days leads to a relative reduction in internal polysaccharide-storing bacteria and the proportion of *Strep. mutans* in plaque. Restitution of the normal diet was accompanied by a return to approximately the original proportions (de Stoppelaar, van Houte & Backer Dirks, 1970). It has been reported in 2 experiments by Jay (1940) and Becks (1950), on *c*. 1600 subjects, that if dietary carbohydrates are limited to 100 g/day for 2 weeks (with a corresponding increase in fat consumption) the lactobacillus counts fall and in many subjects the counts remain low even when the carbohydrates are gradually restored to the diet. The long duration of the effect is difficult to explain, since it would be expected that if any lactobacilli at all remained in the mouth (as they did in about half the subjects) they would rapidly multiply as soon as adequate carbohydrate was made available. In the experiment by Becks (1950), the caries rate was reduced but in Jay's (1940) work only the lactobacillus count was studied. Furthermore, the implantation of streptococci, of oral origin, into the mouth of adult human subjects was facilitated by the increased dietary carbohydrate (Krasse, 1967).

In an experiment still in progress in our laboratory, plaque pH values of 9 subjects were measured before and after 9 days' abstention from tooth cleaning in conjunction with increased amount and frequency of exposure of the plaque to dietary sucrose. In 8 of the 9 subjects, the resting and minimum pH values after test sugar were markedly depressed.

When sucrose in sweets was substituted by either sorbitol or a hydrogenated syrup made from potato starch, less acid was produced *in vivo*. However, it has been suggested that the long term use may, in addition to causing gastrointestinal upsets, lead to an increase in the sorbitol utilizing organisms in the oral flora (Frostell, 1965). The effect of diet on the oral flora has been reviewed by Krasse (1968).

10. Factors Inferred from Presence of Caries

One of the earliest findings on the relation of caries to oral organisms arose from counts of lactobacilli in saliva. It was established during the 1930's and subsequently that counts in caries-free individuals were low or zero compared with 10^4-10^5/ml in saliva from carious mouths. This was often interpreted as evidence for a causal relationship between these organisms and caries although eventually it was realized that, since a low pH value (as in carious cavities) favours lactobacilli, the high count might arise as a result of caries, but the question was never finally decided. More recently, similar results have been quoted for a relationship between *Strep. mutans* and caries (Krasse et al., 1968; Littleton, Kakehashi & Fitzgerald, 1970; Woods, 1971).

This association between caries and these species may be quite non-specific and may apply to the many other species which have not been studied or for which no selective media or reliable method of counting has yet been sought.

The relationship between antibodies in serum and saliva and the intensity of caries is being actively pursued (Challacombe, Guggenheim & Lehner, 1973) but a clear picture has not yet emerged.

11. Effect of Fluoride

The most effective and practicable anticaries agent known is the fluoride ion which decreases the average number of cavities and thereby increases the percentage of subjects who are caries-free. Its main effect is probably exerted by entering the apatite crystals of enamel and converting a proportion of them to the more stable and less soluble fluorapatite. Earlier work suggested that lactobacillus counts were lower in areas with fluoride in the drinking water (Dean et al., 1939) but this difference was not satisfactorily separated from the effect of low counts in the larger proportion of caries-free subjects.

Following the discovery that the dental plaque accumulated fluoride and may contain up to 20 to 30 p/m and even higher concentrations with fluoridated drinking water (Dawes et al., 1965) differences in pH value of plaque have been studied in relation to fluoride. The pH changes *in vitro* after adding sugar to plaques from residents of towns with fluoride in their water are slightly, but consistently, smaller than those from towns without fluoride (Edgar, Jenkins & Tatevossian, 1970). A similar difference was found in plaques from Newcastle residents a few months before and after fluoridation was introduced compared with plaques from a non-fluoridated control. As little difference in caries would be expected over this short period the Newcastle results support an effect on the bacteria rather than effects arising from changes in the number of carious cavities. The effects of fluoride on enamel structure are, however, probably more important for the reduction of caries than those on bacteria.

If these small effects on bacteria are confirmed by future work they could be explained either by differences in bacterial acid production or by changes in the proportions of different species in the plaque. Van Houte et al. (1969) reported that the proportion of bacteria storing intracellular polysaccharides was lower in the plaques of residents of a fluoridated town and this could be associated with differences in change of pH value. More work is needed to co-ordinate chemical and bacteriological studies.

If 1 p/m of fluoride in water influences plaque organisms it would be expected that the much higher concentrations used in mouth rinses (0.05-0.2% of NaF i.e. 250 or 1000 p/m of F) and topical application would have larger effects. Although increased uptake by plaque has been demonstrated following rinses with 0.2% NaF solution, the metabolic effects are not as large as might have been predicted from the observed inhibitions exerted by fluoride on various organisms in pure culture (Bibby & van Kesteren, 1940). Additional work is required by more sensitive methods, but present evidence suggests that even these high concentrations affect enamel rather than bacteria.

12. Conclusions

The inorganic tooth surfaces and the mucosa of the soft tissue serve as selective microbial habitats and the region where they meet, in the gingival crevice, is further modified by crevicular fluid. The salivary secretions *per se*, seem to be suboptimal for supporting the growth, but not metabolism, of the broad spectrum of micro-organisms which comprise the normal oral flora but the limiting factors have not been adequately elucidated. The periodic intake of food by the host provides the micro-organisms with excess exogenous nutrients and at the same time a fluctuating environment from salivary stimulation. The antibacterial factors from saliva and from the normal flora protect the host against contamination by other organisms, which are not normally commensals in the oral cavity. Thorough physical removal of plaque delays recolonization but only temporarily upsets the normal ecological balance.

13. References

BECKS, H. (1950). Carbohydrate restriction in the prevention of dental caries using the *Lactobacillus acidophilus* count as one index. *J.Calif. State Dent. Assoc.* **26,** 53.
BECKS, H. & WAINWRIGHT, W. W. (1943). Human saliva. XIII. Rate of flow of resting saliva of healthy individuals. *J. dent. Res.* **22,** 391.
BERGEIM, O. (1940). Toxicity of intestinal volatile fatty acids from yeast and *Esch. coli*. *J. infect. Dis.* **66,** 222.
BISWAS, S. D. & KLEINBERG, I. (1971). Effect of urea concentration on its utilization, on the pH and the formation of ammonia and carbon dioxide in a human salivary sediment system. *Archs oral Biol.* **16,** 759.
BOWEN, W. H. T. & GILMOUR, M. N. (1961). The formation of calculus-like deposits by pure cultures of bacteria. *Archs oral Biol.* **5,** 145.
BRANDTZAEG, P., FJELLENGER, I. & GJERULDSEN, S. T. (1968). Adsorption of immunoglobulin A onto oral bacteria *in vivo*. *J. Bact.* **96,** 242.
BRAWLEY, R. E. (1935). Studies of the pH of normal resting saliva. I. Variations with age and sex. II. Diurnal variation. III. Effects of vitamins A and D in school children. *J. dent. Res.* **15,** 55, 79, 87.
BIBBY, B. G. & VAN KESTEREN, M. (1940). The effect of fluorine on mouth bacteria. *J. dent. Res.* **19,** 391.
CALDWELL, R. C. & BIBBY, B. G. (1958). The effect of foodstuffs on the pH of dental cavities. *J. Amer. dent. Ass.* **57,** 685.

CARLSSON, J. (1971). Growth of *Streptococcus mutans* and *Streptococcus sanguis* in mixed culture. *Archs oral Biol.* **16**, 963.
CARLSSON, J., GRAHNEN, H., JONSSON, G. & WIKNER, S. (1970). Establishment of *Streptococcus sanguis* in the mouth of infants. *Archs oral Biol.* **15**, 1143.
CHALLACOMBE, S. J., GUGGENHEIM, B. & LEHNER, T. (1973). Antibodies to an extract of *Streptococcus mutans,* containing glucosyltransferase activity, related to dental caries in man. *Archs oral Biol.* **18**, 657.
COURANT, P. (1967). The effect of smoking on the antilactobacillus system in saliva. *Odont. Revy* **18**, 251.
CRITCHLEY, P. (1969). The breakdown of the carbohydrate and protein matrix of dental plaque. *Caries Res.* **3**, 249.
DAVIES, R. M. (1972). General ecology of the commensal microflora of the mouth. In *Host Resistance to Commensal Bacteria.* Ed. T. MacPhee. Edinburgh: Churchill Livingstone.
DAWES, C. (1972). Circadian rhythms in human salivary flow rate and composition. *J. Physiol., Lond.* **220**, 529.
DAWES, C., JENKINS, G. N. & TONGE, C. H. (1963). The nomenclature of the integuments of the enamel surface of teeth. *Brit. dent. J.* **115**, 65.
DAWES, C., JENKINS, G. N., HARDWICK, J. L. & LEACH, S. A. (1965). The relation between the fluoride concentrations in the dental plaque and in drinking water. *Brit. dent. J.* **119**, 164.
DAWES, C. & WOOD, C. M. (1973). The composition of human lip mucous gland secretions. *Archs oral Biol.* **18**, 343.
DEAN, H. T., JAY, P., ARNOLD, F. A. & ELVOVE, E. (1939). Domestic water and dental caries, including certain epidemiological aspects of oral *L. acidophilus. Pub. Health Rep.* **54**, 862.
DE STOPPELAAR, J. D. & GIBBONS, R. J. (1965). Fatty acid inhibition of *Escherichia coli* in the oral cavity. *Internat. Ass. Dent. Res. Abs.*
DE STOPPELAAR, J. D., VAN HOUTE, J. & BACKER DIRKS, O. (1970). The effect of carbohydrate restriction on the presence of *Streptococcus mutans, Streptococcus sanguis* and iodophilic polysaccharide-producing bacteria in human dental plaque. *Caries Res.* **4**, 114.
DOGON, I. L. & AMDUR, B. H. (1970). Evidence for the presence of two thiocyanate-dependent antibacterial systems in human saliva. *Archs oral Biol.* **15**, 987.
DRUCKER, D. B. (1970). Optimum pH values for growth of various plaque streptococci, *in vitro.* In *Dental plaque.* Ed. W. D. McHugh. Edinburgh: Livingstone.
ENGLANDER, H. R., SHKLAIR, I. L. & FOSDICK, L. S. (1959). The effects of saliva on the pH and lactate concentration in dental plaque. I. Caries-rampant individuals. *J. dent. Res.* **38**, 848.
EDGAR, W. M., JENKINS, G. N. & TATEVOSSIAN, A. (1970). The inhibitory action of fluoride on plaque bacteria. Further evidence. *Brit. dent. J.* **128**, 129.
von der FEHR, F. R., LÖE, H. & THEILADE, E. (1970). Experimental caries in man. *Caries Res.* **4**, 131.
FERGUSON, D. B., & FORT, A. (1973). Circadian variations in calcium and phosphate secretion from human parotid and submandibular salivary glands. *Caries Res.* **7**, 19.
FLANAGAN, J. B., CLEMENT, S. C. L. & MOORREES, C. F. A. (1963). The 24-hour pattern of swallowing in man. *Internat. Ass. Dent. Res. Abs.* **165**.
FROSTELL, G. (1965). Substitution of fermentable sugars in sweets. In *Nutrition and Caries-Prevention.* Ed. G. Blix. Uppsala: Almqvist & Wiksell.
GAFFNEY, P. E. (1965). Carbon dioxide effects on glucose catabolism by mixed microbial cultures. *Appl. Microbiol.* **13**, 507.
GEDDES, D. A. M. (1972a). Plaque acids produced during *in vivo* sucrose fermentation. *J. dent. Res.* **51**, 1284.
GEDDES, D. A. M. (1972b). The production of L(+) and D(-) lactic acid and volatile acids by human dental plaque and the effect of plaque buffering acidic strength on pH. *Archs oral Biol.* **17**, 537.
GIBBONS, R. J. & BANGHART, S. B. (1967). Synthesis of extracellular dextran by cariogenic bacteria and its presence in human dental plaque. *Archs oral Biol.* **12**, 11.

GIBBONS, R. J., KAPSIMALIS, B. & SOCRANSKY, S. S. (1964). The source of salivary bacteria. *Archs oral Biol.* **9**, 101.
GIBBONS, R. J. & SOCRANSKY, S. S. (1962). Intracellular polysaccharide storage by organisms in dental plaques. Its relation to dental caries and microbial ecology of the oral cavity. *Archs oral Biol.* **7**, 73.
GIBBONS, R. J., DE STOPPELAAR, J. D. & HARDEN, L. (1966). Lysozyme insensitivity of bacteria indigenous to the oral cavity of man. *J. dent. Res.* **45**, 877.
GILMOUR, M. N. & POOLE, A. E. (1970). Growth stimulation of the mixed microbial flora of human dental plaque by haemin. *Archs oral Biol.* **15**, 1343.
GUNSALUS, I. C. & NIVEN, D. F. (1942). The effect of pH on the lactic acid fermentation. *J. biol. Chem.* **145**, 131.
HANDELMAN, S. L. & MILLS, J. R. (1965). Enumeration of salivary bacterial groups. *J. dent. Res.* **44**, 1343.
HANDLEMAN, S. L. & HESS, C. (1970). Effect of dental prophylaxis on tooth-surface flora. *J. dent. Res.* **49**, 340.
HOERMAN, K. C., ENGLANDER, H. R. & SHKLAIR, J. L. (1956). Lysozyme – its characteristics in human parotoid and submaxillo-lingual saliva. *Proc. Soc. exp. Biol.* **92**, 875.
HOGG, D. McC. & JAGO, G. R. (1970). The antibacterial action of lactoperoxidase. The nature of the bacterial inhibitor. *Biochem. J.* **117**, 779.
HOLMBERG, K. & HALLANDER, H. O. (1973). Production of bactericidal concentration of hydrogen peroxide by *Streptococcus sanguis*. *Archs oral Biol.* **18**, 423.
HURST, V. (1957). Fusiforms in the infant mouth. *J. dent. Res.* **36**, 513.
INWARD, P. W., UPSTONE, D. & VAN HOUTE, J. (1970). Nutritional requirements of oral streptococci. In *Dental Plaque*. Ed. W. D. McHugh. Edinburgh: Livingstone.
JAY, P. (1940). The role of sugar in the aetiology of dental caries. *J. Amer. dent. Ass.* **27**, 393.
JENKINS, G. N. (1970). *The Physiology of the Mouth*, 3rd ed. Oxford: Blackwell.
KELSTRUP, J. & GIBBONS, R. J. (1969). Bacteriocins from human and rodent streptococci. *Archs oral Biol.* **14**, 251.
KELSTRUP, J., RICHMOND, S., WEST, C. & GIBBONS, R. J. (1970). Finger-printing human oral streptococci by bacteriocin production and sensitivity. *Archs oral Biol.* **15**, 1109.
KLEINBERG, I. (1961). Studies on dental plaque, I. The effect of different concentrations of glucose on the pH of dental plaque *in vivo*. *J. dent. Res.* **40**, 1087.
KLEINBERG, I., CHATTERJEE, R., KAMINSKY, F. S., CROSS, H. G., GOLDENBERG, D. J. & KAUFMAN, H. W. (1971). Plaque formation and the effect of age. *J. Periodont.* **42**, 497.
KLEINBERG, I. & JENKINS, G. N. (1964). The pH of dental plaques in the different areas of the mouth before and after meals and their relationship to the pH and rate of flow of resting saliva. *Archs oral Biol.* **9**, 493.
KRASSE, B. (1963). Oral aggregations of microbes. *J. dent. Res.* **42**, 521.
KRASSE, B. (1968). In *Art and Science of Dental Research*. Ed. R. S. Harris. London: Academic Press.
KRASSE, B., EDWARDSSON, S., SVENSSON, I. & TRELL, L. (1967). Implantation of caries-inducing streptococci in the human oral cavity. *Archs oral Biol.* **12**, 231.
KRASSE, B., JORDAN, H. V., EDWARDSSON, S., SVENSSON, I. & TRELL, L. (1968). The occurrence of certain 'caries-inducing' streptococci in human dental plaque material. *Archs oral Biol.* **13**, 911.
LANKE, L. S. (1957). Influence on salivary sugar of certain properties of foodstuffs and individual oral conditions. *Acta odont. Scand.* **15**, Suppl. 23.
LILIENTHAL, B. (1955). An analysis of the buffer systems in saliva. *J. dent. Res.* **34**, 516.
LITTLETON, N. W., KAKEHASHI, S. & FITZGERALD, R. J. (1970). Recovery of specific 'caries-inducing' streptococci from carious lesions in the teeth of children. *Archs oral Biol.* **15**, 461.

LITTLETON, N. W., McCABE, R. M. & CARTER, C. H. (1967). Studies of oral health in persons nourished by stomach tube. II. Acidogenic properties and selected bacterial components of plaque material. *Archs oral Biol.* **12**, 601.

LONGHURST, P. & BERMAN, D. S. (1973). Apples and gingival health. Report of a feasibility study. *Brit. dent. J.* **134**, 475.

McCARTHY, C., SNYDER, M. L. & PARKER, R. B. (1965). The indigenous oral flora of man. I. The newborn to the 1-year-old infant. *Archs oral Biol.* **10**, 61.

MEYNELL, G. G. (1963). Antibacterial mechanism of the mouse gut. III. The fate of *Staphylococcus aureus* in normal and streptomycin-treated mice. *B.J. Exp. Path.* **44**, 625.

MIKX, F. H. M., VAN DER HOEVEN, J. S., KONIG, K. G., PLASSCHAERT, A. J. M. & GUGGENHEIM, B. (1972). Establishment of defined microbial ecosystems in germ free rats. I. Interaction of *Streptococcus mutans* or *Streptococcus sanguis* with *Veillonella alcalescens* on plaque formation and caries activity. *Caries Res.* **6**, 211.

ORAM, J. D. & REITER, B. (1966). The inhibition of streptococci by lactoperoxidase, thiocyanate and hydrogen peroxide. *Biochem. J.* **100**, 373.

PARKER, R. B. (1970). Paired culture interaction of the oral microbiota. *J. dent. Res.* **49**, 804.

PARKER, R. B. & CREAMER, H. R. (1972). Contribution of plaque polysaccharides to growth of cariogenic micro-organisms. *Archs oral Biol.* **16**, 855.

PARKER, R. B. & SYNDER, M. L. (1961). Interactions of the oral microbiota. I. A system for the defined study of mixed cultures. *Proc. Soc. exp. Biol. (N.Y.)* **108**, 749.

PINE, L. (1956). Fixation of carbon dioxide by *Actinomyces* and *Lactobacillus bifidus*. *Proc. Soc. exp. Biol. Med.* **93**, 468.

RITZ, H. L. (1967). Microbial population shifts in developing human dental plaque. *Archs oral Biol.* **12**, 1561.

RITZ, H. L. (1969). Fluorescent antibody staining of *Neisseria, Streptococcus* and *Veillonella* in frozen sections of human dental plaque. *Archs oral Biol.* **14**, 1073.

ROGERS, A. H. (1973). The vitamin requirements of some oral streptococci. *Archs oral Biol.* **18**, 227.

ROGOSA, M. (1964). The genus *Veillonella*. I. General cultural, ecological and biochemical considerations. *J. Bact.* **87**, 162.

SALKIND, A., OSHRAIN, H. I. & MANDEL, I. D. (1971). Bacterial aspects of supragingival and subgingival plaque. *J. Periodont.* **42**, 706.

SAXTON, C. A. & CRITCHLEY, P. (1970). An electron microscope investigation of the effect of diminished protein synthesis on the morphology of the organisms in dental plaque *in vitro*. In *Dental plaque*. Ed. W. D. McHugh. Edinburgh: Livingstone.

SHANNON, I. L. (1967). A formula for human parotid fluid collected without exogenous stimulation. *J. dent. Res.* **46**, 309.

SHILLITOE, E. J. (1972). Immunoglobulins and complement in crevicular fluid. In *Host resistance to Commensal Bacteria*. Ed. T. MacPhee, Edinburgh: Churchill Livingstone.

SIMMONS, N. S. (1952). Studies on the defense mechanisms of the mucous membranes with particular reference to the oral cavity. The bacteriolytic activity of saliva. *Oral Surg.* **5**, 513.

SOCRANSKY, S. S. & MANGANIELLO, S. D. (1971). The oral microbiota of man from birth to senility. *J. Periodont.* **42**, 485.

SOCRANSKY, S. S., MANGANIELLO, A. D., ORAM, V., PROPAS, D., DOGON, I. L. & VAN HOUTE, J. (1971). Development of early dental plaque. *Internat. Ass. Dent. Res. Abs.* 502.

STEPHAN, R. M. (1940). Changes in hydrogen-ion concentration on tooth surfaces and in carious lesions. *J. Amer. dent. Ass.* **27**, 718.

VAN HOUTE, J., BACKER DIRKS, O., DE STOPPELAAR, J. D. & JANSEN, H. M. (1969). Iodophilic polysaccharide-producing bacteria and dental caries in children consuming fluoridated and non-fluoridated drinking water. *Caries Res.* **3**, 178.

VAN HOUTE, J. & SAXTON, C. A. (1971). Cell wall thickening and intracellular polysaccharide in micro-organisms of dental plaque. *Caries Res.* **5**, 30.

WILLIAMS, N. B. & POWLEN, D. O. (1959). Human parotid saliva as a sole source of nutrient for micro-organisms. *Archs oral Biol.* **1**, 48.

WOODS, R. (1971). A dental caries susceptibility test based on the occurrence of *Streptococcus mutans* in plaque material. *Aust. dent. J.* **16**, 116.

WRIGHT, D. E. (1964). The source and rate of entry of leucocytes in the human mouth. *Archs oral Biol.* **9**, 321.

Control of Oral Flora by Hibitane and Other Antibacterial Agents

R. M. DAVIES

Department of Oral Medicine, University of Manchester Dental Hospital, Bridgeford Street, Manchester M15 6FA, England

CONTENTS

1. Introduction 101
2. Antibacterial agents 102
   (a) Theoretical considerations 102
   (b) Antibiotics 102
   (c) Synthetic antibacterial agents 103
      (i) Saliva 104
      (ii) Gingiva and tooth surface 104
3. References 108

1. Introduction

DENTAL PLAQUE, the soft adherent deposit of bacteria and intercellular matrix which forms on the surfaces of teeth, plays an essential role in the aetiology of dental caries and periodontal disease. Bacteria colonize the tooth surface rapidly: between 5 and 20 min after receiving a thorough cleaning 10^6 organisms/cm^2 of tooth surface were counted (Socransky et al., 1971). The predominant organisms in early deposits are Gram positive cocci (Howell, Rizzo & Paul, 1965; Löe, Theilade & Jensen, 1965; Ritz, 1967; Theilade & Theilade, 1970) but after c. 10 days of no oral hygiene, plaque comprises a dense, mixed bacterial population (Löe et al., 1965). Total microscopic count of dental plaque averaged 2.5×10^{11} bacteria/g and total viable counts performed aerobically and anaerobically averaged 2.5×10^{10} and 4.6×10^{10}/g, respectively (Gibbons et al., 1964). The predominant cultivable organisms in plaque were: facultative streptococci, 27%; facultative diphtheroids, 23%; anaerobic diphtheroids, 18%; peptostreptococci, 13%; *Veillonella* spp., 6%; *Bacteroides* spp., 4%; fusobacteria, 4%; *Neisseria* spp., 3% and vibrios, 2%. Certain species of bacteria and particularly *Streptococcus mutans* have been shown to be potentially cariogenic (Guggenheim, 1968; Keyes, 1968; Krasse & Carlsson, 1970) but the relative significance of the different plaque bacteria in periodontal disease has not been established (Socransky, 1970). Therefore, any treatment regime designed to prevent or control periodontal disease and dental caries should ensure the regular removal of all bacterial deposits from the tooth surfaces. Tooth brushing and other mechanical measures will, if performed meticulously, control the formation of plaque (Löe et al., 1965) but the maintenance of a level of oral hygiene compatible with dental health is difficult to achieve by these means (Koch & Lindhe, 1965; Suomi et al., 1971). In an attempt to improve the effectiveness of existing oral hygiene measures antibacterial agents have been evaluated for their ability to prevent or control plaque formation.

2. Antibacterial agents

(a) *Theoretical considerations*

Antibacterial agents could control plaque and dental disease in a number of ways. One possibility is that the agent might suppress the total oral flora to such an extent that bacterial colonization of the tooth surface would not occur (Löe & Rindom Schiøtt, 1970 *a, b*). Other contributors to this symposium, however, have shown that the oral flora is extremely complex and that the nature of the predominant groups of bacteria depends upon the location within the mouth. In other words, the composition of the flora of the tongue, saliva and tooth surface differ from each other. Therefore, suppression of the total oral flora would be unlikely to be obtained in practice. One practical approach might be to try and control plaque formation specifically by interfering with bacteria on the tooth surface. Another possibility might be to reduce the pathogenic potential of plaque by altering its properties and composition. These ideas have been tested in a number of ways by several investigators and the purpose of this paper is to review the results of such studies.

(b) *Antibiotics*

In a typical study (Löe et al., 1967) the teeth of volunteers were scaled and polished and strict oral hygiene measures instituted to ensure that the teeth were clinically free of plaque at the outset. Throughout the experimental period subjects ceased all mechanical oral hygiene procedures and rinsed their mouths, 3 times a day, with one of the following solutions: 0.5% tetracyline, 0.5% vancomycin, 0.25% polymyxin B or distilled water. During the 5 day experimental period the distribution of morphologically distinct groups of bacteria was determined in smear preparations of dental plaque; also samples from the dorsal surface of the tongue were inoculated on Sabouraud agar to detect the presence of yeasts. Plaque accumulated rapidly in the control group and its bacterial composition underwent the sequence of changes described by Löe et al. (1965) and Theilade et al. (1966). There was a proliferation of Gram positive cocci and short rods initially and this was followed by the appearance of

Gram negative cocci and bacilli and filamentous and fusiform organisms. Tetracycline markedly decreased numbers of all plaque bacteria. In contrast, polymyxin depressed Gram negative organisms and the plaque which formed consisted mainly of Gram positive cocci and rods; also, vancomycin depressed the Gram positive flora and smears were characterized by the presence of Gram negative cocci, bacilli, filaments and fusiforms. These observations were supported by the clinical findings: tetracycline reduced the formation of plaque to a greater extent than did either vancomycin or polymyxin. The flora at the gingival margin and on the tongue underwent similar changes to those described on the tooth surface except that the incidence of *Candida albicans* on the tongue increased in the tetracycline group. In all 3 groups using antibiotics the bacterial flora of the tongue became less susceptible to the antibiotic used.

Volpe *et al.* (1969), using a different regime and concentration of vancomycin, confirmed that this antibiotic significantly reduced the Gram positive population at the dento-gingival junction but also found that the incidence of yeasts did not increase. The same workers reported that the macrolide antibiotic, CC 10232, consistently reduced plaque formation when used in a mouth rinse. The antibiotic did not significantly reduce the proportion of Gram positive organisms and no change was observed in the staphylococcal or yeast populations.

In the short term studies described, most antibiotics produced pronounced shifts in the balance of the oral flora and the potential hazards which accompany such changes are well documented (Seelig, 1966; Marples & Kligman 1971; Leyden & Marples, 1973). It is unlikely, therefore, that antibiotics can offer a practical solution to the problem of plaque control in the population although vancomycin (Mitchell & Holmes, 1965) and kanamycin (Loesche *et al.*, 1971) have been used to control plaque formation in mentally retarded patients in institutions. In an effort to overcome some of the problems inherent in the use of antibiotics, a number of synthetic antibacterial agents have been investigated.

(c) *Synthetic antibacterial agents*

Although many compounds, such as cetylypyridinium chloride, benzalkonium chloride and chlorhexidine salts, have been shown to reduce plaque formation (Schroeder, 1969; Volpe *et al.*, 1969; Gjermo, 1972) few studies have assessed their effect on the oral flora.

It has been demonstrated recently that, in the absence of all active oral hygiene measures, 2 daily rinses with a 0.2% solution of chlorhexidine gluconate prevented the formation of plaque and the development of gingivitis (Löe & Rindom Schiøtt, 1970 *a, b*). The effect of chlorhexidine on the oral flora has been determined in some of these studies and the results will now be summarized.

(i) *Saliva*

Samples of saliva were collected throughout the experiments and the number of bacteria/ml was calculated as total aerobes, anaerobes, streptococci and veillonellae (Rindom Schiøtt *et al.* 1970). In some cases the number of enteric bacteria and yeasts also was determined. The average number of bacteria in saliva is shown in Table 1 and the change in the total number of aerobes and anaerobes, expressed as a percentage of day 0 values, is shown in Fig. 1. In the control group cessation of oral hygiene measures over a 22 day period resulted in a 300% increase in the total number of aerobes and anaerobes whereas chlorhexidine caused an 85-90% reduction. Despite this seemingly large reduction the saliva still contained great numbers of organisms and when rinsing stopped the number of bacteria returned to pre-experimental levels within 48h. Similar results were obtained in studies lasting up to 40 days (Davies *et al.*, 1971) and no overgrowth of enteric organisms or yeasts was observed.

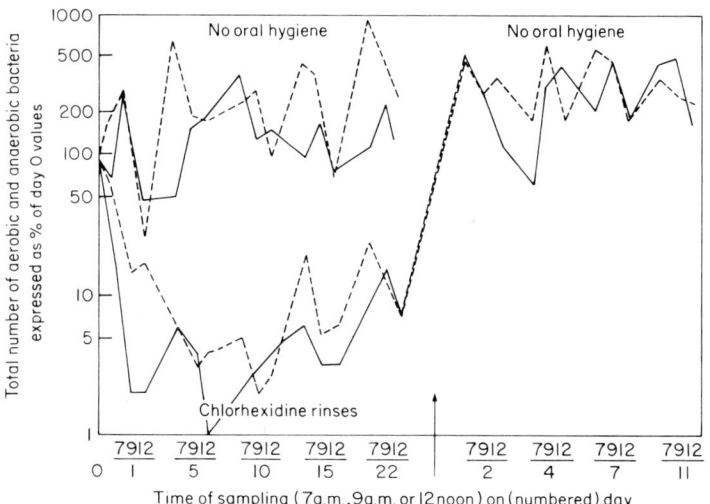

Fig. 1. Average changes in the bacterial content of saliva of one group of subjects using oral chlorhexidine rinses and another group without oral hygiene. Arrow indicates time when the experimental group ceased chlorhexidine treatment. ———, total anaerobic count; – – –, total aerobic count. (Data from Rindom Schiøtt *et al.*, 1970.)

(ii) *Gingiva and tooth surface*

The effect of chlorhexidine rinses on the bacterial flora of the gingiva and tooth surface was studied with the aid of impression preparations. Two impression preparations were taken from the upper right lateral incisor and the adjacent gingiva on each day of examination using the technique described by

Table 1

Average number of bacteria ($\times 10^6$)/ml of saliva in an experimental group of subjects rinsing with chlorhexidine and a control group without oral hygiene (Schiøtt et al., 1970).

	Time of day	Count on day											
		0		1		5		10		15		22	
		Exp.*	Contr.†	Exp.	Contr.	Exp.	Contr.	Exp.	Contr.	Exp.	Contr.	Exp.	Contr.
Total aerobes	7 a.m.	269	510	145	978	14	3162	13	1118	49	2238	62	4072
	9 a.m.	97	110	14	320	3	204	2	310	5	374	11	440
	12 noon	121	258	20	61	4	425	3	227	7	178	8	566
Total anaerobes	7 a.m.	800	1529	150	995	5	769	13	5611	50	1353	60	1553
	9 a.m.	161	192	3	553	6	295	5	240	5	294	23	406
	12 noon	254	258	4	118	3	447	11	357	9	204	20	267
Aerobic streptococci	7 a.m.	207	227	137	415	2	329	8	982	20	578	25	198
	9 a.m.	26	47	1	170	0.6	185	0.9	109	4	518	14	148
	12 noon	90	53	2	52	2	44	1	112	2	55	6	145
Veillonella spp.	7 a.m.	137	342	37	201	1	14	1	1092	1	21	1	205
	9 a.m.	36	66	1	67								
	12 noon	56	22	1	13	1	7	1	22	1	15	1	82

* Experimental group
† Control group

Löe et al. (1965); one preparation was stained with gentian violet and the other with Gram stain. The numbers and morphological types of bacteria on the marginal and attached gingiva and the tooth surface were assessed.

At the outset the gingiva of all subjects was composed of Gram positive organisms. On cessation of oral hygiene the gingival flora of control (untreated) subjects proliferated and its composition closely resembled that of the tooth surface. Chlorhexidine reduced the flora, particularly at the gingival margin, to such an extent that large areas of epithelium were devoid of bacteria. In some areas, however, certain changes were observed in the composition of the gingival flora. The most conspicuous change was a temporary increase in the proportion of Gram negative bacilli many of which exhibited gliding motility (Davies et al., 1972).

A sparse flora of Gram positive cocci and short bacilli was present on the tooth surface at the start of the experiment. On cessation of all active oral hygiene measures bacterial colonization occurred rapidly in the control subjects and the bacterial composition of the developing plaque underwent the sequence of changes already described (Theilade et al., 1966). Chlorhexidine prevented the formation of plaque completely in most subjects and few, if any, bacteria were observed on the smooth surfaces of the teeth for up to 40 days. In 4 students, however, a thin line of plaque developed at the gingival margin of the canine and premolar teeth after 21 days rinsing with chlorhexidine. The numbers of different colony types of streptococci were determined and related to the total number of streptococci isolated on *mitis salivarius* agar: the predominant streptococci in all plaque samples were *Streptococcus sanguis* (Davies, Schiøtt & Löe, 1973). Finally, when chlorhexidine rinses were stopped, the bacterial composition of the plaque which subsequently accumulated was similar to that of the control group.

These experiments were performed initially to test the hypothesis that suppression of the total oral flora at regular intervals would reduce the bacterial colonization of the tooth surfaces. Although rinsing, twice daily, with a 0.2% solution of chlorhexidine reduced the salivary flora by 85-90%, the saliva still contained large numbers of organisms. The reduction in plaque formation, however, was far greater than might have been accounted for by the decrease in salivary organisms, unless there was a specific effect preventing salivary bacteria attaching to teeth and gingiva. In order to test whether the important site of action of chlorhexidine was the tooth surface, a 2% solution of chlorhexidine was applied topically, once a day for 15 days to the teeth of 6 volunteers. This method of application did not reduce the salivary flora but the tooth surfaces remained practically free of bacteria (Davies et al., 1970). These observations suggested, therefore, that the plaque-preventing effect of chlorhexidine was primarily the result of a local antibacterial action at the tooth surface.

Rolla, Löe & Rindom Schiøtt (1970, 1971) concluded from *in vitro*

Table 2
Selection of mutants resistant to chlorhexidine (Hennessey, 1973).

Organism		Number of successive subcultures					
		0	5	10	16	23	40
Streptococcus mutans		0.78*	1.56	1.56	6.25	1.56	3.12
Streptococcus sanguis		0.78	0.78	3.12	3.12	N.T.	N.T.
'Plaque streptococcus' { 1		0.19	3.12	3.12	3.12	3.12	6.25
2		3.12	6.25	25	6.25	25	6.25
3		1.56	3.12	3.12	3.12	1.56	6.25
4		3.12	12.5	12.5	12.5	12.5	25

* Figure in Table is the concentration of chlorhexidine (μg/ml) tolerated by the organisms.
N.T. Not tested.

experiments that hydroxyapatite, teeth and proteins adsorb chlorhexidine and acquire antibacterial activity in the process. The process of adsorption was reversible and these workers suggested that, during a mouthrinse or topical application, reservoirs of chlorhexidine are formed on the teeth and/or mucous membranes. Subsequently, the slow release of the agent prevented plaque formation for a few hours. It may be noted that Jensen & Christensen (1971) detected antibacterial concentrations of chlorhexidine in saliva for $c.$ 5h after a single mouthrinse.

There is some concern that the prolonged use of chlorhexidine in the oral cavity might lead to proliferation of either resistant organisms or new opportunist pathogens. Such changes might reduce the clinical efficacy of the agent or produce undesirable side-effects. In order to predict possible changes in the susceptibility of the oral flora during the potential use of chlorhexidine, various strains of streptococci were passaged in the presence of sub-inhibitory concentrations of the agent (Hennessey, 1973). Althouth the susceptibility of streptococci appeared to decrease (Table 2) the changes were of doubtful significance because minor fluctuations were frequently encountered during these experiments. Rindom Schiøtt & Löe (1972) examined the sensitivity of oral streptococci isolated from persons who had used chlorhexidine regularly for 6 months. They reported that although there was a slight decrease in sensitivity the change was relatively inconspicuous.

In conclusion, these studies with chlorhexidine have demonstrated that plaque formation might be controlled by selectively inhibiting bacterial growth on the tooth surface. The potential value of such an agent in the prevention of dental caries and periodontal disease justifies further research into the use of antibacterial compounds. Whether the long term use of chlorhexidine in the oral cavity will prove to be an effective and safe procedure must await the results of extended clinical trials.

3. References

DAVIES, R. M., JENSEN, S. B., RINDOM SCHIØTT, C. & LÖE, H. (1970). The effect of topical application of chlorhexidine on the bacterial colonization on the teeth and gingiva. *J. periodont. Res.* **5,** 976.

DAVIES, R. M., LÖE, H., SCHIØTT, C. R. & JENSEN, S. B. (1971). Chlorhexidine in the prevention of gingivitis. In *The Prevention of Periodontal Disease.* Eds J. E. Eastoe, D. C. A. Picton & A. G. Alexander. London: Henry Kimpton.

DAVIES, R. M., JENSEN, S. B., RINDOM SCHIØTT, C., LÖE, H. & THEILADE, J. (1972). Anaerobic gliding bacteria isolated from the oral cavity. *Acta Path. Microbiol. Scand. Section B.* **80,** 397.

DAVIES, R. M., SCHIØTT, C. R. & LOE, H. (1973). Streptococci isloated from plaque in subjects rinsing with chlorhexidine. *Archs oral Biol.* **18,** 297.

GIBBONS, R. J., SOCRANSKY, S. S., de ARAUJO, W. C. & van HOUTE, J. (1964). Studies of the predominant cultivable microbiota of dental plaque. *Archs oral Biol.* **9,** 365.

GJERMO, P. (1972). Chemical cleaning of teeth. In *Oral Hygiene.* Ed. A. Frandsen. Copenhagen: Munksgaard.

GUGGENHEIM, B. (1968). Streptococci of dental plaques. *Caries Res.* **2**, 147.
HENNESSEY, T. D. (1974). Some antibacterial properties of chlorhexidine. *J. periodont. Res.* **8**, Suppl. 12, 61.
HOWELL, A., RIZZO, A. & PAUL, F. (1965). Cultivable bacteria in developing and mature human dental calculus. *Archs oral Biol.* **10**, 307.
JENSEN, J. & CHRISTENSEN, F. (1971). A study of the elimination of chlorhexidine from the oral cavity using a new spectrophotometric method. *J. periodont. Res.* **6**, 306.
KEYES, P. (1968). Research in dental caries. *J. Am. Dent. Assoc.* **76**, 1357.
KOCH, G. & LINDHE, J. (1965). The effect of supervised oral hygiene on the gingiva of children. The effect of toothbrushing. *Odont. Revy* **16**, 327.
KRASSE, B. & CARLSSON, J. (1970). Various types of streptococci and experimental caries in hamsters. *Archs oral Biol.* **15**, 25.
LEYDEN, J. J. & MARPLES, R. R. (1973). Ecologic principles and antibiotic therapy in chronic dermatoses. *Archs Dermatol.* **107**, 208.
LÖE, H., THEILADE, E. & JENSEN, S. B. (1965). Experimental gingivitis in man. *J. Periodont.* **36**, 177.
LÖE, H., THEILADE, E., JENSEN, S. B. & RINDOM SCHIØTT, C. (1967). Experimental gingivitis in man. III. The effect of antibiotics on gingival plaque development. *J. periodont. Res.* **2**, 282.
LÖE, H. & RINDOM SCHIØTT, C. (1970a). The effect of suppression of the oral microflora upon the development of dental plaque and gingivitis. In *Dental Plaque*. Ed. W. D. McHugh. Edinburgh: Livingstone.
LÖE, H. & RINDOM SCHIØTT, C. (1970b). The effect of mouthrinses and topical application of chlorhexidine on the development of dental plaque and gingivitis in man. *J. periodont. Res.* **5**, 79.
LOESCHE, W., GREEN, E., NAFE, D. & KENNY, E. (1971). The effect of topical kanamycin on plaque accumulation. *I.A.D.R. 49 Gen. Meet. Chicago. Abstr. No. 281*, p. 123.
MARPLES, R. R. & KLIGMAN, A. M. (1971). Ecological effects of oral antibiotics on the microflora of human skin. *Archs Dermatol.* **103**, 148.
MITCHELL, D. F. & HOLMES, L. A. (1965). Topical antibiotic maintenance of oral health. *J. oral Ther.* **4**, 83.
RINDOM SCHIØTT, C., LÖE, H., JENSEN, S. B. KILIAN, M., DAVIES, R. M. & GLAVIND, K. (1970). The effect of chlorhexidine mouthrinses on the human oral flora. *J. periodont. Res.* **5**, 84.
RINDOM SCHIØTT, C. & LÖE, H. (1972). The sensitivity of oral streptococci to chlorhexidine. *J. periodont. Res.* **7**, 192.
RITZ, H. L. (1967). Microbial population shifts in developing human dental plaque. *Archs oral Biol.* **12**, 1561.
ROLLA, G., LÖE, H. & RINDOM SCHIØTT, C. (1970). The affinity of chlorhexidine for hydroxyapatite and salivary mucins. *J. periodont. Res.* **5**, 90.
ROLLA, G., LÖE, H. & RINDOM SCHIØTT, C. (1971). Retention of chlorhexidine in the human oral cavity. *Archs oral Biol.* **16**, 1109.
SCHROEDER, H. E. (1969). *Formation and Inhibition of Dental Calculus*. Berne: Hans Huber.
SEELIG, M. S. (1966). The role of antibiotics in the pathogenesis of *Candida* infections. *Am. J. Med.* **40**, 887.
SOCRANSKY, S. S. (1970). Relationship of bacteria to the aetiology of periodontal disease. *J. dent. Res.* **49**, 203.
SOCRANSKY, S. S., MANGANIELLO, A. D., ORAM, V., PROPAS, D., DOGON, I. L. & van HOUTE, J. (1971). Development of early dental plaque. *I.A.D.R. 49 Gen. Meet. Chicago. Abstr. No. 502*, p. 178.
SUOMI, J. D., GREENE, J. C., VERMILLION, J. R., DOYLE. J., CHANG, J. J. & LEATHERWOOD, E. C. (1971). The effect of controlled oral hygiene procedures on the progression of periodontal disease in adults. Results after third and final year. *J. Periodont.* **42**, 152.

THEILADE, E., WRIGHT, W. H., JENSEN, S. B. & LÖE, H. (1966). Experimental gingivitis in man. II. A longitudinal clinical bacteriological investigation. *J. periodont. Res.* **1**, 1.

THEILADE, E. & THEILADE, J. (1970). Bacteriological and ultrastructural studies of developing dental plaque. In *Dental Plaque.* Ed. W. D. McHugh. Edinburgh: Livingstone.

VOLPE, A. R., KUPCZAK, L. J., BRANT, J. H., KING, W. J., KESTENBAUM, R. C. & SCHLISSEL, H. J. (1969). Antimicrobial control of bacterial plaque and calculus and the effects of these agents on the oral flora. *J. dent. Res.* **48**, 832.

Structural and Ecological Aspects of Dental Plaque

H. N. NEWMAN AND D. F. G. POOLE

*M.R.C. Dental Unit, The Dental School,
Lower Maudlin Street, Bristol BS1 2LY, England*

CONTENTS

1. Introduction	.	111
2. Plaque formation	.	112
3. The acquired pellicle	.	113
4. The attachment of plaque to the tooth	.	114
5. The plaque flora	.	115
6. Plaque structure	.	116
7. Interbacterial contacts	.	118
8. Filament formation and plaque	.	118
9. The dynamics of plaque formation	.	120
10. Plaque and other natural microbial films	.	120
11. Plaque and disease	.	124
12. References	.	127

1. Introduction

TWO OF THE COMMONEST of human diseases are, between them, responsible for the generalized tooth loss which is a feature of advancing age in so-called civilized society. Dental caries, the familiar tooth decay, and chronic inflammatory periodontal disease, which gradually erodes the supporting tissues of the teeth, are responsible for virtually all human tooth loss and, therefore, an enormous loss in man-hours of work, due both to local and systemic effects and to the current methods of treatment of these conditions. Furthermore, the pain which is the outstanding symptom of dental caries, and the suffering which is an almost inevitable concomitant of present treatment methods, provide ample reason for current intensive research into causes, mode of onset and possible methods of prevention of these diseases. Many of these studies are concerned with dental plaque, the natural bacterial layer present on the erupted surfaces of human teeth. For it is believed that at least some of the bacteria of plaque are responsible for the production of acid which results in caries, and for the destruction of dental supporting tissues which occurs in chronic inflammatory periodontal disease.

It should, perhaps, be emphasised at the outset that the mere presence of bacteria on the natural tooth surface is not, in itself, indicative of disease. To cite the example of caries, plaque bacteria may be found on the enamel surface, and even in natural pits in that surface (Plates 1 & 2), without there being any evidence of carious attack of the underlying enamel crystallites (Newman & Poole, 1973).

It is now over eighty years since this microbial film was first described by Miller (1890), who postulated that both bacteria and dietary carbohydrate were needed to produce dental caries, and that toxic metabolites of oral bacteria were involved in chronic inflammatory periodontal disease. While some organisms may contribute to caries, by being more acidogenic and aciduric than others, there is no more definite evidence today than there was in the time of Miller (1890) that either caries or chronic inflammatory periodontal disease is caused by a single bacterial strain or species. Furthermore, investigations of the pathogenicity of plaque have, so far, been little concerned with possible beneficial effects of this microbial film in relation to oral and systemic health. Nor has plaque been compared from an ecological standpoint with other naturally occurring microbial films.

The present review is concerned with describing the formation, growth and structure of natural human dental plaque and with considering some of the similarities between plaque and other natural microbial films. At the same time, an attempt has been made to correlate some of these features of plaque biology with the role of this portion of man's natural microbial flora in disease and health.

2. Plaque Formation

How does plaque form? Current theories suggest that, as the tooth enters the mouth, there is present on the surface of the crystallites, of which the tooth enamel surface is composed, a very thin organic cuticle, the remains of the formative epithelium of the tooth crown (Plate 3). It is this cuticle that is initially colonized by plaque bacteria (Newman, 1973a, b), sometimes accompanied by intervening material which has been shown by a number of workers to consist of salivary glycoprotein (Plate 4).

Staining of the crowns of erupting teeth reveals the distribution of plaque over the tooth surface. Plaque bacteria are found in the fissures in the biting surfaces of the teeth, and on the smooth surfaces. The organic integument of the smooth enamel surface is a zoned film. Two of these zones, which appear to be composed primarily of the formative epithelium of the enamel, are, in health, covered by the gum or gingiva. The third zone is present on that portion of the tooth crown exposed in the mouth, and is composed of plaque and salivary pellicle. Its topography is determined by the cleansibility of the surface, that is, the ability of such agencies as tongue and toothbrush to limit plaque deposition and growth. Plaque is, therefore, most abundant in the most stagnant sites on the surface, especially around the area where adjacent teeth are in contact. The contact area itself is, however, almost free of plaque (Plate 5). The natural enamel

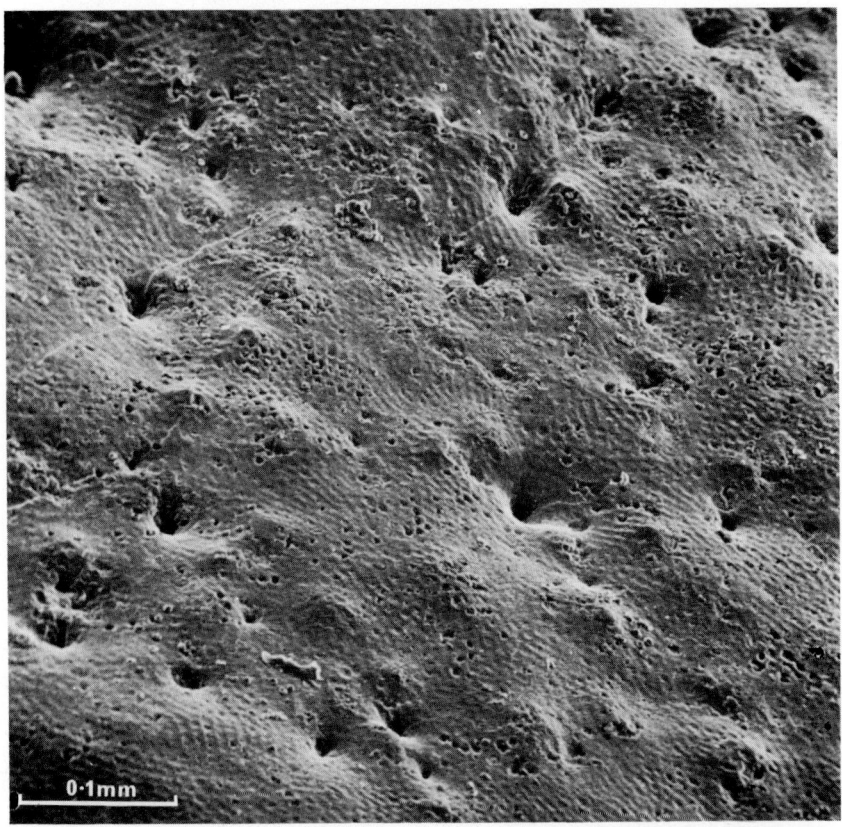

Plate 1. Scanning electron micrograph (SEM). The natural human enamel surface is marked by a variety of irregularities in which bacteria may lodge when the tooth enters the mouth.

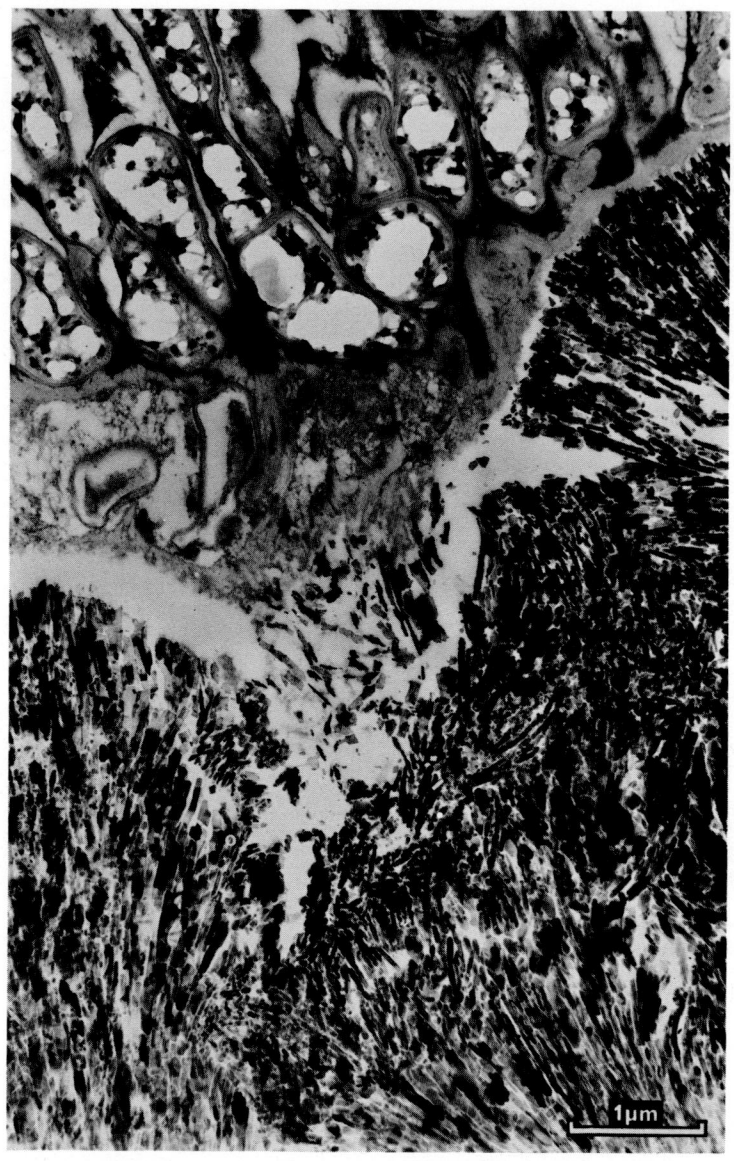

Plate 2. Transmission electron micrograph (TEM). Bacteria may be present in a natural surface pit without there being any evidence of carious demineralization of the underlying enamel crystallites.

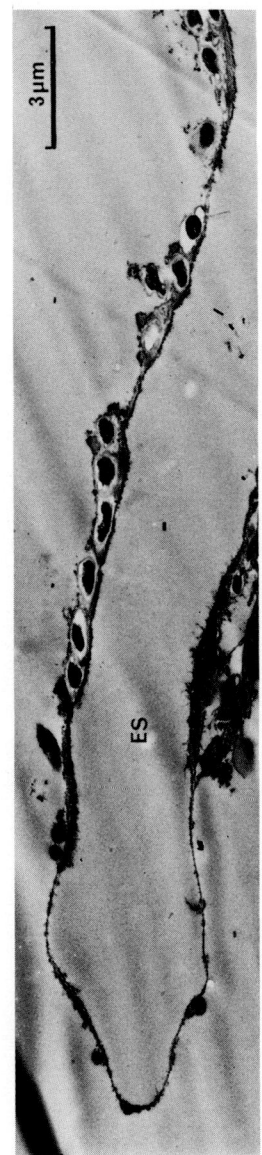

Plate 3. TEM. Dental plaque bacteria are initially deposited on a thin cuticle derived from the formative epithelium of the enamel, as seen on this demineralized specimen. ES = enamel space.

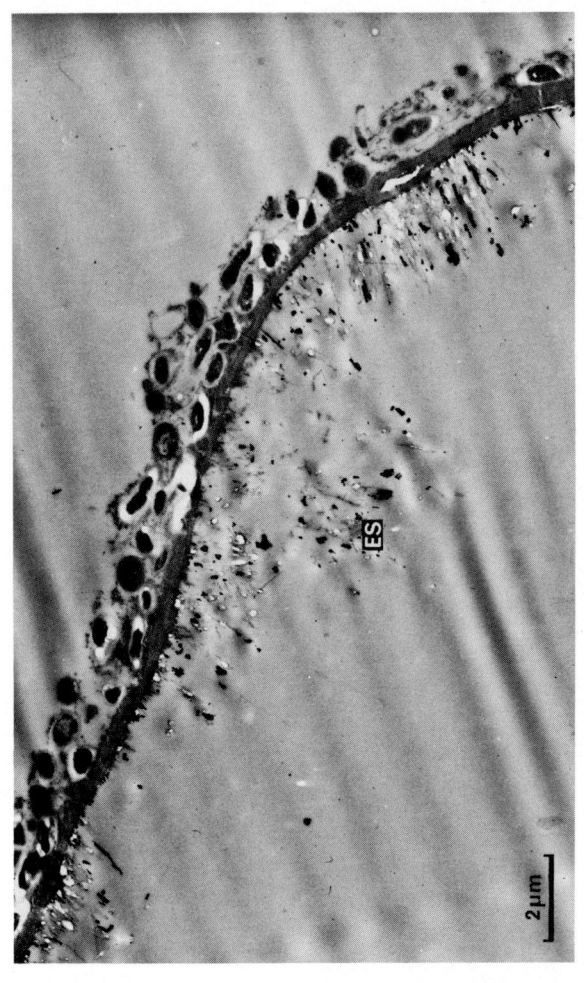

Plate 4. TEM. Amorphous material of salivary origin may contribute to this layer, forming the acquired pellicle.

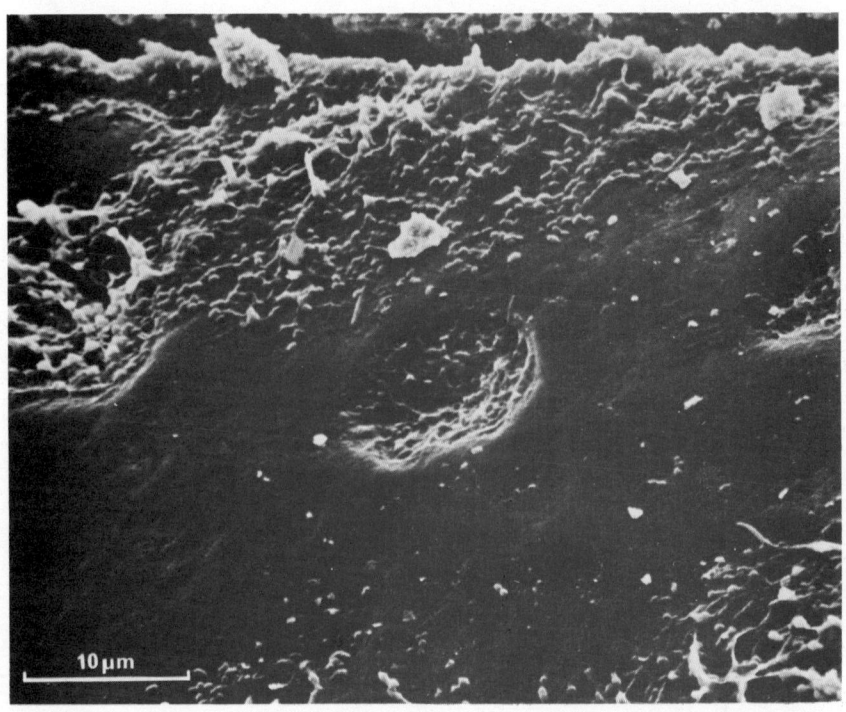

Plate 5. SEM. The contact area is relatively free from plaque, except for deposits in surface pits.

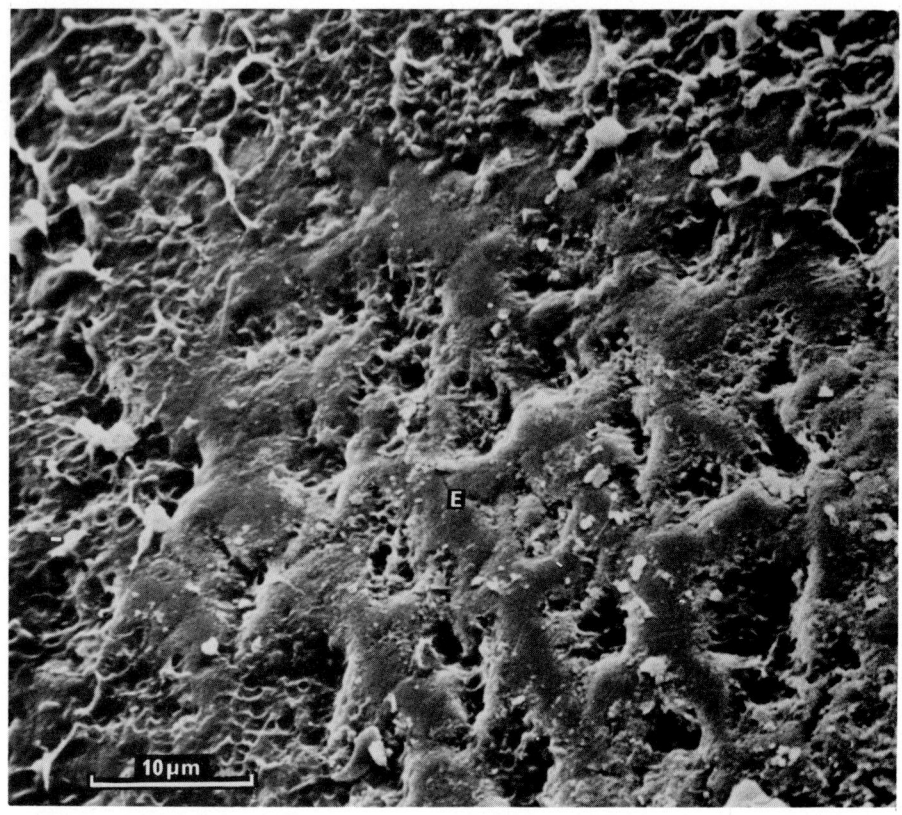

Plate 6. SEM. Separation of gingiva from tooth in the early stages of chronic inflammatory periodontal diseases results in the penetration of the crevice region by micro-organisms. E = enamel.

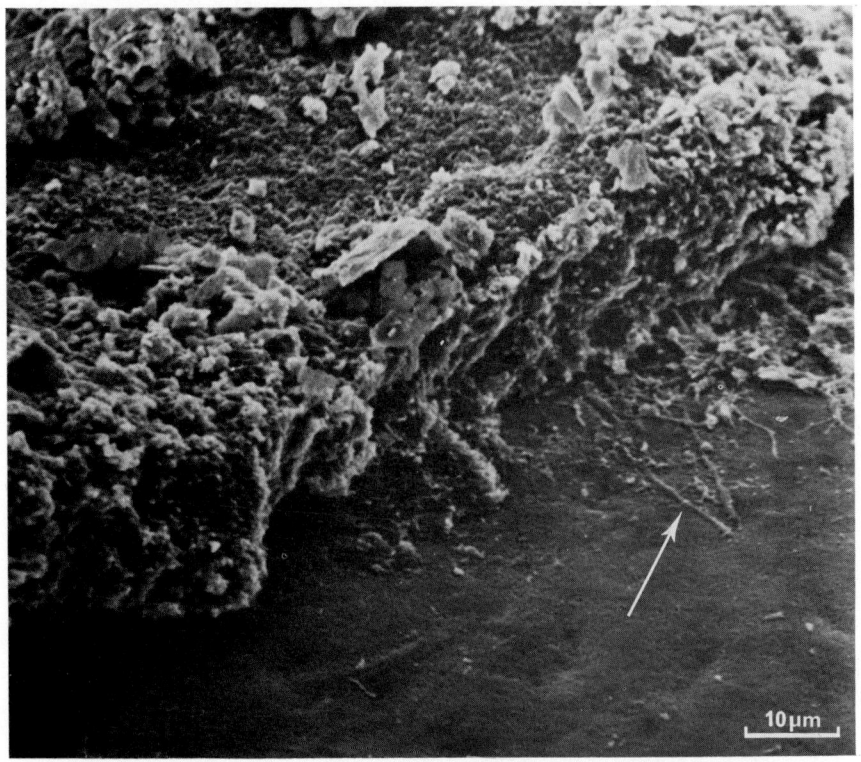

Plate 7. SEM. In health, the ingress of organisms into the crevice is prevented by the firm apposition of gingiva to tooth. Here (arrow) bacterial penetration into the crevice is in its initial stages.

Plate 8. TEM. In this undemineralized specimen much of the plaque matrix, even near its salivary surface is occupied by polysaccharide or other carbohydrate-containing macromolecules, disclosed by the osmium thiosemicarbazide method.

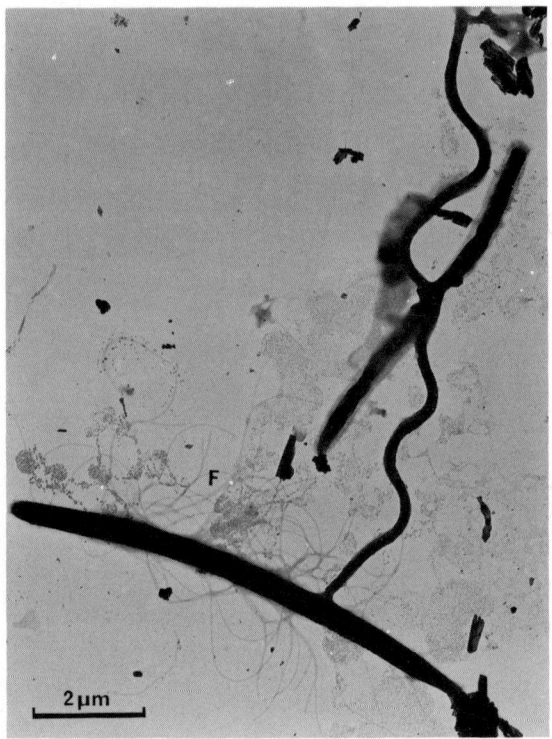

Plate 9. TEM. Spirochaetes are among the plaque organisms not easily observed in sections. In this ultrasonicated specimen a spirochaete is attached to a fusiform bacillus which is itself in relation to fibrillar material (F), possibly of matrix origin.

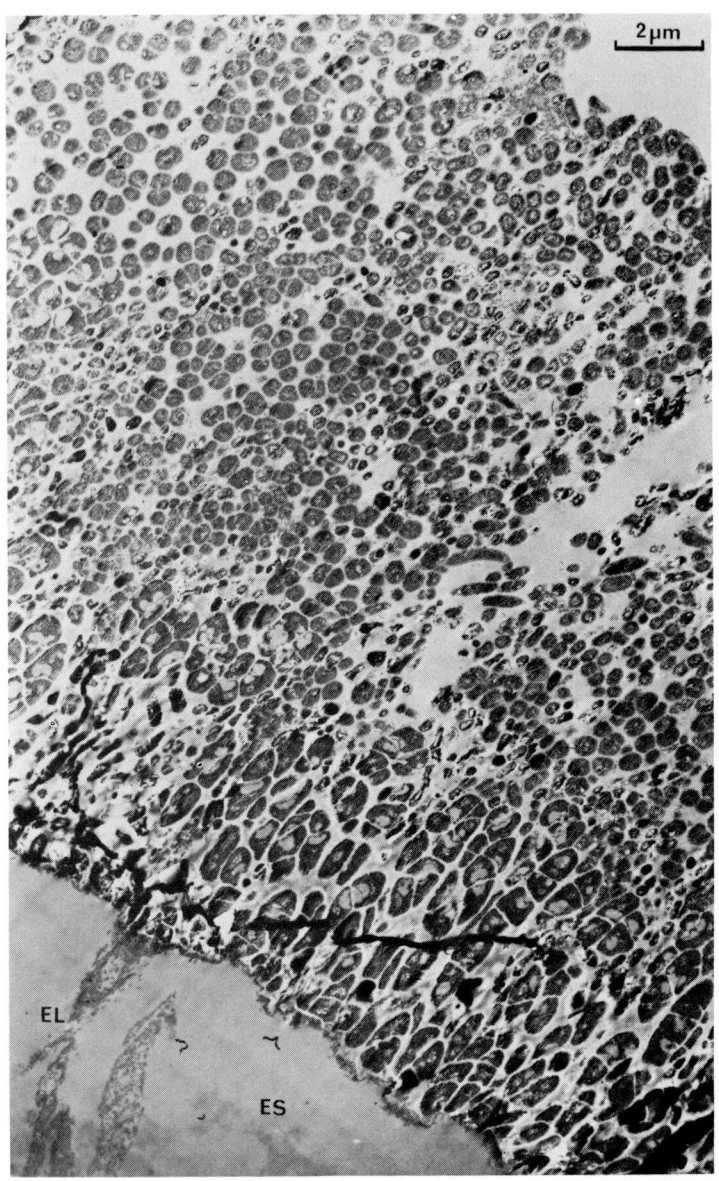

Plate 10. TEM. Organisms at the deep surface of the plaque have become aligned in parallel. EL = enamel lamella.

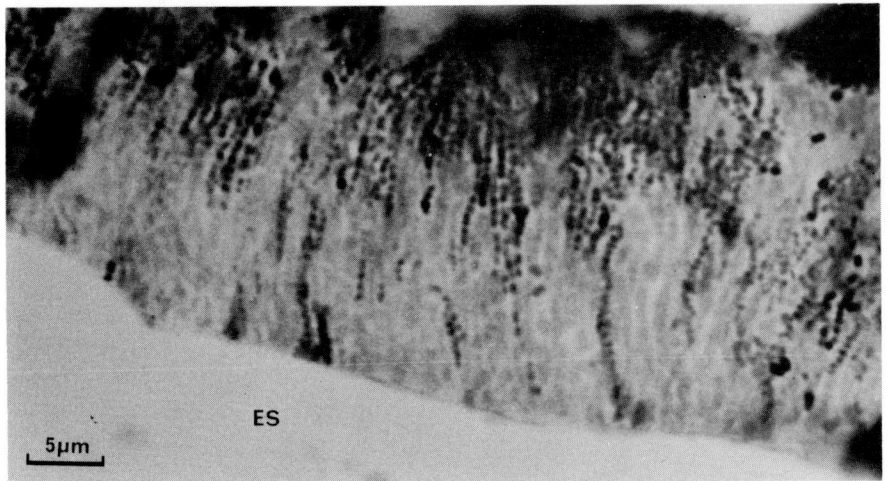

Plate 11. Paraffin section. Parallel chains of Gram positive cocci are present in this section.

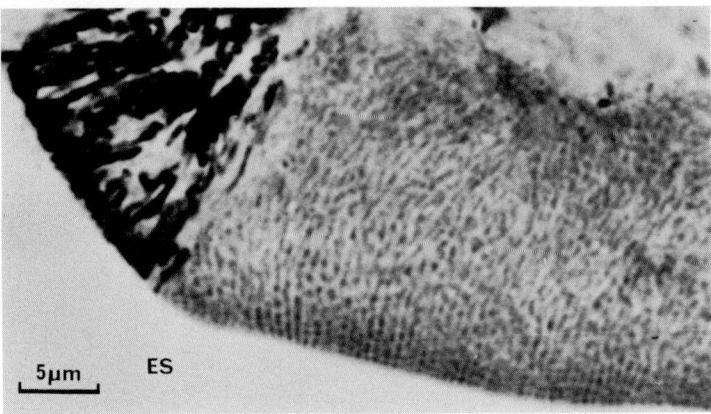

Plate 12. Paraffin section. Palisades of Gram negative organisms adjoin a similar colony of Gram positive forms.

Plate 13. Freeze-etch micrograph (FEM). Parallel filaments near salivary surface of plaque. Many cells are in direct contact via their cell walls. A = antifreeze solution and matrix.

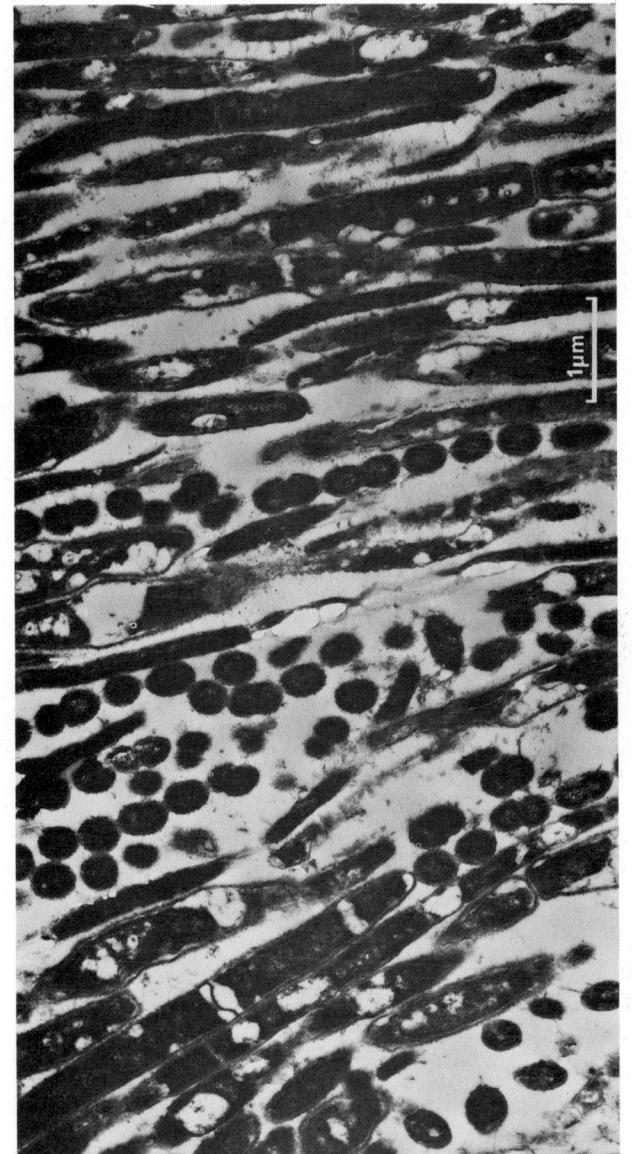

Plate 14. TEM. Note alignment of cocci in chains between parallel septate filamentous forms.

Plate 15. FEM. Adjoining palisades of filaments and coccoid and coccobacillary forms. Freeze-etching avoids much of the shrinkage artefacts of TEM preparation, and thus reveals that organisms in the superficial plaque are less closely packed than their counterparts nearer the tooth surface. Palisades of filaments are to the left and of coccal forms to the right of the micrograph.

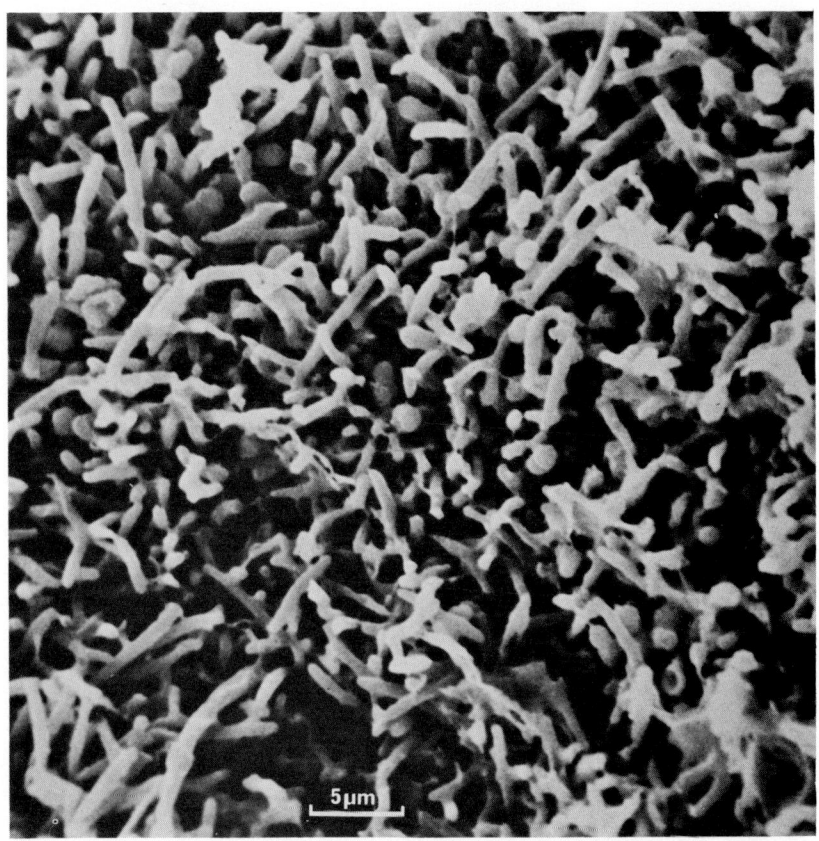

Plate 16. SEM. Mixture of filaments, coccal and bacillary forms in the superficial plaque.

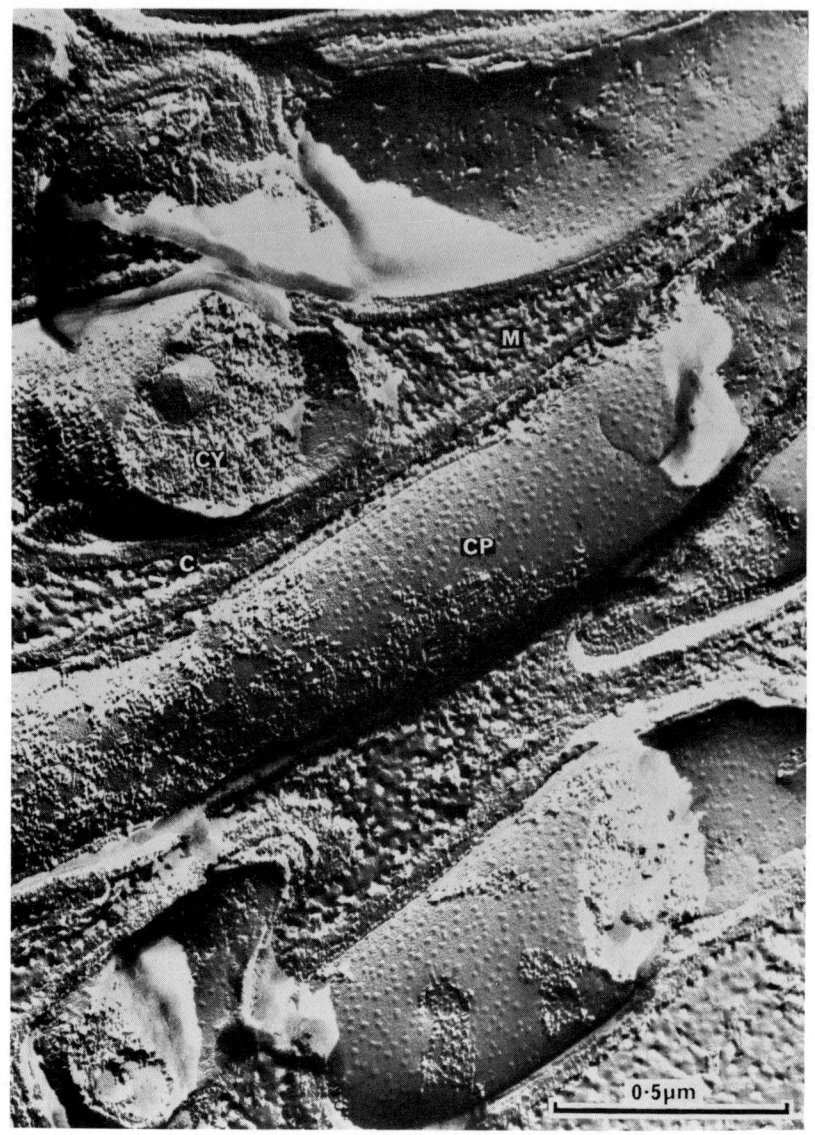

Plate 17. FEM. Note thick cell walls (CW) of parallel forms in deep plaque. Organisms in this region may be in contact with one another as at (C), or separated from each other by narrow zones of matrix (M). CP = cytoplasmic membrane; CY = cytoplasm.

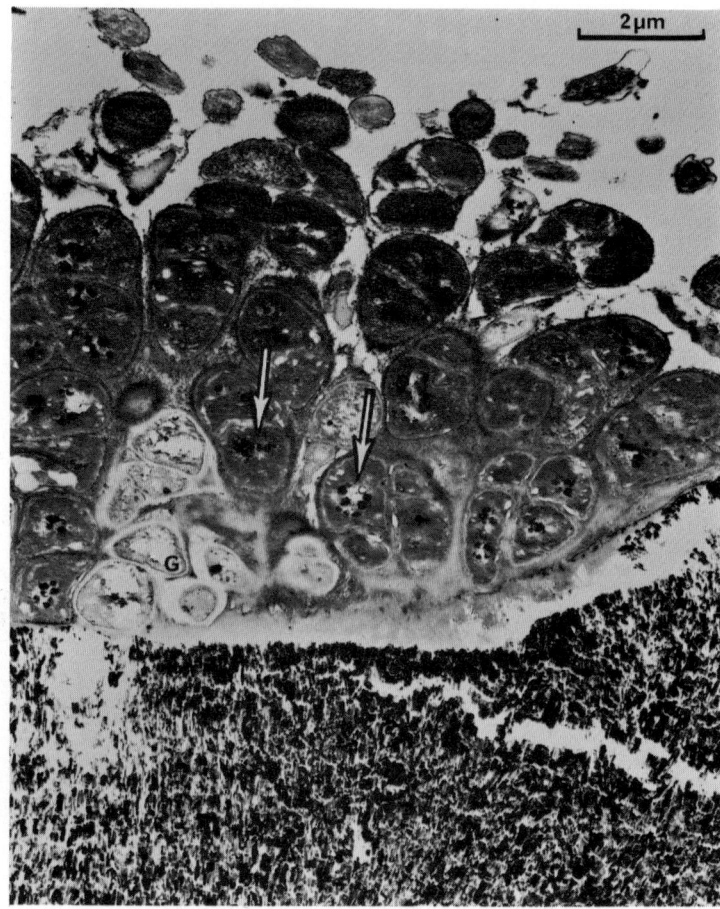

Plate 18. TEM. Note overgrowth of cell ghosts (G) near tooth surface by colony of organisms containing granules of extracellular polysaccharide (arrows).

Plate 19. SEM. Coated filaments of 'corn cob' configurations in the superficial plaque.

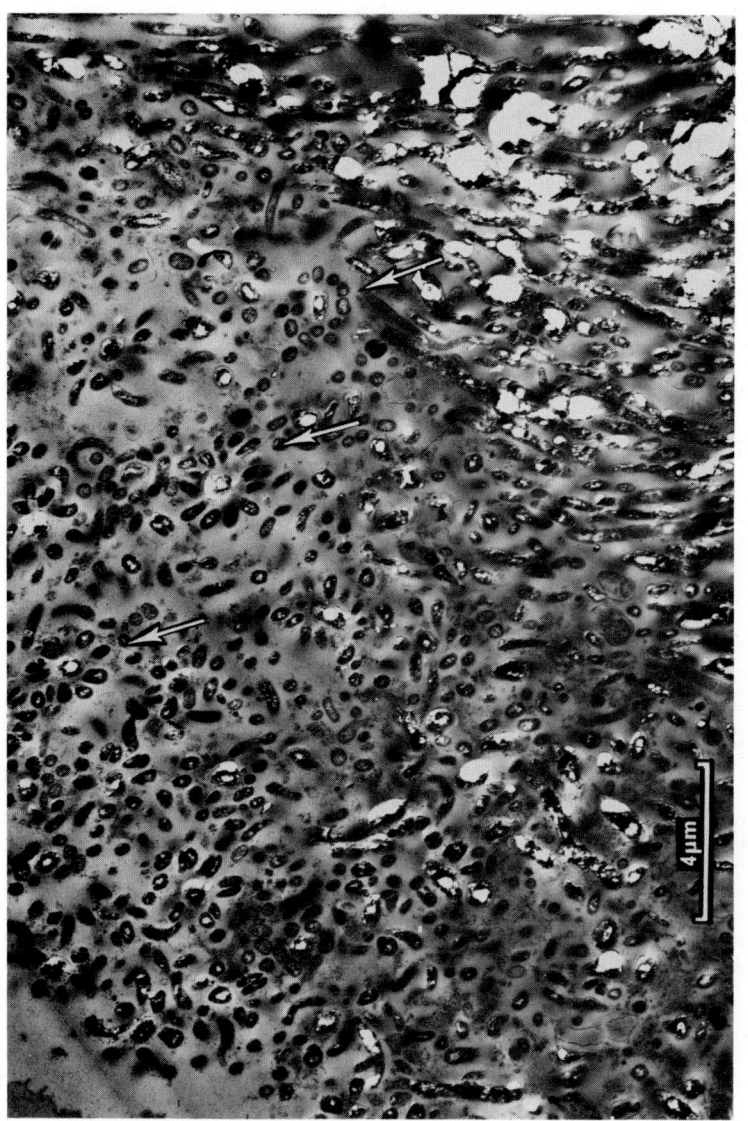

Plate 20. TEM. In section, in the superficial plaque, the corn cob configuration appears as a rosette (arrows).

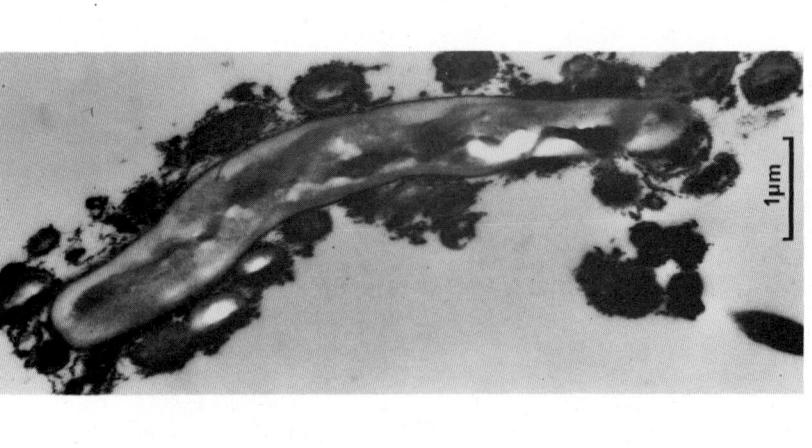

Plate 22.

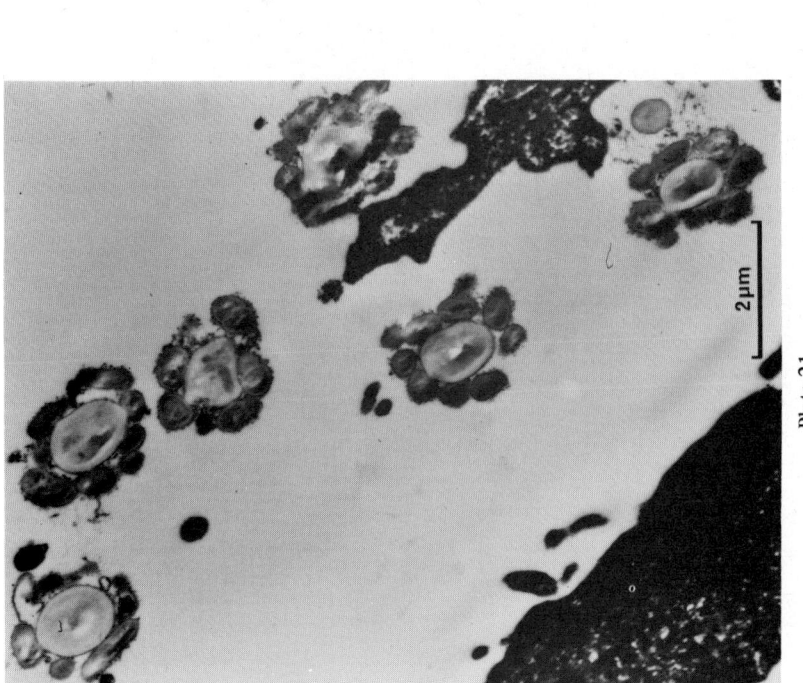

Plate 21.

Plate 21. TEM. The configuration consists usually of Gram variable or electron-dense coating organisms around a Gram negative or electron-lucent central filament.

Plate 22. TEM. In longitudinal section the real nature of this configuration becomes apparent. Coating organisms may be directly attached

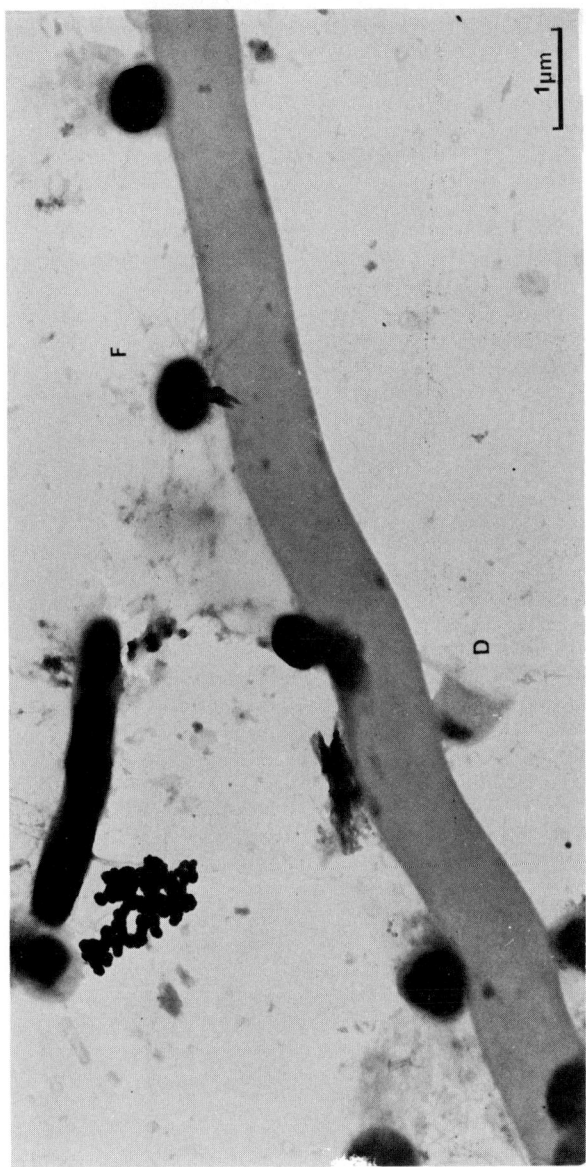

Plate 23. TEM. This attachment is not easily disrupted, even by prolonged ultrasonication, as in this specimen where one coating cell has been disrupted and lost its contents (D) while other coccal forms remain attached, some by means of fibrils (F).

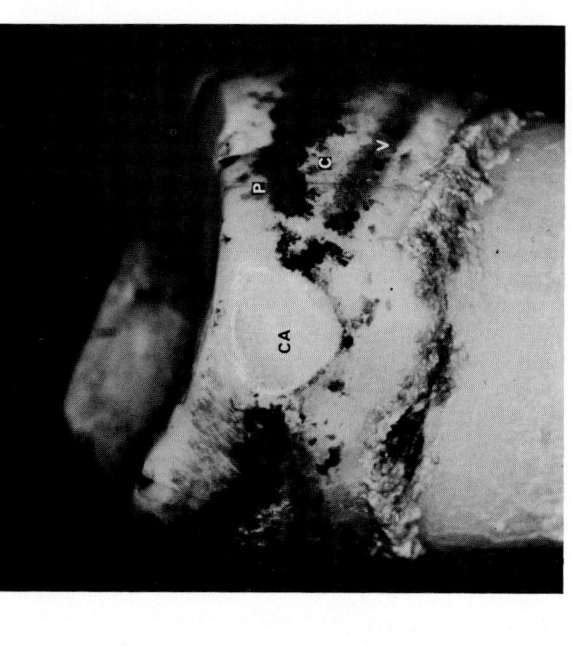

Plate 25.

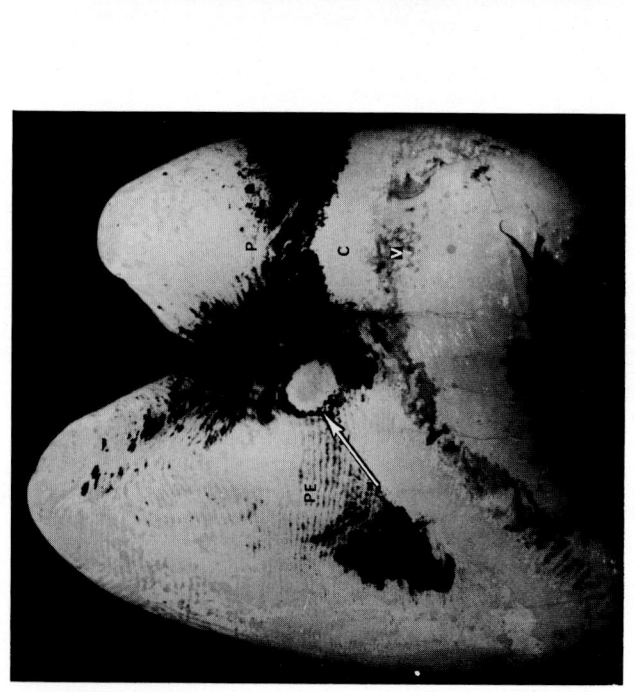

Plate 24.

Plate 24. The zoned organic integument of human enamel, as present on the approximal surface of a child's premolar tooth. P = dental plaque, C = cuticle, V = cellular remains of enamel organ. Note the distribution of plaque around the contact area (arrow) and in the troughs of the wave-like enamel markings of perikymata (PE).

Plate 25. Approximal surface of third molar from adult North American Eskimo. The contact area (CA) is now a marked wear facet and is almost free of plaque around its borders. Note also the considerable wear of the occlusal surface compared with the unworn state of the specimen depicted in Plate 24.

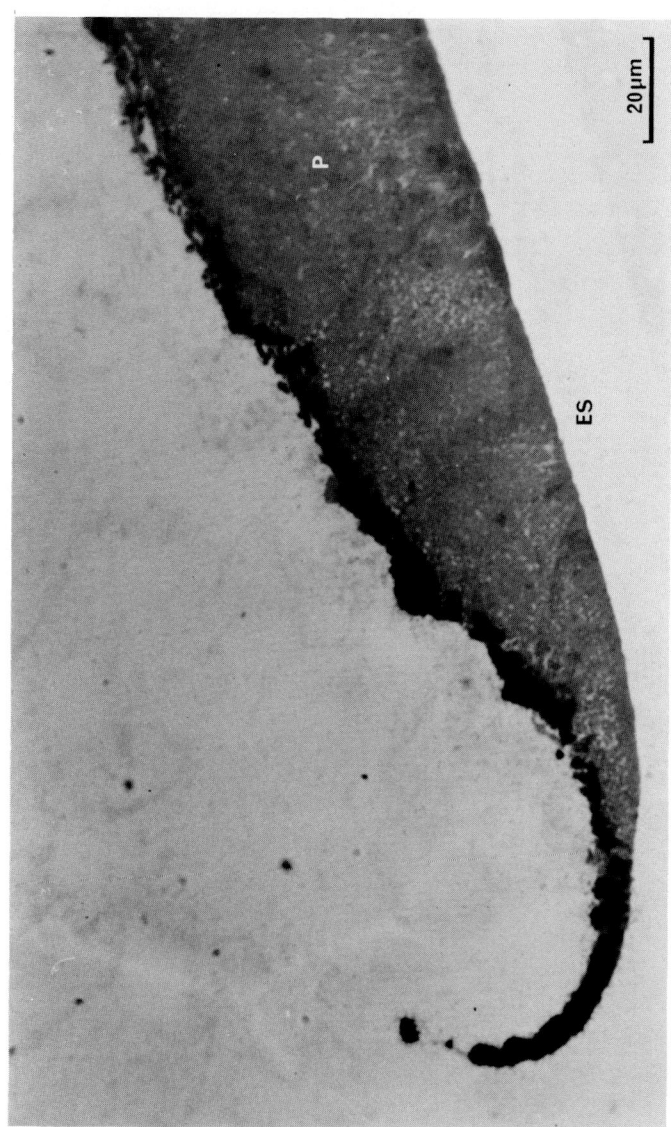

Plate 26. Autoradiograph of natural human dental plaque. 0.5 μm section of specimen incubated for 4 h in saliva containing ^{3}H-glucose, showing the pattern of labelling in the region near the edge of the plaque. Colonies of organisms at the outer surface are densely labelled, but organisms in the deeper plaque show no evidence of incorporation of ^{3}H-glucose. Strained crystal violet. P = plaque, ES = enamel space. Courtesy of Mr. R. P. Shellis, MRC Dental Unit, Bristol.

Plate 27. Specimen as in Plate 26, but showing a thick portion of plaque. A labelled colony (arrow) is present in the plaque interior, which is otherwise unlabelled, except for a few grains associated with filamentous organisms (F) at the outer surface. The pattern of ^{3}H-glucose incorporation by natural plaque, as depicted in these two plates, is evidently heterogeneous, and could be the result of several factors, including diffusion restriction and variations in rate and type of metabolism in different areas of plaque. Courtesy of Mr. R. P. Shellis, MRC Dental Unit, Bristol.

surface possesses a number of wave-like ridges known as perikymata: the troughs of these 'waves' act as minor stagnation sites and, together with other smaller surface irregularities, become foci for plaque deposition.

With the onset of chronic inflammatory periodontal disease, plaque will invade the crevice between gum and tooth (Plate 6). Until this happens, the firm approximation of gum to tooth prevents the ingress of bacteria into the crevice between them, and sharply delineates the plaque border in this region (Plate 7) (Newman, 1972a, b; 1973a, b). It is now considered that adhesive forces rather than structural interconnections maintain the tooth-epithelium relationship in the crevice region (Selvig, 1971), by which the ingress of bacteria is prevented.

3. The Acquired Pellicle

The amorphous layer on the tooth surface often found between the fine pre-eruptive cuticle and the plaque, and also on the cleansible, almost bacteria-free, surfaces, is known as the acquired pellicle. It consists principally of salivary glycoprotein, although it also contains bacterial cell wall material (Leach & Saxton, 1966; Armstrong, 1967). Thus the attachment of plaque to the tooth is invariably by means of an organic layer. In considering the physico-chemical nature of this adhesion, it is necessary to study two aspects, namely, the adhesion of plaque to pellicle/cuticle, and of the latter to the tooth. The acquired pellicle has been found to be connected to the tooth by forces of adhesion, in experiments using cleaned surfaces exposed subsequently to saliva. It has been suggested that carboxyl groups in the glycoprotein molecules adsorb to calcium ions in the enamel hydroxyapatite (Rølla, 1967, 1969, 1971; McGaughey & Stowell, 1967, 1971). Also, calcium ions enhance the adsorption of salivary proteins to hydroxyapatite by promoting protein adsorption (McGaughey & Stowell, 1971). Bacteria do not appear to be essential for the formation of the acquired pellicle (Mayhall, 1970).

It has also been suggested that specific parotid (Armstrong, 1971) and sublingual (Rølla et al., 1969) salivary glycoproteins are selectively adsorbed by hydroxyapatite. However, Selvig (1970) observes that plaques are frequently, but not always, attached to the tooth by means of an organic layer, and that bacteria may be seen in contact with either this layer or the tooth (Frank & Brendel, 1966; van Houte, Hillman & Gibbons, 1970).

It would seem that the acquired pellicle aids the attachment of plaque to the tooth surface by physical means, by virtue of its intimate connection with the pre-eruptive layer and the enamel surface crystallites (Meckel, 1965; Newman, 1973b). Rølla (1971) suggests that cariogenic streptococci are aided in their adhesion to the tooth surface by the secretion of extracellular polysaccharides, which possess an affinity for the glycoprotein molecules present in the acquired pellicle.

There are two possible modes of formation of the acquired pellicle. The first suggests that terminal compounds, such as sialic acid, are split off from salivary glycoproteins, either by salivary or bacterial enzymes, with the resultant precipitation of the 'degraded' protein on to the tooth surface (Dawes & Jenkins, 1963; Dawes; 1964; Leach, 1967; Leach & Hayes, 1968; Leach & Melville, 1970). A second possibility is that intact salivary proteins are precipitated on to the tooth without such degradation taking place (Armstrong, 1967; Ericson, 1967; Leach et al., 1967; McGaughey & Stowell, 1967, 1971; Rølla, 1967; Jenkins, 1968; Bowen, 1969; Hay, 1969, 1970; Mayhall, 1970; Schroeder, 1970).

Having considered the attachment of acquired pellicle to enamel, it is now opportune to survey present concepts of the mode of attachment of plaque to pellicle or pre-eruptive cuticle. More will be said later concerning the function of such organic layers generally in mediating the attachment of bacteria to a variety of surfaces in nature.

4. The Attachment of Plaque to the Tooth

Rølla (1971) believes that bacteria can be directly adsorbed on to tooth mineral, but McGaughey & Stowell (1971) observe that the instances in which a fresh enamel surface is exposed to the saliva are few, and Schroeder (1970) states that, when plaque begins to develop, the sound enamel surface is already covered with organic material derived from the crevice epithelium, the salivary glycoproteins, or both. Selvig (1970), as noted previously, suggests that plaques are frequently, but not always, attached to the tooth by means of organic cuticle. Leach & Hayes (1968) observe that bacteria become embedded in pellicle. Gibbons & Spinell (1970) postulate that salivary induced agglutination of plaque bacteria may contribute to the intermicrobial adhesion which exists between microorganisms in plaque. The importance of adhesion has also been stressed by Frank & Brendel (1966), Schroeder (1969, 1970) and van Houte, Hillman & Gibbons (1970). Meckel (1971) notes that the absence of a cuticular layer appears to delay the attachment of bacteria to a tooth for one day, that is, the time taken, in his experiments, to form a cuticle on a previously cleaned tooth specimen.

Silverman & Kleinberg (1967) state that several factors promote bacterial aggregation and attachment to the tooth. They suggest that the lowering of pH favours cell aggregation, since it reduces the negative electrostatic charges present on cell membranes, thus allowing the cells to move closer together. In addition, they suggest that these charges are reduced by increasing ionic strength, and especially by divalent cations which act in the formation of intercellular bridges. They point out that various acellular components which bind calcium, lower the isoelectric point of cell-protein combinations, enabling the binding of macromolecules to cell membranes.

Divalent cations may also be important in the formation of an intercellular matrix in plaque itself. Recent work (Grant *et al.,* 1973) suggests that such ions are required in the formation of links between polysaccharide chains. The presence of intercellular matrix at the outer surface of the plaque, obscuring details of individual bacteria on intact specimens (Boyde & Lester, 1968; Newman, 1972*b*), contrasts with the relative absence of such matrix on conventional transmission electron microscopy specimens when those specimens are removed from enamel with the calcium complexing agent EDTA. This indicates that superficial plaque matrix, at least, may contain polysaccharides linked by calcium bridging. Preliminary results using the osmium thiosemicarbazide method for polysaccharide do suggest, in fact, that polysaccharide matrix occurs frequently in the superficial plaque on undemineralized sections (Plate 8).

5. The Plaque Flora

Most workers are agreed that the first bacteria to be deposited on a tooth surface are cocci which form a monolayer (Frank & Brendel, 1966; McGaughey & Stowell, 1967; Ritz, 1967, 1969, 1970; Armstrong & Hayward, 1968; Frank & Houver, 1970; Hayward & Armstrong, 1970; Löe, 1970; Schroeder & de Boever, 1970; Børglum Jensen & Löe, 1971; Newman, 1972 *a, b*). Aerobic nocardial species are possibly also involved in initial plaque formation (Ritz, 1963). Filamentous forms are not usually present in significant numbers in the early stages of plaque formation (Schwartz & Massler, 1969). They appear, however, to increase rapidly in the first week, and less so in the second week (Lynch, Crowley & Ash, 1969). Löe (1969) states that filaments appear in two to three days, as do fusiforms, and that vibrios and spirochaetes can be found in ten-day-old plaque. Meckel (1968) notes that veillonellae and fusobacteria increase in numbers with the age of the plaque. Ritz (1967) observes similar increases in the counts of actinomycetes, veillonellae and corynebacteria in a 9-day-old plaque. The general consensus would appear to be that aerobic organisms predominate initially, and that more or less anaerobic forms increase in number with the age and thickness of the plaque. Most of these results are based on studies of model plaques grown over short periods of time in the oral cavity. The actual state of affairs pertaining *in vivo* in naturally established plaque, and the relationships between organisms in the developing plaque remain somewhat ill-defined.

The biochemical interrelationships between these bacteria remain almost unknown (Carlsson, 1971 *a, b,* Mikx *et al.,* 1972). Poole, Gilmour & Mills (1967) suggest that intermicrobial reactions occur in plaque which affect its biochemical capabilities, as indicated by the close relationships between filaments on the one hand, and cocci and rods on the other, a matter discussed elsewhere in this review. Ritz (1967, 1969, 1970) has asserted that the growth of anaerobes in plaque is dependent upon establishment and growth of aerobic and facultative forms. He notes (Ritz, 1967) that fusobacteria are favoured in their growth by neisseriae in this way. The importance of understanding the collective biological activities of the plaque bacteria has been emphasised by Bowen (1971). As yet, however, relatively few of these relationships between species have been defined. To cite a few examples, Fosdick & Wessinger (1940) have noted a symbiotic effect between mouth yeast and what was then called *Lactobacillus acidophilus,* possibly due to a yeast phosphatase being utilized by lactobacilli, and a reductase produced by the latter being used by the yeast; Loesche (1968) has observed commensal relationships between *Treponema dentium, Bacteroides melaninogenicus* and *Veillonella,* and Carlsson (1971a) has described a parasitic dependence of *Streptococcus mutans* on *Streptococcus sanguis.* It should be emphasised that only limited conclusions can be drawn from these experiments since they are based on laboratory rather than oral conditions.

Many studies have been carried out in the field of plaque microbiology, in attempts to implicate various bacterial 'species' (Cowan, 1970) or combinations of micro-organisms in the causation of dental disease. What has emerged is a varied and complex catalogue of the plaque flora (Burnett & Scherp, 1968). Some idea of the complexity may be gained from the review presented by Hardie & Bowden at this symposium, and from the findings by other workers of the existence of many different microbial strains in plaque, especially streptococci (Guggenheim, 1968) and filaments (Baboolal, 1969). Plaque contains many morphological types: cocci, rods, filaments, vibrios and spirochaetes (Børglum Jensen & Löe, 1971). Although cocci appear to predominate numerically in viable count studies, filaments are regarded as occupying a large proportion of mature plaque *in situ* on the tooth surface (Burnett & Scherp, 1968; Löe, 1969; Schroeder & de Boever, 1970), and as forming the framework within which smaller forms are distributed (Ennever, Robinson & Kitchen, 1951; Thoma & Goldman, 1960; Jones, 1971, 1972; Poole & Newman, 1971; Newman, 1972a). In spite of the variety of morphological types in plaque, only coccal, bacillary, filamentous and yeast forms are easily recognizable *in situ.* The presence of others, including vibrios, spirochaetes (Plate 9) and viruses, is generally revealed only by culture and isolation techniques.

6. Plaque Structure

In its early stages, plaque appears to have no structural organization, its constituent organisms being randomly arranged in relation to one another. In areas inaccessible to cleansing forces such as tongue or toothbrush, plaque increases in thickness. As this occurs, organisms at the tooth surface become aligned in parallel and appear to grow out from that surface towards the superficial aspect of the film (Plate 10). This palisading is a feature of both filaments and cocci (Plates 11-13), and the individual palisades form the outstanding structural components of plaque. Indeed, smaller forms are often constrained to grow in parallel chains between palisaded filaments (Plate 14) (Poole & Newman, 1971; Newman, 1972a, 1973b).

Plaque may show either a heterogeneous or colonial type of substructure. The heterogeneous type is more commonly associated with palisaded regions of the film, especially in the deeper layers, or with regions of new plaque formation near the gum margin or the limits of plaque distribution on the tooth surface. Colony-type plaque regions display a variety of forms arranged in microcolonies within the body of the plaque. These may be randomly distributed or orientated in palisades (Plates 15 & 16).

The close packing of organisms is especially noticeable in the deeper layers and results in a distortion of individual cell outlines. This may be responsible for the organization of plaque from deep surface outwards. Other features associated particularly with organisms in the deeper layers of plaque include: a greater proportion of cells to matrix (Plate 15), the presence of thick cell walls, including some with fine fibrillar projections (Plate 17), and a fibrillar matrix, the latter feature being associated especially with degenerate forms. This matrix often contains more polysaccharide than that of the superficial plaque. Cell ghosts (Plate 18) are also more common in deeper plaque, as are structures resembling vesicular-type mesosomes which may be associated, by virtue of their large size (Rogers, 1970; Freese, 1973), with other features of cells with a slow rate of turnover (Chung, 1971). As the functions of mesosomes remain unclear at the present time, any explanation of their frequent occurrence, especially in deep plaque, must await the result of further experimental work on these organelles (Reusch & Burger, 1973).

Conventional histological examination does not always allow a precise definition of biological inter-relationships. However, several features observed may be of significance in plaque biology. One of these is the coating of the terminal, superficially-sited portions of some filaments with coccal or bacillary forms (Plate 19). The coating cells are usually Gram variable or electron-dense, the central filament Gram negative or electron-lucent. Central filament and coating cells may be either in direct contact or separated by amorphous extracellular material. In section this configuration has a rosette-like appearance (Newman, 1972a) (Plates 20-23). Work at the turn of the century by Goadby (1900) and much more recently by Newman & McKay (1973) suggests that the coating cells are sometimes streptococci or lactobacilli; the central filament has not yet been identified, but may be a yeast (Vreven & Frank, 1973) in which case the central organism must be moribund, since yeasts are strongly Gram positive. Yeast forms have been found occasionally in relation to the configurations (McKay, personal communication). The significance of the relationship has yet to be demonstrated, although a nutritional function seems likely. While the terminal portion of the filament is frequently expanded, its proximal portion often shows evidence of collapse and loss of cytoplasmic contents, which indicates that coating cells may utilize elements of the central organism.

Palisading too, remains to be explained, although it is possible that this arrangement permits transport of nutrients to organisms in the deeper layers of plaque (Poole & Newman, 1971); in this respect it is interesting to note the alignment of small forms between parallel filaments. Parallelism always commences at the deep surface of the film next to the tooth enamel, proceeds outwards, and appears to be a function of plaque thickness, being present even in plaque regions only a few cells deep. Organisms in the palisade units are separated from each other by narrow zones of matrix, as a rule, although sometimes their cell walls are in direct contact, and may even be linked by fine fibrils. Similar palisading of organisms in pure culture has been noted by Henneberg (1964). It is evident that palisading results in an efficient organization of plaque bacteria in terms of ratio of cell surface area to volume. At the same time it appears to be a function of the planes of division of individual cells, since the majority of cell division planes in plaque occur parallel to the tooth surface. In spite of the well-known rigidity of bacterial cell walls, the close relationship of organisms to one another in palisade regions of plaque produces a distortion of the outlines of individual cells: thus cell division would seem to occur in a plane parallel to the tooth surface, that is, a plane perpendicular to the direction of least physical resistance to growth.

7. Interbacterial Contacts

Apart from relationships between different morphological types there are also several forms of structural contact between plaque bacteria (Newman, 1972 *a, b, c, d*; Newman & Britton, in press). Wall-to-wall junctions have been observed between filaments in the superficial plaque and between smaller forms in the deeper layers. These contacts may be continuous or intermittent along the length of the cell wall. The fibrils associated with the thick-walled cells frequently appear to mediate intercellular connections in plaque. In other instances adjacent cell walls may be separated by narrow zones of matrix. As might be expected, the close packing of cells in the deeper layers reduces the distance between contiguous cell walls.

8. Filament Formation and Plaque

It has been mentioned that filaments predominate in intact plaque *in situ* on the tooth surface, whereas coccal forms are far more common in viable count experiments. A variety of stimuli, including heat (Ferroni & Innis, 1973), lack of nutrients (Brock, 1966) and a number of antibiotics (O'Grady & Greenwood, 1973) can stimulate an organism to grow in filamentous fashion. Some experiments indicate that filament formation is, in fact, a mode of bacterial adaptation to unfavourable growth circumstances. A coccus is limited, by a diminishing ratio of cell surface area to volume in its ability to assimilate nutrients at a rate commensurate with adequate growth and division. In the case of a filament, however, the organism merely elongates, and the surface area : volume ratio remains the same. In this manner a filament can, as it were, grow towards an adequate source of nutrients (Brock, 1966). Such an organism is clearly at an advantage in regions of plaque where substrates, especially protein and nucleic acid components essential for growth and division, are in short supply, as indicated by the various features of deep plaque associated with cells with a slow rate of division and growth. In this respect it is interesting to note that the addition of polysaccharides to pure cultures of plaque *Nocardia* favours the production of a filamentous mycelium, and inhibits cell division and the onset of filament fragmentation (Erikson, 1954).

Some organisms in plaque may be unable to grow because they are too closely packed together. Thus, Hurst (1950) isolated coccal forms present at the deep surface of the plaque, but found that they grew in pure culture in the filamentous mode and were, in fact, species of *Actinomyces*. It would appear that very few plaque filaments grow as such in pure culture: however, in contrast, Snyder, Bullock & Parker (1967) suggested that only *Leptotrichia* species grew in filament form on isolation from plaque.

Adaptability must clearly be an essential property of an organism if it is to be established successfully in a given ecological habitat. The examination of even large numbers of plaque specimens by a variety of histological methods can fail to give an adequate impression of plaque as a functioning biological community. Even in one section, the variety in form and structure observed can be most marked. When examining such sections it must be remembered that what one is observing is merely a very small portion of a very large bacterial community as it existed at a fixed point in time. The organisms observed are those that have succeeded in establishing themselves in relation to one another at that particular site on the tooth surface. It is, unfortunately, only by *in vitro* experiments, in which antagonistic and synergistic relationships between organisms can be investigated in a practical manner (Scrivener *et al.* 1950; Rosebury, Gale & Taylor, 1954; Parker, 1970; Holmberg & Hallander, 1972; Donoghue, 1972, 1973) that one realizes that the establishment and growth of organisms in relation to one another in plaque is not simply a passive process. In terms of adaptability it is interesting to note, for instance, that very few plaque commensals are strictly aerobic or anaerobic: the vast majority are facultative; and that organisms which fail to survive are often clearly evident in the form of groups of cell ghosts between established cells.

9. The Dynamics of Plaque Formation

It is, perhaps, opportune at this point to consider the factors involved in the establishment of plaque structure as at present understood. Several workers (van Houte, Gibbons & O'Gara, 1971; van Houte, Gibbons & Pulkkinen, 1971) have suggested that certain species have the ability to adhere selectively to a given site on the tooth surface. These workers have postulated that the possession of this attribute is a partial explanation of the involvement of certain streptococci in the cause of dental caries. However, only certain areas of teeth are successfully colonized by bacteria in the natural situation in the mouth, as described elsewhere in this study, namely, where normal cleansing mechanisms do not restrict the thickness of the bacterial film. Organisms on cleansible portions of the tooth surface are largely removed no matter what special mechanisms of adherence they may possess. The numbers of different strains present in the mouth at any one time evidently vary, and this must affect the numbers of any one strain actually impinging on a tooth surface at that given time. The successful establishment of that strain must also depend upon the presence of a favourable nutritional environment and upon its success in dealing with synergistic or antagonistic interactions with organisms already on the tooth, or which arrive subsequently. It should be noted that the growth of more facultative or anaerobic forms in older, thicker plaque is probably not due to later arrival of these organisms on the tooth but, rather, to suitable conditions for their growth having been created by such factors as an increase in plaque thickness.

To summarize, the structural appearances and microbial configurations observed in plaque specimens indicate that adjacent organisms had succeeded in establishing themselves in those relationships at the time of sampling of the specimen, to form, as it were, a number of micro-environments.

10. Plaque and Other Natural Microbial Films

Organisms are not often found in direct contact with the intact enamel surface. They are usually separated from it by an organic layer. Such a layer has, indeed, been found by other workers to mediate the attachment of bacteria to a wide variety of surfaces, notably in the case of marine (Corpe, 1970; Marshall, Stout & Mitchell, 1971 *a, b*), pond (Jones, Roth & Sanders, 1969) and leaf (Last & Warren, 1972) organisms. This is but one instance of a characteristic common to many natural microbial layers. In fact, dental plaque is not unique: the natural adsorption of organic matter, both biotic and abiotic, to mineral surfaces is a relatively common physicochemical phenomenon by means of which living cells are brought into contact with essential nutrient organic matter (Brock, 1966). It is not yet clear how such organic matter may mediate interactions between contiguous organisms which colonize it and the mineral host surface. In the human mouth, however, it seems clear that saliva plays an important biological role in helping to maintain the integrity of the exposed tooth surface (Poole & Newman, 1971; Poole & Silverstone, 1973). Such function is necessary, even in the absence of dental caries, since enamel, which is unique amongst biological tissues in being composed almost entirely of mineral (principally calcium phosphate — hydroxyapatite), possesses no physiological repair capacity of its own, and since mineral may be lost from the tooth surface even under normal conditions of diet and resultant wear or attrition. Saliva is always supersaturated with respect to hydroxyapatite (McCann, 1968) and this implies that mineral lost from the tooth surface may be replenished in this way from the saliva. In practical terms, the hardness of enamel, and even dentine, at the biting surfaces of the teeth is maintained in this manner once the surface is exposed to saliva, be it damaged as a result of caries or natural wear. It is also possible that salivary glycoprotein coating the tooth surface is more easily retained by virtue of the general irregularities of that surface (Newman & Poole, 1973) and that it provides a protective layer between the teeth and the mouth environment, as indicated by the work of Stack & Fletcher (1970).

There are other similarities between dental plaque and other natural floras, apart from the frequent presence of an organic cuticle or pellicle between colonizing organisms and the host surface. In fact, most surfaces so far examined possess plaques of their own which live in intimate contact with their particular hosts in either a commensal or symbiotic relationship (Marshall, Stout & Mitchell, 1971a, b). In certain instances, as in the case of both plant leaves and roots, such commensal organisms may actually invade the underlying plant cells without any apparent untoward effects on the host (Last & Warren, 1972; Rovira, personal communication). Natural microbial films have been observed in a variety of situations apart from those mentioned, including seawater surfaces (Ferguson Wood, 1958; Corpe, 1970; Marshall, Stout & Mitchell, 1971a, b; Fletcher & Floodgate, 1973), slimes in stagnant liquids (Jones, Roth & Sanders, 1969; Parsons & Dugan, 1971), sand grains (Meadows & Anderson, 1966) and rocks (Brock, 1966; Silverman & Munoz, 1970) and the skin and gut epithelial surfaces of a variety of animals, including man (Nissle, 1916; Duguid, 1959; Meynell, 1963; Meynell & Subbaiah, 1963; Duguid, Anderson & Campbell, 1966; Hentges, 1967, 1970; Savage, 1970, 1972; Fuller & Turvey, 1971; Wistreich & Baker, 1971; Fuller, 1972; Pedersen, Frøholm & Bøvre, 1972; Selwyn & Ellis, 1972; Ward & Watt, 1973). In this light, dental plaque may be considered as only one example of a coating microbial film, in this case present on an organic-coated mineral surface composed principally of the calcium phosphate salt, hydroxyapatite.

These films or plaques are found most abundantly in stagnation sites, for instance, in stagnant liquids (Jones, Roth & Sanders, 1969: Parsons & Dugan,

1971), in crevices in sand grains (Meadows & Anderson, 1966) and in the stomatal antechambers of leaves (Campbell, 1972). In the case of the flora on leaves and roots as in the case of that on teeth the deposition of smaller forms is followed by that of filaments, and smaller forms may be found coating larger ones (Campbell, personal communication). A further similarity is the presence of polysaccharides, especially the extracellular variety, in these films. Polysaccharides currently attract considerable attention amongst dental plaque researchers, since they are thought to be involved in the causation of caries by virtue of their stickiness, their diffusion-limiting effects, and their capacity to be broken down to produce acid. It is, however, more probable that they are produced by bacteria as a response to environmental conditions within the plaque, as indicated by the work of Erikson (1954). Polysaccharides are, in fact, present at the outer surface of most cells (Rambourg, 1972) and fulfil many essential functions in terms of bacterial growth and survival (Wilkinson, 1958; Parsons & Dugan, 1971; Rambourg, 1972; Pessac & Defendi, 1972). Several floras in addition to plaque on teeth possess organisms in which fibrillar structures mediate the attachment of cells to each other or to the host surface, for instance, intestinal epithelium (Duguid, 1959; Duguid, Anderson & Campbell, 1966; Wistreich & Baker, 1971; Pedersen, Frøholm & Bøvre, 1972; Ward & Watt, 1973; Fuller, personal communication). But perhaps the most important similarities between these separate microbial plaques lie in their common ability to exclude alien organisms from the living hosts they colonize and, in many instances, to aid in host nutrition (Nissle, 1916; Meynell, 1963; Meynell & Subbaiah, 1963; Bowden & Rovira, 1961; Hentges, 1970; Savage, 1970, 1972; Çetin et al., 1971; Fuller & Turvey, 1971; Hawksworth, Drasar & Hill, 1971; Fuller, 1972; Rovira, 1972; Selwyn & Ellis, 1972).

While the nutritional functions of natural floras have been the object of research for many years, it is only relatively recently that efforts have been renewed to understand the role of these floras in contributing otherwise to the health of their hosts. This is especially surprising since the antibiotic properties of commensal bacteria were discovered at the same period as the involvement of bacteria in the cause of infectious disease. In 1877, Pasteur & Joubert (Florey, 1946) noted that growth of the anthrax bacillus in urine was inhibited by 'common' micro-organisms, and they suggested that the latter might be used to treat anthrax. This was followed by the work of Cantani, who in 1885 (Florey, 1946) attempted to replace tubercle bacilli in human lungs with a mixture of non-pathogens, and by that of Emmerich (Florey, 1946) who in 1887 reported on his attempts to prevent the spread of anthrax by deliberately infecting subjects with beta-haemolytic streptococci. Basing his efforts on these studies, Florey (1946) made several efforts to turn these natural antibiotic properties to therapeutic use, unfortunately, without success. It is, perhaps a realization of the limited usefulness of antibiotics (Sanders, 1969; Richmond, 1972) that has brought about a resurgence of interest in interbacterial antagonisms. Among the more interesting results of post-war studies has been the discovery that various commensal enterobacteria could produce antibacterial substances lethal to *Proteus sp.* (Sieburth et al., 1952) and the demonstration by Freter (1955) that guinea pigs could be infected with *Vibrio cholera* only after reduction of the normal gastrointestinal flora by starvation and streptomycin.

Many studies of bacterial antagonisms have been carried out on the flora of the mouth and nasopharynx. Thompson & Shibuya (1946) succeeded in demonstrating that the antibacterial action of saliva noted by earlier workers was, in fact, due to inhibitory organisms present in it. They showed, for example, that *Streptococcus mitis* completely inhibited the growth of diphtheria bacilli. There followed a series of such studies, as a result of which it has become clear that many oral and nasopharyngeal commensals, especially viridans and anhaemolytic streptococci, play a most important role in the protection of the human host against a variety of known pathogens. Myers (1959), Dineen (1960) and Dubos & Schaedler (1962) have shown how staphylococcal infection may be controlled by the commensal flora. Indeed, Myers (1959) was able to demonstrate that both neisseria and viridans streptococci suppressed staphylococcal growth, and he suggested that the 'hospital strain' of *Staphylococcus aureus* had probably been created by excessive use of antibiotics resulting in an imbalance of the natural flora. Bartels, Blechman & Lories (1960) showed that beta-haemolytic streptococci were inhibited *in vitro* by an oral alpha-haemolytid streptococcus. Sprunt & Redman (1968) demonstrated that viridans streptococci were potent inhibitors of Gram negative pathogenic bacteria, for instance, *Escherichia coli, Aerobacter* and *Pseudomonas aeruginosa*. Their work also indicated that antibiotics could potentiate the growth of various coliform organisms, and that prolonged antibiotic treatment might eliminate the natural antagonisms of the commensal flora. Sanders (1969) observed that viridans streptococci were also highly active against coagulase-positive staphylococci and meningococci, and Johanson et al. (1970) noted a similar inhibitory effect of viridans streptococci on pneumococcal strains. In fact, Johanson and co-workers (1970) suggested that the pathogenicity of carrier pneumococcal strains might in part be due to their resistance to these commensals of the nasopharynx.

The principal agents mediating this antibacterial activity of natural floras on animals appear to be fatty acids (Meynell, 1963; Hentges, 1967, 1970; Savage & McAllister, 1970; Galbraith et al., 1970; Fuller, 1972), effective especially in the more anaerobic sites in the gut and at a pH of 5.5-6.0 (Hentges, 1967, 1970). These acids are thought to produce their effect by inhibiting substrate transport into cells (Freese, Sheu & Galliers, 1973). Such information provides a somewhat novel, if incomplete, interpretation of the functions of dental plaque acids in biological terms: until now these acids have been considered solely as causative factors in dental caries. Plaque bacteria possess additional antibiotic properties of possible benefit to the host which are not, apparently, acid in nature (Kelstrup & Gibbons, 1969; Çetin et al., 1971; Donoghue, 1972, 1973; Holmberg & Hallander, 1973). Analogous agents have been found to mediate the antibacterial activity of floras as varied as those of human skin (Selwyn & Ellis, 1972), gut (Meynell, 1963), marine forms (Ferguson Wood, 1958) and pine needles (Campbell, 1972). However, attempts to characterize these other antibacterial agents chemically have so far met with little success. Sanders (1969) has suggested that successful antagonists may produce their effect in one of several ways. They may compete successfully with pathogens for essential substrates; they may elaborate a variety of antibiotic substances, or they may create a restrictive physiological environment, for example, by altering pH or redox potential. Little more is known about the various antibiotic substances other than acid except that some of them are easily denatured, especially by heat (Myers, 1959).

In summary, evidence currently available indicates a need for further investigations into the possibilities of controlling pathogens by natural floras. In the present context it may be noted that commensals of the mouth and nasopharynx, especially the viridans and anhaemolytic streptococci, may play a significant role in preventing the establishment in the human host of a variety of pathogens. It is, therefore, becoming increasingly evident that future possible methods of prevention of both caries and chronic inflammatory periodontal disease will have to avoid the suppression of the local natural flora. The reconsideration of the role of plaque in disease which follows may help to explain that suppressive treatment of the local flora is not only contra-indicated, but does not take into account the nature of the mechanisms which make plaque pathogenic.

11. Plaque and Disease

It is evident from the aspects of plaque biology encompassed in this review that plaque is one of a multitude of naturally occurring microbial films. It is not yet clear, however, in what manner such a film has come to be involved in the cause of diseases, and those among the commonest affecting man. The siting of organisms on mineral surfaces and the use of that mineral as a substrate are natural phenomena by no means confined to the mouth. The use of acid for such demineralization is widespread (Silverman & Munoz, 1970; Ness, 1971). Furthermore, the acids involved in caries are not specific toxins, but the natural products of carbohydrate metabolism. Another feature worth emphasising is that organisms in acid milieus other than those on teeth have been found to produce extracellular polysaccharides which help to raise the pH in relation to the producer organisms (Dugan, MacMillan & Pfister, 1970a, b).

Similarly, many of the agents involved in the cause of chronic inflammatory periodontal disease are relatively simple substances such as ammonia and urea, the products of protein catabolism (Rizzo, 1967; Frostell & Söder, 1970; Cole & Longton, 1972; Hayes & Hyatt, 1972; Howden, 1972. Relatively simple substances such as these may constitute part of the mechanisms whereby bacteria control their environment and exert antagonistic or synergistic effects on neighbouring organisms (Rosebury, Gale & Taylor, 1954; Parker, 1970; Donoghue, 1972, 1973; Holmberg & Hallander, 1973). The problem remains, however, to explain how these and other chemical agents are maintained in contact with dental and periodontal tissues so that disease supervenes. There is, as yet, no definitive answer. But it would appear that the dietary changes that have occurred within the last few hundred years, a relatively inconsequential period in an evolutionary context, have had a very unfavourable effect on the biology of human dental plaque. The most obvious of these changes has been the consumption of increasingly larger quantities of fermentable carbohydrates, especially sucrose, with the resultant production of excessive amounts of acid implicated in caries. Sucrose has a further damaging effect since it stimulates the production in the plaque intercellular matrix of large amounts of extracellular polysaccharide which (Bowden, 1969) have been shown to restrict diffusion within plaque, this causing the retention of acid against tooth surfaces. An equally significant dietary change has been the alteration in the texture of the foods we consume (Poole & Newman, 1971). In sum, their fibre content and, therefore, the degree of mastication needed to render them suitable for swallowing, has been drastically reduced. Recent evidence (Burkitt, 1973) indicates that this reduction in dietary fibrous texture may be largely responsible for a variety of gastro-intestinal and other disorders. It now seems clear that a further consequence of this change in diet has been the increased accumulation of plaque deposits on the teeth and around the gums (Ainamo, 1972) which has, in turn, let to an increased incidence of both caries and chronic inflammatory periodontal disease. This reduction in masticatory function manifests itself most clearly by a lack of wear or attrition of both the biting and contiguous surfaces of 'civilized' as compared with naturally worn teeth (Plates 24 & 25). Lack of wear on the biting surfaces leads to the persistence of surface pits and fissures which act as stagnation sites and where the interaction of trapped bacteria and sugar results in decay. Lack of wear of the contiguous surfaces indicates a restricted degree of movement of the teeth in relation to one another. This may be seen from such specimens as are available (Newman, 1972*b*) to have resulted in an unnatural build-up of plaque in the site which is most stagnant and, therefore, most at risk, namely, that portion of the tooth surface below the contact areas of contiguous teeth. In such a region saliva can not reach the surface, and the normal protective mechanism by which salivary calcium and phosphate replenish lost tooth surface mineral is insufficient to compensate for carious dissolution (Poole & Silverstone, 1973).

Indications such as these mitigate against one particular organism or group of organisms being the direct cause of either condition, in the accepted sense of the term infectious disease. Nonetheless, plaque contains organisms belonging to several genera and, since the original correlation of oral bacteria with caries and periodontal disease, considerable efforts have been made to find whether or not one or more of these genera contained strains more pathogenic than others. There is, unfortunately, as much confusion with regard to the taxonomy of the oral flora as to that of skin, soil, plant or any other complex natural microbiota. In view of the difficulties of applying conventional taxonomic methods, in this as in other fields, several workers (Bleiweis et al., 1971; Coykendall et al., 1971; Baboolal, 1972; Drucker, 1972; Drucker et al., 1972, 1973; Stack et al., 1972, 1973) are concerned with trying to develop new techniques which could be used to identify strains for experimental purposes. Results have so far shown the deficiencies inherent in assigning precise species names to complex groups of organisms, for instance, the streptococci of plaque (Guggenheim, 1968), a group of intereest to dentists primarily because of its involvement in dental caries and secondarily for its role as the cause of bacterial endocarditis in patients with valvular heart lesions. The organisms making up this group are very numerous and show a wide range of shared characteristics. However, it is clear from many studies that organisms of the *mutans* streptococcus group are primarily involved in caries because of their ability to produce and survive in large amounts of acid and extracellular polysaccharide and their relative tolerance of large amounts of dietary sucrose. A similar situation pertains in the crevice between gum and tooth where plaque bacteria are in contact with protein in the form of desquamating oral epithelial cells, degenerating leucocytes and serum proteins in the crevicular exudate. Here, too, it seems likely that organisms which can tolerate the presence of a relatively alkaline environment and which have a high proteolytic potential are more likely to survive (Burnett & Scherp, 1968; Loesche, 1968; Thonard, 1968; Berry, 1970; Ishikawa & Kinoshita, 1972; Loesche, Hockett & Syed, 1972). It would appear that it is not primary bacterial infection but, rather, the changes in plaque formation and distribution produced by modern diet which have modified bacterial metabolism in plaque and resulted in disease, and such is the premise on which much current work is based. The work presented in this review has derived from studies of tooth enamel formation, structure and pathology. There is clearly a limit to the usefulness of structural studies by themselves in interpreting biological function. However this work is being used as the basis for different approaches to the problem of understanding the role of plaque in health and disease. These include autoradiography (Plates 26 & 27), diffusion studies, taxonomy of plaque bacteria and investigations of the *in vivo* and *in vitro* inter-relationships between plaque organisms, which are being carried out by the various members of the MRC Dental Unit in Bristol. Together they form part of a much more widespread effort to understand plaque in biological and ecological terms, to which the following references may provide a guide (McHugh, 1970; Bowen, 1971; Carlsson, 1971 a, b; Gibbons, 1971; van Houte & Saxton, 1971; Ikeda & Sandham, 1971: Tanzer, Wood & Krichevsky, 1971; Ainamo, 1972; Cole & Longton, 1972; Duany, Zinner & Jablon, 1972; van der Hoeven et al., 1972; Holmberg & Hallander, 1972; Jenkins, 1972, 1973; Kelstrup & Funder-Nielsen, 1972; Mäkinen, 1972; Mäkinen & Scheinin, 1972; Mikx et al., 1972; Regolati, Guggenheim & Mühlemann, 1972; Sims, 1972; Tanzer, 1972; Tanzer, Chassy & Krichevsky, 1972; Bowden, Nash & Spiers, 1973; Braverman, 1973; Montgomery, 1973).

We would like to thank Mr. M. S. Gillett and Mr. A. B. Britton for their skilled assistance, Dr. H. D. Donoghue for her helpful comments and Mrs. K. Peat for her kind cooperation in the preparation of the manuscript.

12. References

AINAMO, J. (1972). Relationship between occlusal wear of the teeth and periodontal health. *Scand. J. dent. Res.* **80,** 505.
ARMSTRONG, W. G. (1967). The composition of the organic films formed on human teeth. *Caries Res.* **1,** 89.
ARMSTRONG, W. G. (1971). Characterization studies on the specific human salivary proteins adsorbed *in vitro* by hydroxyapatite. *Caries Res.* **5,** 24.
ARMSTRONG, W. G. & HAYWARD, A. F. (1968). Acquired organic integuments of human enamel: a comparison of analytical studies with optical, phase contrast and electron microscope examinations. *Caries Res.* **2,** 294.
BABOOLAL, R. (1969). Cell wall analysis of oral filamentous bacteria. *J. gen. Microbiol.* **58,** 217.
BABOOLAL, R. (1972). A study of the enzyme patterns of some oral filamentous bacteria by starch-gel electrophoresis. *Archs oral Biol.* **17,** 691.
BARTELS, H. A., BLECHMAN, H. & LORIES, D. (1960). The *in vitro* inhibition of beta streptococci by an oral streptococcus. *J. dent. Res.* **39,** 687.
BERRY, W. C. (1970). Increased antibody production to *Bacteroides melaninogenicus* in acute gingival disease. *J. dent. Res.* **49,** 167.
BLEIWEIS, A. S., CRAIG, R. A., ZINNER, D. D. & JABLON, J. M. (1971). Chemical composition of purified cell walls of cariogenic streptococci. *Infection and Immunity,* **3,** 189.
BØRGLUM JENSEN, S. & LÖE, H. (1971). The effect of dextranase on plaque and gingivitis in man. In *The Prevention of Periodontal Disease.* Eds J. E. Eastoe, D. C. A. Picton & A. G. Alexander. London: Henry Kimpton.
BOWDEN, G. H. (1969). The components of the cell walls and extracellular slime of four strains of *Staphylococcus salivarius* isolated from human dental plaque. *Archs oral Biol.* **14,** 685.
BOWDEN, G. H., NASH, R. & SPIERS, R. L. (1973). The localization and retention of ^{32}P and ^{45}Ca within surface deposits of *Streptococcus sanguis* and the influence of such deposits on the release of these isotopes from enamel. *Caries Res.* **7,** 185.
BOWDEN, G. D. & ROVIRA, A. D. (1961). The effects of micro-organisms on plant growth. I. Development of roots and root hairs in sand and agar. *Pl. Soil,* **15,** 166.

BOWEN, W. H. (1969). The gingival environment. In *Biology of the Periodontium*. Eds A. H. Melcher & W. H. Bowen. London and New York: Academic Press.

BOWEN, W. H. (1971). The effects of calcium, magnesium and manganese on dextran production by a cariogenic streptococcus. *Archs oral Biol.* **16**, 115.

BOYDE, A. & LESTER, K. S. (1968). Scanning electron microscopy of carious cavity plaque after ethylene diamine treatment. *Archs oral Biol.* **13**, 1413.

BRAVERMAN, C. M. (1973). Effect of sugar concentrations on growth of certain oral streptococci. *J. dent. Res.* **52**, abs. 5, p. 60 (Amer. Div., Int. Ass. Dent. Res.)

BROCK, T. D. (1966). *Principles of Microbial Ecology*. Englewood Cliffs, New Jersey: Prentice-Hall Inc.

BURKITT, D. P. (1973). Some diseases characteristic of modern Western civilization. *Brit. med. J.* No. 5848, **1**, 274.

BURNETT, G. W. & SCHERP, H. W. (1968). In *Oral Microbiology and Infectious Diseases*. Baltimore: Williams and Wilkins Co.

CAMPBELL, R. (1972). Electron microscopy of the epidermis and cuticle of the needles of *Pinus nigra* var. *maritima* in relation to infection by *Lophodermella sulcigena*. *Ann. Bot.* **36**, 307.

CARLSSON, J. (1971a). Bacteriological populations associated with the periodontium. In *The Prevention of Periodontal Disease*. Eds J. E. Eastoe, D. C. A. Picton & A. G. Alexander. London: Henry Kimpton.

CARLSSON, J. (1971b). Growth of *Streptococcus mutans* and *Streptococcus sanguis* in mixed culture. *Archs oral Biol.* **16**, 963.

ÇETIN, E. T., AŇG, Ö., TÖRECI, K. & BERKITEN, R. (1971). Investigations on aerobic oral and nasal flora of university students. *Path. Microbiol.* **37**, 185.

CHUNG, K. L. (1971). Thickened cell walls of *Bacillus cereus* growth in the presence of chloramphenicol: their fate during cell growth. *Canad. J. Microbiol.* **17**, 1561.

COLE, J. S. & LONGTON, R. W. (1972). Influence of small metabolites from bacteria on cultured oesophageal epithelial cells. *J. dent. Res.* **51**, abs 341, p. 133 (Amer. Div., Int. Ass. dent. Res.)

CORPE, W. A. (1970). Attachment of marine bacteria to solid surfaces. In *Adhesion in Biological Systems*. Ed. R. S. Manly. New York and London: Academic Press.

COWAN, S. T. (1970). Heretical taxonomy for bacteriologists. *J. gen. Microbiol.* **61**, 145.

COYKENDALL, A. C., DAILY, D. P., KRAMER, M. J. & BEATH, M. E. (1971). DNA-DNA hybridization of *Streptococcus mutans*. *J. dent. Res.* **50**, 131.

DAWES, C. (1964). Is acid-precipitation of salivary proteins a factor in plaque formation? *Archs oral Biol.* **9**, 375.

DAWES, C. & JENKINS, G. N. (1963). Studies related to the formation of dental plaque. *J. dent. Res.* **42**, abs. 362, p. 126 (Br. Div., Int. Ass. dent. Res.).

DINEEN, P. (1960). Effect of reduction of bowel flora on experimental staphylococcal infection in mice. *Proc. Soc. Exp. Biol. Med.* **104**, 760.

DONOGHUE, H. D. (1972). Antagonisms among bacteria isolated from dental plaque. *J. dent. Res.* **51**, 1239.

DONOGHUE, H. D. (1973). Effect of glucose and sucrose on antagonist production by plaque bacteria *in vitro*. *J. dent. Res.* (Abs.) (in press).

DRUCKER, D. B. (1972). The identification of streptococci by gas-liquid chromatography. *Microbios* **51**, 109.

DRUCKER, D. B., GRIFFITHS, C. J. & MELVILLE, T. H. (1972). Effect of changes in growth conditions on the fatty acid profiles of *Streptococcus mutans*. *J. dent. Res.* **51**, 1276.

DRUCKER, D. B., GRIFFITHS, C. J. & MELVILLE, T. H. (1973). Application of gas-liquid chromatography to the identification of oral streptococci. *J. dent. Res.* (Abs.) **52**, 964.

DUANY, L. F., ZINNER, D. D. & JABLON, J. M. (1972). Epidemiologic studies of caries-free and caries-active students: II. Diet, dental plaque and oral hygiene. *J. dent. Res.* **51**, 727.

DUBOS, R. & SCHAEDLER, R. W. (1962). Some biological effects of the digestive flora. *Amer. J. Med. Sci.* **244**, 265.

DUGAN, P. R., MACMILLAN, C. B. & PFISTER, R. M. (1970a). Aerobic heterotrophic bacteria indigenous to pH 2.8 acid mine water; microscopic examination of acid streamers. *J. Bact.* **101**, 973.

DUGAN, P. R., MACMILLAN, C. B. & PFISTER, R. M. (1970b). Aerobic heterotrophic bacteria indigenous to pH 2.8 acid mine water; predominant slime-producing bacteria in acid streamers. *J. Bact.* **101**, 982.

DUGUID, J. P. (1959). Fimbriae and adhesive properties in *Klebsiella* strains. *J. gen. Microbiol.* **21**, 271.

DUGUID, J. P., ANDERSON, E. S. & CAMPBELL, I. (1966). Fimbriae and adhesive properties in salmonellae. *J. Path. Bact.* **92**, 107.

ENNEVER, J., ROBINSON, H. B. G. & KITCHIN, P. C. (1951). *Actinomycetes* and the dentobacterial plaque. *J. dent. Res.* **30**, 88.

ERICSON, T. (1967). Adsorption to hydroxylapatite of proteins and conjugated proteins from human saliva. *Caries Res.* **1**, 52.

ERIKSON, D. (1954). Factors promoting cell division in a 'soft' mycelial type of *Nocardia: Nocardia turbata* n. sp. *J. gen. Microbiol.* **11**, 198.

FERGUSON WOOD, E. J. (1958). The significance of marine microbiology. *Bact. Rev.* **22**, 1.

FERRONI, G. D. & INNISS, W. E. (1973). Thermally caused filament formation in the psychrophile *Bacillus insolitus. Canad. J. Microbiol.* **19**, 581.

FLETCHER, M. & FLOODGATE, G. D. (1973). An electron-microscopic demonstration of an acidic polysaccharide in the adhesion of a marine bacterium to solid surfaces. *J. gen. Microbiol.* **74**, 325.

FLOREY, H. W. (1946). The use of micro-organisms for therapeutic purposes. *Yale J. Biol. Med.* **19**, 101.

FOSDICK, L. S. & WESSINGER, G. D. (1940). Carbohydrate degradation by mouth organisms. II. Yeast. *J. Amer. dent. Ass. and Dent. Cosmos.* **27**, 203.

FRANK, R. M. & BRENDEL, A. (1966). Ultrastructure of the approximal dental plaque and the underlying normal and carious enamel. *Archs oral Biol.* **11**, 883.

FRANK, R. M. & HOUVER, G. (1970). An ultrastructural study of human supragingival dental plaque formation. In *Dental Plaque.* Ed. W. D. McHugh. Edinburgh and London: E. and S. Livingstone.

FREESE, E. B. (1973). Unusual membranous structures in cytochrome A-deficient mutants of *Bacillus subtilis. J. gen. Microbiol.* **75**, 187.

FREESE, E., SHEU, C. W. & GALLIERS, E. (1973). Function of lipophilic acids as antimicrobial food additives. *Nature, Lond.* **241**, 321.

FRETER, R. (1955). The fatal enteric cholera infection in the guinea pig achieved by inhibition of normal enteric flora. *J. Infect. Dis.* **97**, 57.

FROSTELL, G. & SÖDER, P-O. (1970). The proteolytic activity of plaque and its relation to soft tissue pathology. *Int. dent. J.* **20**, 436.

FULLER, R. (1972). Bacteria that stick in the gut. *New Scientist* 30th Nov. 506.

FULLER, R. & TURVEY, A. (1971). Bacteria associated with the intestinal wall of the fowl. *(Gallus domesticus). J. appl. Bact.* **34**, 617.

GALBRAITH, H., MILLER, T. B., PATON, A. M. & THOMPSON, J. K. (1971). Antibacterial activity of long chain fatty acids and the reversal with calcium, magnesium, ergocalciferol and cholesterol. *J. appl. Bact.* **34**, 803.

GIBBONS, R. J. (1971). Microbial ecology and its relation to dental disease. (U.S. Dept. of Health, Education and Welfare), *Dental Research in the United States and Canada 1971*, p. 1-19.

GIBBONS, R. J. & SPINELL, D. M. (1970). Salivary-induced aggregation of plaque bacteria. In *Dental Plaque*, Ed. W. D. McHugh. Edinburgh and London: E. & S. Livingstone.

GOADBY, K. W. (1900). Micro-organisms in dental caries. *Dent. Cosmos* **42**, 210.

GRANT, G. T., MORRIS, E. R., REES, D. A., SMITH, P. J. C. & THOM, D. (1973). Biological interactions between polysaccharides and divalent cations. *FEBS Letters* **32**, 195.

GUGGENHEIM, B. (1968). Streptococci of dental plaques. *Caries Res.* **2**, 147.

HAWKSWORTH, G., DRASAR, B. S. & HILL, M. J. (1971) Intestinal bacteria and the hydrolysis of glycosidic bonds. *J. med. Microbiol.* **4**, 451.
HAY, D. I. (1969). Some observations on human saliva proteins and their role in the formation of the acquired enamel pellicle. *J. dent. Res.* **48**, 806.
HAY, D. I. (1970). Isolation and partial characterization of salivary proteins adsorbed by hydroxyapatite. *J. dent. Res.* **49**, 71.
HAYES, M. L. & HYATT, A. T. (1972). A possible relationship between ammonia in the dental plaque and gingival inflammation. *J. dent. Res.* **51**, 1259.
HAYWARD, A. F. & ARMSTRONG, W. G. (1970). Parallel electron microscope and analytical investigations of enamel integuments. In *Dental Plaque*. Ed. W. D. McHugh. Edinburgh and London: E. & S. Livingstone.
HENNEBERG, G. (1964). (ed.) *Bildatlas pathogener Microorganismen. (Pictorial Atlas of Pathogenic Microorganisms)* Stuttgart: Gustav Fischer Verlag.
HENTGES, D. J. (1967). Influence of pH on the inhibitory activity of formic and acetic acids for *Shigella. J. Bact.* **93**, 2029.
HENTGES, D. J. (1970). Enteric pathogen – normal flora interactions. *Amer. J. Clin. Nutrition.* **23**, 1451.
VAN DER HOEVEN, J. S., MIKX, F. H. M., PLASSCHAERT, A. J. M. & KÖNIG, K. G. (1972). Methodological aspects of gnotobiotic caries experimentation. Preliminary investigation into the microbial ecology of dental plaque. *Caries Res.* **6**, 203.
HOLMBERG, K. & HALLANDER, H. O. (1972). Interference between Gram-positive micro-organisms in dental plaque. *J. dent. Res.* **51**, 588.
HOLMBERG, K. & HALLANDER, H. O. (1973). Production of bactericidal concentrations of hydrogen peroxide by *Streptococcus sanguis. Archs oral Biol.* **18**, 423.
HOWDEN, G. F. (1972). The effect of some dental plaque metabolites on HeLa cells in tissue culture. *J. dent. Res.* **51**, 1258.
HURST, V. (1950). Morphologic instability of actinomycetes associated with enamel. *J. dent. Res.* **29**, 571.
IKEDA, T. & SANDHAM, H. J. (1971). Prevalence of *Streptococcus mutans* in various tooth surfaces in negro children. *Archs oral Biol.* **16**, 1237.
ISHIKAWA, I. & KINOSHITA, S. (1972). Identification of proteolytic activity in the periodontal pocket. *J. dent. Res.* **51**, 1301.
JENKINS, G. N. (1968). The mode of formation of dental plaque. *Caries Res.* **2**, 130.
JENKINS, G. N. (1972). Current concepts concerning the development of dental caries. *Int. dent. J.* **22**, 350.
JENKINS, G. N. (1973). Dental plaque. *Indent.* **1**, 4.
JOHANSON, W. G., BLACKSTOCK, R., PIERCE, A. K. & SANDFORD, J. P. (1970). The role of bacterial antagonism in pneumococcal colonization of the human pharynx. *J. Lab. Clin. Med.* **75**, 946.
JONES, H. C., ROTH, I. L. & SANDERS, W. M. (1969). Electron microscopic study of a slime layer. *J. Bact.* **99**, 316.
JONES, S. J. (1971). Natural plaque on tooth surfaces: a scanning electron microscope study. *Apex* **5**, 93.
JONES, S. J. (1972). A special relationship between spherical and filamentous micro-organisms in mature human dental plaque. *Archs oral Biol.* **17**, 613.
KELSTRUP, J. & GIBBONS, R. J. (1969). Bacteriocins from human and rodent streptococci. *Archs oral Biol.* **14**, 251.
KELSTRUP, J. & FUNDER-NIELSEN, T. D. (1972). Molecular interactions between the extracellular polysaccharides of *Streptococcus mutans. Archs oral Biol.* **17**, 1659.
LAST, F. T. & WARREN, R. C. (1972). Non-parasitic microbes colonizing green leaves: their form and function. *Endeavour* **31**, 143.
LEACH, S. A. (1967). The acquired integuments of the teeth. *Brit. dent. J.* **122**, 537.
LEACH, S. A., CRITCHLEY, P., KOLENDO, A. B. & SAXTON, C. A. (1967). Salivary glycoproteins as components of the enamel integuments. *Caries Res.* **1**, 104.
LEACH, S. A. & HAYES, M. L. (1968). A possible correlation between specific bacterial enzyme activities, dental plaque formation and cariogenicity. *Caries Res.* **2**, 38.

LEACH, S. A. & MELVILLE, T. H. (1970). Investigation of some human oral organisms capable of releasing the carbohydrates from salivary glycoproteins. *Archs oral Biol.* **15**, 87.

LEACH, S. A. & SAXTON, C. A. (1966). An electron microscopic study of the acquired pellicle and plaque formed on the enamel of human incisors. *Archs oral Biol.* **11**, 1081.

LÖE, H. (1969). Present day status and direction for future research on etiology and prevention of periodontal disease. *J. Period. Perioc.* **40**, 678.

LÖE, H. (1970). A review of the prevention and control of plaque. In *Dental Plaque*. Ed. W. D. McHugh. Edinburgh and London: E. & S. Livingstone.

LOESCHE, W. J. (1968). Importance of nutrition in gingival crevice microbial ecology. *Periodontics* **6**, 245.

LOESCHE, W. J., HOCKETT, R. N. & SYED, S. A. (1972). The predominant cultivable flora of tooth surface plaque removed from institutionalized subjects. *Archs oral Biol.* **17**, 1311.

LYNCH, M., CROWLEY, M. C. & ASH, M. (1969). Correlation between plaque and bacterial flora. *J. Period. Perioc.* **40**, 634.

MÄKINEN, K. K. (1972). The role of sucrose and other sugars in the development of dental caries: a review. *Int. dent. J.* **22**, 363.

MÄKINEN, K. K. & SCHEININ, A. (1972). The effect of various sugars and sugar mixtures on the activity and formation of enzymes of dental plaque and oral fluid. *Acta odont. Scand.* **30**, 259.

MARSHALL, K. C., STOUT, R. & MITCHELL, R. (1971a). Selective sorption of bacteria from seawater. *Canad. J. Microbiol.* **17**, 1413.

MARSHALL, K. C., STOUT, R. & MITCHELL, R. (1971b). Mechanism of the initial events in the sorption of marine bacteria to surfaces. *J. gen. Microbiol.* **68**, 337.

MAYHALL, C. W. (1970). Composition and source of acquired enamel pellicle of human teeth. *J. dent. Res.* **49**, abs. 92, p. 71 (Amer. Div., Int. Ass. dent. Res.).

McCANN, H. G. (1968). In *Art and Science of Dental Caries Research*. Ed. R. S. Harris. Chicago: University Press.

McGAUGHEY, C. & STOWELL, E. C. (1967). The adsorption of human salivary proteins and porcine submaxillary mucin by hydroxyapatite. *Archs oral Biol.* **12**, 815.

McGAUGHEY, C. & STOWELL, E. C. (1971). Adsorption of salivary proteins by hydroxyapatite: relations between the effects of calcium ions, hydrogen ions, temperature, and exposure time. *J. dent. Res.* **50**, 542.

McHUGH, W. D. (1970). *Dental Plaque*. Edinburgh and London: E. & S. Livingstone.

MEADOWS, P. S. & ANDERSON, J. G. (1966). Micro-organisms attached to marine and freshwater sand grains. *Nature, Lond.* **212**, 1059.

MECKEL, A. H. (1965). The formation and properties of organic films on the teeth. *Archs oral Biol.* **10**, 585.

MECKEL, A. H. (1968). The nature and importance of organic deposits on dental enamel. *Caries Res.* **2**, 104.

MECKEL, A. H. (1971). The importance of organic cuticles to plaque formation on enamel. *J. dent. Res.* **50**, abs. 654, p. 216 (Amer. Div., Int. Ass. dent. Res.).

MEYNELL, G. G. (1963). Antibacterial mechanisms of the mouse gut. II. The role of E_h and volatile fatty acids in the normal gut. *Br. J. exptl. Path.* **44**, 209.

MEYNELL, G. G. & SUBBAIAH, T. V. (1963). Antibacterial mechanisms of the mouse gut. I. Kinetics of infection by *Salmonella typhi-murium* in normal and streptomycin-treated mice studied with abortive transductants. *Br. J. exptl. Path.* **44**, 197.

MIKX, F. H. M., VAN DER HOEVEN, J. S., KONIG, K. G., PLASSCHAERT, A. J. M. & GUGGENHEIM, B. (1972). Establishment of defined microbial ecosystems in germ free rats. I. The effect of the interaction of *Streptococcus mutans* or *Streptococcus sanguis* with *Veillonella alcalescens* on plaque formation and caries activity. *Caries Res.* **6**, 211.

MILLER, W. D. (1890). *The Micro-organisms of the Human Mouth, The Local and General Diseases which are caused by Them*. Philadelphia: The S. S. White Mfg. Company.

MONTGOMERY, P. C. (1973). Immunobiology of the oral cavity. *J. dent. Res.* **52**, abs. 1, p. 59 (Amer. Div., Int. Ass. dent. Res.).

MYERS, D. M. (1959). An antibiotic effect of viridans streptococci from the nose, throat and sputum, and its inhibitory effect on *Staphylococcus aureus*. *Am. J. clin. Path.* **31,** 332.
NESS, A. R. (1971). The natural history of enamel caries. *Apex* **5,** 15.
NEWMAN, H. N. (1972a). Clinical significance of micro-organisms and organic films on human enamel. *Proc. roy. Soc. Med.* **65,** 908.
NEWMAN, H. N. (1972b). Structure of approximal human dental plaque as observed by scanning electron microscopy. *Archs oral Biol.* **17,** 1445.
NEWMAN, H. N. (1972c). Freeze-etching and dental research. *J. periodont. Res.* **7,** 91.
NEWMAN, H. N. (1972d). Bacteriological inter-relationships in dental plaque. *J. dent. Res.* **51,** 1235.
NEWMAN, H. N. (1973a). Zone demarcation of organic films present on human enamel surfaces *in vivo*. *Brit. dent. J.* **134,** 273.
NEWMAN, H. N. (1937b). The organic films on enamel surfaces. *Brit. dent. J.* **135,** 64 & 106.
NEWMAN, H. N. & McKAY, G. S. (1973). An unusual microbial configuration in naturally occurring human dental plaque. *Microbios* **8,** 117.
NEWMAN, H. N. & POOLE, D. F. G. (1974). Human surface enamel and its natural integuments. *Helv. odont. Acta* **7,** 58.
NISSLE, (1916). In: HENTGES, D. J. (1970). Enteric infection – normal flora interactions. *Amer. J. Clin. Nutrition* **23,** 1451.
O'GRADY, F. W. & GREENWOOD, D. (1973). Antibiotic-induced damage in bacteria. *The Glaxo Volume* **38,** 5.
PARKER, R. B. (1970). Paired culture interaction of the oral microbiota. *J. dent. Res.* **49,** 804.
PARSONS, A. B. & DUGAN, P. R. (1971). Production of extracellular polysaccharide matrix by *Zooglea ramigera*. *Appl. Microbiol.* **21,** 657.
PEDERSEN, K. B., FRØHOLM, L. O. & BØVRE, K. (1972). Fimbriation and colony type of *Moraxella bovis* in relation to conjunctival colonization and development of keratoconjunctivitis in cattle. *Acta path. microbiol. Scand.* section B, **80,** 911.
PESSAC, B. & DEFENDI, V. (1972). Cell aggregation: role of acid mucopolysaccharides. *Science* **175,** 898.
POOLE, A. E., GILMOUR, M. N. & MILLS, J. R. (1967). The microbial content of natural and membrane plaques. *Caries Res.* **1,** 239.
POOLE, D. F. G. & NEWMAN, H. N. (1971). Dental plaque and oral health. *Nature, Lond.* **234,** 329.
POOLE, D. F. G. & SILVERSTONE, L. M. (1973). Remineralization of enamel. In *Hard Tissue Growth, Repair and Remineralization* (CIBA Foundation Symposium 11. Amsterdam: ASP (Elsevier, Excerpta Medica, North-Holland).
RAMBOURG, A. (1972). Morphological and histochemical aspects of glycoproteins at the surface of animal cells. *Int. Rev. Cytol.* **31,** 57.
REGOLATI, B., GUGGENHEIM, B. & MÜHLEMANN, H. R. (1972). Synergisms and antagonisms of two bacterial strains superinfected in conventional Osborne-Mendal rats. *Helv. odont. Acta* **16,** 84.
REUSCH, V. M. & BURGER, M. M. (1973). The bacterial mesosome. *Biochim. biophys. Acta* **300,** 79.
RICHMOND, M. H. (1972). Some environmental consequences of the use of antibiotics: or 'What goes up must come down'. *J. appl. Bact.* **35,** 155.
RITZ, H. L. (1963). Immunofluorescent studies of plaque *Nocardia*. *J. dent. Res.* **42,** abs. 12, p. 35 (Amer. Div., Int. Ass. dent. Res.).
RITZ, H. L. (1967). Microbial population shifts in developing human dental plaque. *Archs oral Biol.* **12,** 1561.
RITZ, H. L. (1969). Fluorescent antibody staining of *Neisseria. Streptococcus* and *Veillonella* in frozen sections of human dental plaque. *Archs oral Biol.* **14,** 1073.
RITZ, H. L. (1970). The role of aerobic neisseriae in the initial formation of dental plaque. In *Dental Plaque*. Ed. W. D. McHugh. Edinburgh and London: E. & S. Livingstone.

RIZZO, A. A. (1967). Rabbit corneal irrigation as model system for studies on the relative toxicity of bacterial products implicated in periodontal disease. The toxicity of neutralized ammonia solutions. *J. Periodont.* **38**, 491.
RØLLA, G. (1967). Adsorption of sialic acid-containing glycoprotein to human enamel *in vivo. J. periodont. Res.* **2**, 243.
RØLLA, G. (1969). The adsorption of radioactive salivary glycoprotein to hydroxylapatite *in vitro. J. periodont. Res.* **4**, 165.
RØLLA, G. (1971). Adsorption of dextran to saliva-treated hydroxyapatite. *Archs oral Biol.* **16**, 527.
RØLLA, G., CHRISTENSEN, T., MATHIESEN, P. & POVATONG, L. (1969). The properties of a human sublingual glycoprotein in the analytical ultracentrifuge and its adsorption of hydroxylapatite *in vitro. Caries Res.* **3**, 211.
ROVIRA, A. D. (1972). Studies on the interactions between plant roots and microorganisms. *J. Austral. Inst. Agr. Sci.* **38**, 90.
ROGERS, H. J. (1970). Bacterial growth and the cell envelope. *Bact. Rev.* **34**, 194.
ROSEBURY, T., GALE, D. & TAYLOR, D. F. (1954). An approach to the study of interactive phenomena among micro-organisms indigenous to man. *J. Bact.* **67**, 135.
SANDERS, E. (1969). Bacterial interference. I. Its occurrence among the respiratory tract flora and characterization of inhibition of Group A streptococci by viridans streptococci. *J. infect. Dis.* **129**, 698.
SAVAGE, D. C. (1970). Associations of indigenous micro-organisms with gastrointestinal epithelia. *Amer. J. Clin. Nutrition* **23**, 1495.
SAVAGE, D. C. (1972). Survival in mucosal epithelia, epithelial penetration and growth in tissues of pathogenic bacteria. In *Microbial Pathogenicity in Man and Animals* (Symposia Soc. Gen. Microbiol. No. 22).
SAVAGE, D. C. & McALLISTER, J. S. (1970). Microbial interactions at body surfaces and resistance to infectious diseases. In *Resistance to Infectious Disease.* Eds R. H. Dunlop & H. W. Moon. Saskatoon: Saskatoon Modern Press.
SCHROEDER, H. E. (1969). *Formation and Inhibition of Dental Calculus.* Bern, Stuttgart, Vienna: Hans Huber Publishers.
SCHROEDER, H. E. (1970). The structure and relationship of plaque to the hard and soft tissues: electron microscopic interpretation, *Int. dent. J.* **20**, 353.
SCHROEDER, H. E. & DE BOEVER, J. (1970). The structure of microbial dental plaque. In *Dental Plaque.* Ed. W. D. McHugh. Edinburgh and London: E. & S. Livingstone.
SCHWARTZ, R. S. & MASSLER, M. (1969). Tooth accumulated materials: a review and classification. *J. Periodont.* **40**, 407.
SCRIVENER, C. A., MYERS, A. I., MOORE, N. A. & WARNER, B. W. (1950). Some antagonistic activity of bacteria from the human oral cavity. *J. dent. Res.* 29, 784.
SELVIG, K. A. (1970). Attachment of plaque and calculus to tooth surfaces. *J. periodont. Res.* **5**, 8.
SELVIG, K. A. (1971). Interfaces between epithelium and soft and hard connective tissue. In *The Prevention of Periodontal Disease.* Eds J. E. Eastoe, D. C. A. Picton & A. G. Alexander. London: Henry Kimpton.
SELWYN, S. & ELLIS, H. (1972). Skin bacteria and skin disinfection reconsidered. *Brit. med. J.* No. 5793, **1**, 136.
SIEBURTH, J. M., McGINNIS, J. & SKINNER, C. E. (1952). Effect of terramycin on the antagonism of certain bacteria against species of *Proteus. J. Bact.* **64**, 163.
SILVERMAN, G. & KLEINBERG, I. (1967). Fractionation of human dental plaque and characterization of its cellular and acellular components. *Archs oral Biol.* **12**, 1387.
SILVERMAN, M. P. & MUNOZ. E. F. (1970) Fungal attack on rock; solubilisation and altered infra-red spectra. *Science* **169**, 985.
SIMS, W. (1972). The concept of immunity in dental caries. II. Specific immune responses. *Oral Surg.* **34**, 69.
SNYDER, M. L., BULLOCK, W. W. & PARKER, R. B. (1967). Morphology of Gram-positive filamentous bacteria identified in dental plaque by fluorescent antibody technique. *Archs oral Biol.* **17**, 1269.

SPRUNT, K. & REDMAN, W. (1968). Evidence suggesting importance of role of interbacterial inhibition in maintaining balance of normal flora. *Ann. Int. Med.* **68**, 579.
STACK, M. V., DONOGHUE, H. D., TYLER, J. E. & MARSHALL, M. (1972). Identification of oral bacteria by pyrolysis gas chromatography. *J. dent. Res.* **51**, 1238.
STACK, M. V., DONOGHUE, H. D., TYLER, J. E. & MARSHALL, M. (1973). Identification of oral streptococci using standardised data from pyrolysis gas chromatography. *J. dent. Res.* **52**, 969.
STACK, M. V. & FLETCHER, R. P. (1971). Inhibition of enamel reactivity by silanes and surfactants. *J. dent. Res.* **50**, 693.
TANZER, J. M. (1972). Studies on the fate of the glucosyl moiety of sucrose metabolized by *Streptococcus mutans*. *J. dent. Res.* **51**, 415.
TANZER, J. M., CHASSY, B. M. & KRICHEVSKY, M. I. (1972). Sucrose metabolism by *Streptococcus mutans*, SL – 1. *Biochim. Biophys. Acta (G)* **261**, No. 2, 379.
TANZER, J. M., WOOD, W. I. & KRICHEVSKY, M. I. (1971). Linear growth kinetics of plaque-forming streptococci in the presence of sucrose. *J. gen. Microbiol.* **58**, 125.
THOMA, K. H. & GOLDMAN, H. M. (1960). In *Oral Pathology* (5th ed.) St. Louis: C. V. Mosby Co.
THOMPSON, R. & SHIBUYA, M. (1946). The inhibitory action of saliva on the diphtheria bacillus: the antibiotic effect of salivary streptococci. *J. Bact.* **51**, 671.
THONARD, J. C. (1968). The microbiology of periodontal disease. *Ala. J. med. Sci.* **5**, 302.
VAN HOUTE, J., GIBBONS, R. J. & O'GARA, M. M. (1971). Adherence of oral streptococci to dental plaques, the tongue and vestibular mucous membranes in humans. *J. dent. Res.* **50**, 180.
VAN HOUTE, J., GIBBONS, R. J. & PULKKINEN, A. J. (1971). Adherence as an ecological determinant for streptococci in the human mouth. *Archs oral Biol.* **16**, 1131.
VAN HOUTE, J., HILLMAN, J. D. & GIBBONS, R. J. (1970). The sorption of bacteria to human enamel power. *J. dent. Res.* **49**, abs. 158, p. 88. (Amer. Div., Int. Ass. dent. Res.).
VAN HOUTE, J. & SAXTON, C. A. (1971). Cell wall thickening and intracellular polysaccharide in micro-organisms of the dental plaque. *Caries Res.* **5**, 30.
VREVEN, J. & FRANK, R. M. (1973). Cytochimie ultrastructurale de la phosphatase acide et de la pyrophosphatase dans la plaque dentaire humaine. *J. Biol. Buccale* **1**, 63.
WARD, M. E. & WATT, P. J. (1973). How infectious is gonorrhoea? *Brit. med. J.* No. 5851, **1**, 485.
WILKINSON, J. F. (1958). The extracellular polysaccharides of bacteria. *Bact. Rev.* **22**, 46.
WISTREICH, G. A. & BAKER, R. F. (1971). The presence of fimbriae (pili) in three species of *Neisseria*. *J. gen. Microbiol.* **65**, 167.

Aerial Dispersal of Micro-organisms from the Human Respiratory Tract

O. M. LIDWELL

*Central Public Health Laboratory,
Colindale Avenue, London NW9 5HT, England*

CONTENTS

1. Introduction	135
2. Droplet dispersal from the respiratory tract	135
3. Salivary streptococci in the air	140
4. Some examples of the dispersal of specific pathogens	143
(a) β-haemolytic streptococci	143
(b) Mycobacteria	144
(c) *Staphylococcus aureus*	146
(d) Respiratory viruses	149
5. Conclusion	150
6. Summary	152
7. Acknowledgements	152
8. References	152

1. Introduction

THE EARLY HISTORY of bacteriology is intimately associated with the ubiquitous presence of micro-organisms in the air. As the science developed in the course of the nineteenth century the germ theory of disease progressed from a plausible guess to a well evidenced phenomenon. At the same time many workers were convinced that airborne dispersal was responsible for the epidemic spread of zymotic disease. If this was so then the methods by which the specific virus became airborne and the distance to which the infective agent was carried were of major interest. A summary of the situation as it appeared at the turn of the century is given by Gordon (1904). Morbific virus, he states, apart from direct contact or inoculation, may either be ingested or inhaled. Control over food and water has to a considerable extent been achieved but there is no bacteriological test for the pollution of air by material given off from the human body. On the basis of the large numbers of streptococci, in particular 'streptococcus brevis' (a short chain streptococcus of the *viridans* group) found in normal 'saliva' he goes on to suggest that the presence or absence of this organism in the air might be used as an index of respiratory pollution.

2. Droplet Dispersal from the Respiratory Tract

The projection of secretion by the acts of sneezing, coughing and talking is one of the obvious facts of life. When it became apparent that this might constitute a method of spreading disease many bacteriological studies were made of these activities, of which the best known are those of Flügge and his co-workers (1897, 1899). Koeniger (1900) and Gordon (1904) also contributed. However, the technical equipment of all these investigators was inadequate to provide a comprehensive and quantitative description.

The use of a high intensity short duration flash gives a vivid visual demonstration, which may be recorded photographically, of the actual projection of particles of secretion (Bourdillon & Lidwell, 1941; Jennison, 1942) and an exhaustive analysis of the numbers of particles expelled and their size distribution was made by Duguid (1945, 1946). The photographs show clearly that even during sneezing the particles are projected mainly from the mouth although occasionally strings of mucus may be ejected from the nostrils.

Duguid examined the size distribution of the particles of saliva. The larger droplets, $> c.$ 100 μm diam, fall fairly rapidly to the ground. The distance they may have been projected varies with their size and the velocity with which they are expelled but is limited to a few metres in a forward direction. These are the 'projectiles' of Flügge and others. Smaller droplets will dry out before reaching the ground and since the solid content of saliva is normally a little $< 1\%$, the diameters of these 'droplet nuclei', as they were christened by W. F. Wells (1934), are only $c.$ $\frac{1}{2}$th of the diameters of the original droplets. The settling velocities of such small particles are only a few cm/sec and, since the turbulent air movements in a room are usually several times greater than this, these small particles are carried with the air currents and may remain suspended in the air for a considerable time. They do, however, settle and are removed from the air in this way as well as by ventilation. Duguid (1945, 1946) also pointed out that the probability that a particle of saliva would contain a viable micro-organism was dependent both on the size of the particle and the number of micro-organisms contained in the liquid dispersed. Hid results have been recalculated in graphical form (Lidwell, 1967). Fig. 1 shows the result for the particles expelled during coughing or talking. Unless the density of the original suspension exceeds 10^7/ml, the size distribution of the resulting air-suspended particles carrying an organism is largely independent of the suspension density and has a median around 10-15 μm. A vigorous cough might produce between 10^3 and 10^4 droplets. A similar number resulted from speaking loudly some 2000 words although the numbers dispersed on talking vary widely from person to person (*Report,* 1948), a fact of which everyone is probably aware. The particles expelled by sneezing are in general somewhat smaller but the difference is not very great (Fig. 2). The most striking, and obvious, fact is the very large numbers of particles produced, $c.$ 10^6 by one average sneeze. The size distribution of airborne particles carrying salivary streptococci found in occupied places has been found to agree quite well with these calculations, (Fig. 3). It should be pointed out that the size distributions are very broad. The quartiles fall at

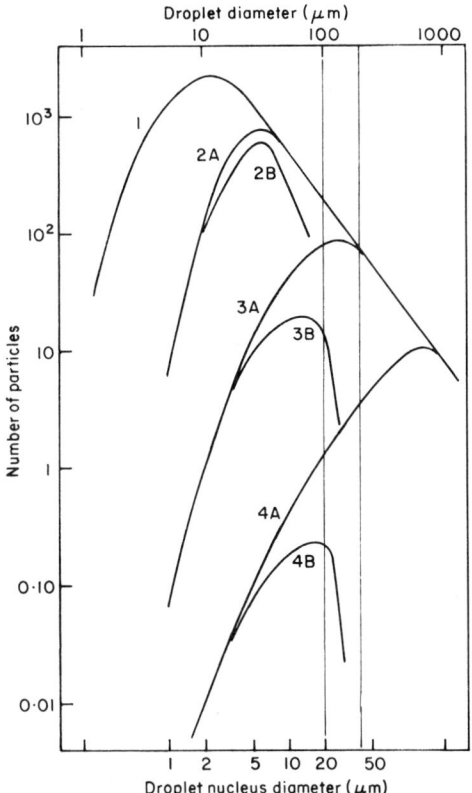

Fig. 1. The size distribution of the particles of secretion expelled by coughing or talking and of the resulting 'droplet nuclei' in a room 10 ft high ventilated at 4 air changes/h. 1, The numbers and size distribution of the expelled droplets produced by 2 coughs or speaking loudly 4000 words (10^4 droplets). 2A, The numbers and size distribution of those droplets carrying one or more viable units for a level of infection in the secretion of 10^8 viable units/ml. 2B, The relative exposure to the 'droplet nuclei' of different sizes formed by evaporation of the droplets described by the curve 2A. 3A, 3B, 4A, 4B, The same distributions as illustrated in 2A and 2B but resulting from levels of infection in the secretion of 10^6 viable units/ml for 3A and 3B, and 10^4 viable units/ml for 4A and 4B. The figure is drawn on a double log scale so that a log-normal distribution of particle size would appear as a symmetrical curve with approximately linear tails. The scale at the top of the figure gives the sizes of the original droplets; the bottom scale gives the resulting droplet nuclei (curves B), assuming evaporation to a residuum of 1.25%. The curves are based on the data given by Duguid (1946).

diameters 2-3 times the median and at corresponding fractions.

The particle diameters bear no relation to the size of the micro-organisms themselves so that similar size distribution would be expected for virus carrying particles also although high titres ($>10^8$/ml) would lead to rather smaller

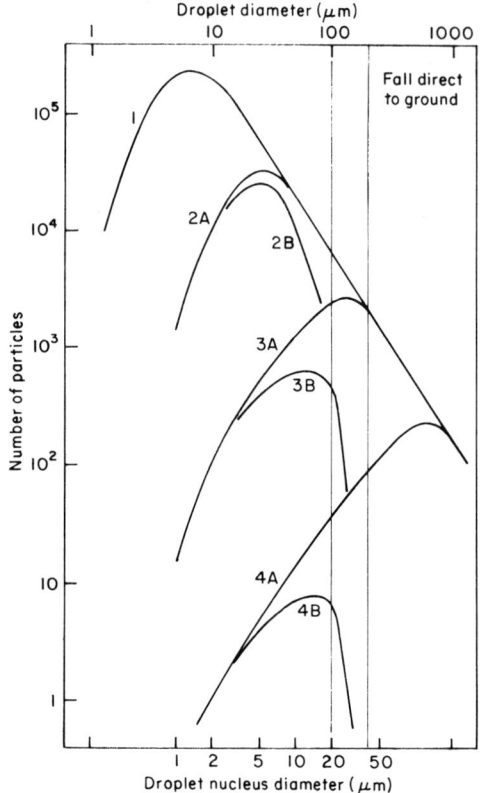

Fig. 2. The size distribution of the particles of secretion expelled by sneezing and of the resulting 'droplet nuclei'. 1, The numbers and size distribution of the expelled droplets produced by one average sneeze (10^6 droplets). 2A, 3A, 4A, The numbers and size distribution of those droplets carrying one or more viable units for levels of infection in the secretion of 10^8, 10^6 and 10^4 viable units/ml. 2B, 3B, 4B, The relative exposure to the 'droplet nuclei' of different sizes formed by evaporation of the droplets described by the curves 2A, 3A, 4A. The curves are based on the data given by Duguid (1946).

median diameters. Possibly more important, if absolute filtration or determination of high levels of air 'sterility' is required, is the absence for virus particles of the absolute cut-off at a minimum particle diameter above 1 μm.

Buckland & Tyrrell (1964) explored dispersal from the respiratory tract when the point of origin of the infected material was in the nose. They introduced artifically small quantities of a spore suspension and determined the amounts discharged into a plastic bag by talking, coughing and sneezing. The contamination levels found in the throat were down to one log unit lower than those in the nose while those in the saliva might be as much as 4 log units down and

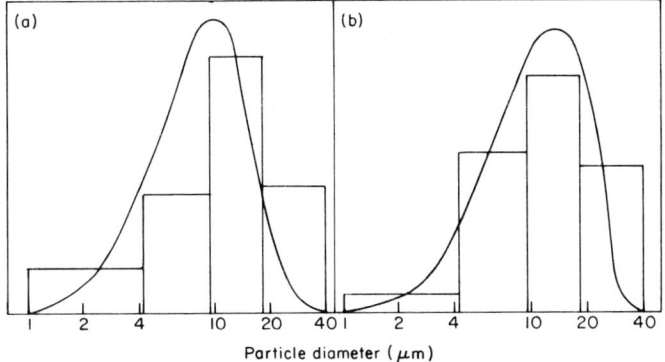

Fig. 3. The size distribution of airborne particles, presumptively of mouth origin, carrying streptococci. (a), The histogram shows the size distribution of all the mouth streptococci. The full line is the size distribution of the infected droplet nuclei produced by talking or coughing for a level of infection in the secretion of 10^7 viable units/ml (comparable to curves 2B, 3B and 4B in Fig. 2). (b), The histogram shows the size distribution of those particles carrying *Streptococcus salivarius*. The full line is the size distribution of the infected droplet nuclei produced by talking or coughing for a level of infection in the secretion of 10^6 viable units/ml (the same curve allowing for changes of scale, as 3B in Fig. 2). The scale of particle size is logarithmic. All the curves have been normalized to enclose the same area. The data for the airborne streptococcus-carrying particles are taken from observations made in a group of clerical offices (Kingston, Lidwell & Williams, 1962).

were usually more than 2 log units below that in the nose (Table 1). The relatively small size of the bag leads to rapid deposition of all but the smallest particles on the walls but it is clear that coughing was less effective than talking on producing small particles. The difference between sneezing and talking was similar to that reported by Duguid (1945, 1946) for the total material dispersed, a ratio of 1.5×10^4 in place of 4×10^3 for an average sneeze compared with speaking 100 words.

Table 1
*Dispersal of spores placed in the nose**

Activity	Spores found	
	On wall of bag	In air of bag
Declaim 10 lines of Shakespeare	380	40
Deep coughing	380	0
Sneeze	5.7×10^6	1.9×10^3

* 10^8 spores of *Bacillus mycoides* in 0.1 ml of water were placed in the nose.

3. Salivary Streptococci in the Air

Duguid's work (1945, 1946) demonstrates the regular and continuous dispersal of droplets from the mouth and thus substantiates Gordon's (1904) suggestion that the presence of salivary streptococci in the air would be an indication of air pollution.

Several investigators have made use of this and attempted to relate air contamination measured in this way to the risk of the spread of upper respiratory tract disease and to the efficacy of environmental measures, such as ventilation, in reducing one or both.

An extensive study was made during the years 1947-1949 in a group of schools in the borough of Southall, London (*Air Hygiene Committee,* 1954; Williams *et al.,* 1956). Air samples were taken regularly in the class rooms and estimates made of the numbers of streptococci (Tables 2 & 3). Care was taken to distinguish between those types of streptococci presumptively derived from the saliva of the children and other similar organisms found in the samples (Williams *et al.,* 1956). When this was done, a good correlation was found between the numbers of those types present in the air and the amount of talking in the class-room at the time thus confirming the direct oral origin of these bacteria (Table 4). Of much greater interest is the close correlation found between the measles secondary attack rate in the class-rooms and the numbers of streptococci in the air of those class-rooms (Fig. 4) (Reid *et al.,* 1956).

A somewhat similar series of observations was made over the years 1951-1957

Table 2
Bacterial species comprised in general flora of school classrooms
(based on a total of 930 colonies from 29 air samples)

Bacteria	No. of organisms/ft^3 (based on colony count)	Count as a percentage of total count
Micrococci*	35.04	81.6
Diphtheroid organisms	2.94	6.8
Aerococci	1.41	3.3
Streptococci	1.34	3.1
Coliform organisms	1.15	2.7
Aerobic spore-bearing organisms	0.42	1.0
Staphylococcus aureus	0.12	0.3
Others	0.54	1.3
Total flora	42.96	100.1

* The term 'micrococci' is used for all Gram positive cocci that are not streptococci, coagulase positive staphylococci, or aerococci. The last-named is a Gram positive coccus, not a streptococcus, distinguished from the micrococci chiefly by its ability to grow in the presence of 1 : 500,000 of crystal violet, and to produce greening of blood agar (Williams *et al.,* 1953)

Table 3
Varieties of streptococci in schoolroom air (based on a total of 7390 colonies from 100 air samples)

Type of streptococci	Count of organisms/ft^3 (based on colony count)	Count as percentage of total count
Viridans-type streptococci	0.51	44.7
Streptococcus salivarius	0.29	25.5
Enterococci: α-haemolytic	0.17 ⎫	14.9 ⎫
β-haemolytic	0.01 ⎬ 0.23	0.9 ⎬ 20.2
Non-haemolytic	0.05 ⎭	4.4 ⎭
Non-haemolytic streptococci not *Strep. salivarius*	0.08	7.0
β-haemolytic streptococci	0.03	2.6
Total streptococci	1.14	100.0

The mean total count on the selective plates from which these streptococci were isolated was 7.16 colonies/ft^3 of which 4.95/ft^3 were aerococci, and the remainder, apart from the streptococci, were other micrococci, coliforms and aerobic spore-bearing organisms.

Table 4
Coefficients of the standard partial regression equations for bacterial count in school classrooms on activity, talking and ventilation

Organism	Activity*	Talking*	Ventilation*
Streptococcus salivarius	0.15	0.42	−0.17
Mouth streptococci	0.05	0.33	−0.06
General flora	0.17	0.22	−0.28
Aerococci	0.30	0.14	−0.26
Enterococci	0.13	0.10	−0.01

* The standard errors of these coefficients are all similar in magnitude and lie between 0.06 and 0.08.

in a group of Government offices at Newcastle-on-Tyne (Lidwell & Williams, 1961 *a, b*; Kingston *et al.*, 1962). Air samples were examined in a similar way to those from the schools and streptococci, presumptively of oral origin, were consistently found although in smaller numbers than in the class-rooms (Table 5). However, in this environment it was not possible to show any correlation between the numbers of streptococci in the air and the incidence of respiratory disease.

A consideration of these 2 studies shows something of both the possibility of using the numbers of airborne streptococci as an index of human respiratory

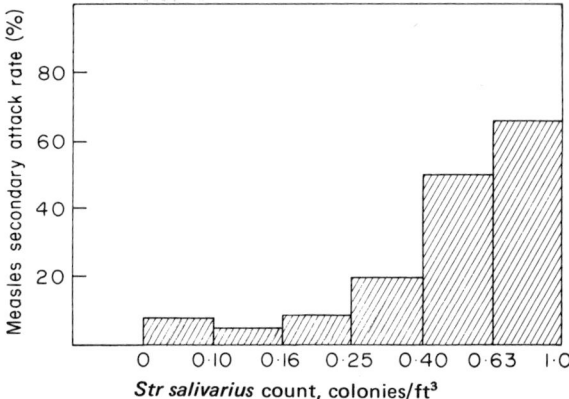

Fig. 4. Counts of airborne *Strep. salivarius* and measles cross-infection rates in classrooms. Adapted from Reid *et al.* (1956).

Table 5

Salivary streptococci in the air of offices and the incidence of colds

Year	Treatment	Mouth streptococci (no./m^3)	*Streptococcus salivarius* (no./m^3)	No. of colds	
				observed	expected
1955-56	Control	6.7	1.7	132	130
	Dummy treatment	12.0	2.0	184	191
	High ventilation	4.9	0.8	182	182
	Chemical disinfection	4.9	1.3	67	61
1956-57	Control	12.4	0.8	141	136
	Dummy treatment	9.9	1.2	194	195
	High ventilation	4.6	0.7	170	175
	UV irradiation	3.5	0.4	62	61

* Based on the overall average for the period, standardized for the age and sex distribution in the individual groups.

tract pollution and the limitations of attempting to derive from this any useful guidance on the risk of infection within the environment.

The investigations of infection by measles within the schools were able, because of the immunologically simple nature of the disease, to identify both the introduction of a primary case into a class-room and any secondary cases resulting from it, as well as the number of susceptibles exposed. Within this limited context the risk of cross infection could be seen to be directly related to the level of air contamination. However, there was only an insignificant reduction in the overall incidence of measles in those schools provided with UV irradiation of the class-room air, although this resulted in a reduction in the numbers of salivary streptococci in the air of these rooms of between 3 and 4-fold. Other investigators have reported similar results (Wells et al., 1942; Perkins et al., 1947). Irradiation of the class-rooms sometimes prolonged the duration of an epidemic but had little if any effect on the total number of cases. Two, essentially similar, factors combine to bring about this result. The children are exposed to risk of infection in other places and they may be repeatedly exposed in the class-room so that, although the risk of acquiring the disease on any one occasion is reduced, the majority of the susceptibles eventually succumb. Similar factors would also have been operative in the office situation. Here, however, the immunological situation and the problems of disease identity among the many minor upper respiratory infections described as 'colds' made it impossible to attempt any real assessment of secondary attack rates within the office environment. In contrast to measles, where susceptible children are all more or less equally susceptible to infection, susceptibility to 'colds' is probably highly variable. This means that reducing the infectious dose is likely to have a disproportionately small effect on the incidence of infection i.e. a very large reduction in dose is needed to effect a moderate reduction in disease (Lidwell, 1963). All the methods available for reducing the bacterial content of the air of occupied places, however, e.g. increased ventilation, UV irradiation, disinfecting vapours, have only a very moderate effect, producing reductions which rarely exceed 75% in practice (Lidwell & Williams, 1961b).

The limitations of the use of the numbers of airborne salivary streptococci as an index of air purity are then apparent. Compared with e.g. the *E. coli* test for water supplies, the volumes of air inhaled are very large, 5-10,000 l over an 8 h working day, the sources of supply are numerous and essentially difficult to control, the human effluent is discharged directly into the supply close to the point of consumption and, finally, the methods of purification are inefficient applied to this situation.

4. Some Examples of the Dispersal of Specific Pathogens

(a) β-haemolytic streptococci

A study of dispersal of β-haemolytic streptococci was made by Hamburger & Robertson (1948). The numbers dispersed by talking and coughing were too few for any detailed analysis. The distribution of organisms recovered as a result of a series of sneezes showed clearly, however, the dominant effect of the concentration of the organism in the saliva (Fig. 5). The numbers dispersed were approximately proportional to this and apparently independent of the level of nose or throat carriage. The limited data obtained from the coughing experiments showed the highest dispersal from a carrier with only moderate levels of β-haemolytic streptococci in the saliva. Further confirmation that the material dispersed by coughing was not derived from the mouth was found in the absence of dispersal of α-streptococci by this activity. As was the case after sneezing, most of the larger droplets, which settled within a few minutes, were projected <2 m.

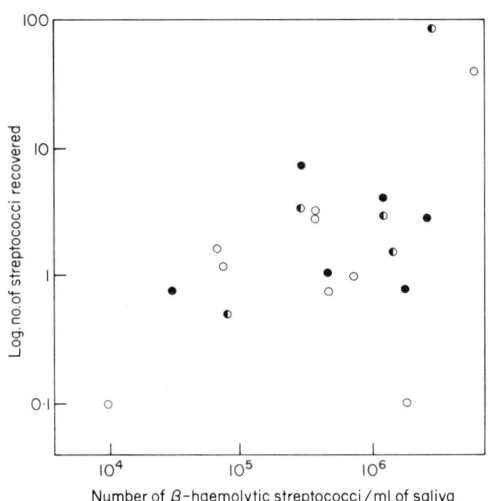

Fig. 5. Dispersal of β-haemolytic streptococci by sneezes. The vertical scale is the sum of the numbers collected in bubbler samples plus 1/10 of the numbers found on settle plates. ●, heavy nose and throat carrier; ◐, heavy throat carrier (nose weak); ○, weak nose and throat carrier. After Hamburger et al. (1948).

(b) *Mycobacteria*

Among the earliest attempts to study in detail the mode of transmission of a specific pathogen were studies of the dispersal of tubercle bacilli by patients suffering from pulmonary tuberculosis. During the closing years of the 19th century Flügge and his co-workers at Breslaw extended their work on the dispersal by talking, coughing and sneezing of saliva artificially contaminated with *B. prodigiosus (Serratia marcescens)* to explore the actual dissemination of virulent organisms from patients (Flügge et al., 1899); Gordon (1904) in discussing these results points out that although droplets containing the bacillus were not found further than 1-2 m from the patient they appeared to be present in the air of the room for several hours after the act of dispersal and that guineapigs kept in a ventilating shaft at the Brompton Hospital acquired tuberculosis. This last observation was expanded into a major experiment by Riley and others (Sultan, et al., 1960; Riley & O'Grady, 1961; Riley et al.,

1962) who examined the dispersal of tubercle bacilli by patients in a 6 bed isolation ward by passing the ventilating air from the rooms over cages containing susceptible guineapigs. The first result of their studies was the demonstration that extensive dispersal was limited to a small minority of patients (Fig. 6). Over ⅔ of the animal infections could be related to only 3

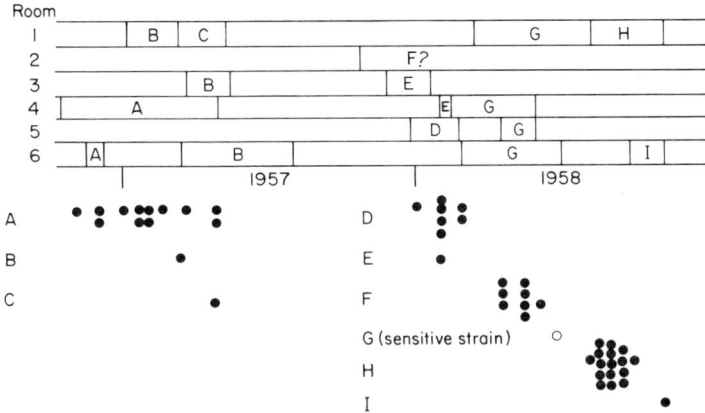

Fig. 6. Patient sources of tuberculosis infection in guineapigs. Each horizontal strip represents the sequence of events in a single patient room with the vertical lines indicating the periods of time during which the room was occupied by a patient carrying the strain denoted by the letters placed between them. The air extracted from the 6 patient rooms by the ventilation plant was passed through a chamber in which there were cages containing up to 180 susceptible guineapigs. These were tuberculin tested at monthly intervals and any positive reactors autopsied. The circles in the lower part of the figure show the times at which guineapigs were found to be infected with strains of mycobacteria identical with those carried by patients. Each circle corresponds to a single guineapig infected with the strain indicated by the letter shown to the left of each sequence of circles. After Sultan *et al.* (1960).

among the 77 patients who occupied the ward during the study. While the average patient dispersed only *c.* 30 infectious particles a day into the ward the most prolific generated 50 times as many. The second observation related the infectiveness of the room air for the animals to the experience of the nurses. The average time for tuberculin conversion for a susceptible nurse was about one year. During her time on duty during this period she would have inhaled *c.* 750 m^3 of air. The guineapigs developed one microscopically detected tubercule for every 400 m^3 of air they breathed. Although the numbers of infectious particles generated may seem small it must be remembered that the average person breathes nearly a cubic metre of air/h so that these numbers are quite adequate to explain the explosive epidemics which have been frequently reported from class-rooms and other close communities.

The site of infection is in the lung and it is the coughed up sputum which contains the organism. The generation from such viscous material of droplets small enough to reach the susceptible site in the deep lung, i.e. no more than 5 μm diam, requires an active process. Dispersal into the saliva of the mouth might make this possible. Direct projection on coughing might also achieve it but would be very inefficient. An interesting association with singing has been reported in several instances (e.g. Bates, Potts & Lewis, 1965), including a widespread outbreak due to the activities of pop-groups in the Netherlands (*Medical World News*, 1965). It is suggested that prolonged vibration of the vocal cords leads to the production of droplets and that these may then be effectively dispersed through the clear passage ways associated with singing.

Leprosy frequently attacks the nasal cartilage and large numbers of the bacillus may be present in nasal secretion. Similar experiments to those of Flügge (1897, 1899) were conducted by Schaeffer with leprosy patients. Leprosy bacilli were recovered in a similar spatial distribution.

(c) Staphylococcus aureus

The normal site of growth of *Staph. aureus* is in the anterior nares. There appears to be very little direct dispersal of this organism from the upper respiratory tract. For example, in a study of surgical masks, Shooter, Smith & Hunter (1959) found that carriers dispersed only very small numbers of *Staph. aureus* on to plates exposed in a small chamber around the head (Table 6) even when talking. Where skin sites are colonized, as in patients with infected eczemas or in perineal carriers, extensive dispersal occurs. This appears, however, to be only a more dramatic example of the common route of dispersal on desquamating skin which has been contaminated from the upper respiratory tract (Davies & Noble, 1962). The difference in this respect between *Staph. aureus* and e.g. the salivary streptococci, is presumably to be found in the tolerance or susceptibility of the different species to the fatty acids of the skin.

Table 6
Dispersal on talking, by carriers of Staphylococcus aureus

	Average no. of colonies on 20 plates*	
	All species	*Staph. aureus*
Dummy	15	–
Silent	30	0.05
Talking	50	0.14

* Exposed during 15 min quiet conversation in a box of 13 ft^3 capacity.

There is, however, a study of staphylococcal dispersal by babies in an infant nursery which suggests that in some circumstances a very different situation may arise (Eichenwald et al., 1960). During a series of epidemics of staphylococcal infection with type 80/81 the dispersal of the organism from several hundred infants was examined by placing them clad only in a paper napkin within a small cubicle. The results (Fig. 7) showed that while all those infants with skin infections dispersed substantial numbers of staphylococci there was also a substantial group of dispersers among the infants with no detectable infected lesions. It also appeared that, in contrast to other dispersers, the organisms

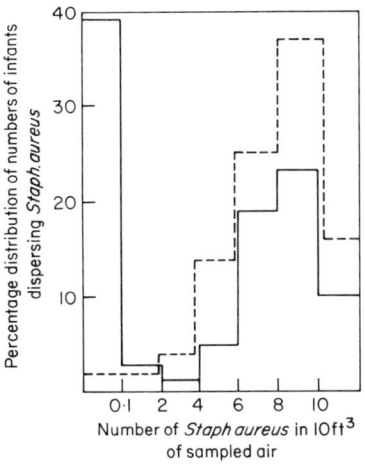

Fig. 7. Dispersal of *Staphylococcus aureus* by babies. The percentage distribution of the numbers of infants dispersing the organism between defined levels is given. Dispersal was estimated by sampling 100 ft^3 of air from a small cubicle of 150 ft^3 capacity in which the baby, clothed only in a clean napkin, lay asleep. The sampling took c. 45 min and was done after an infant had lain undisturbed in the cubicle for 1 h; ——, symptomless carriers; - - - -, babies with clinical pyoderma. After Eichenwald et al. (1960).

dispersed by these 'cloud-babies' were carried on particles the great majority of which were <5 μm in diam. By enclosing some of the infants in polythene bags tied around the neck it was clear that this dispersal was taking place from around the head (Fig. 8). Dispersal from most of the infants with pyoderma was greatly reduced by enclosure in the bag but that from the 'cloud-babies' was unaffected. It then appeared that transfer of an infant who was not himself a staphylococcal carrier (who had been on novobiocin since birth) from a nursery containing 'cloud-babies' to another where there were none such, although there were non-disseminating carriers of the staphylococcus present, could result in the appearance of 'cloud-babies' in this second nursery. Viral studies on nasal and pharyngeal swabs from 8 'cloud-babies' showed a close association between virus recovery and staphylococcal dispersal (Fig. 9). Similar studies on 5 disseminating

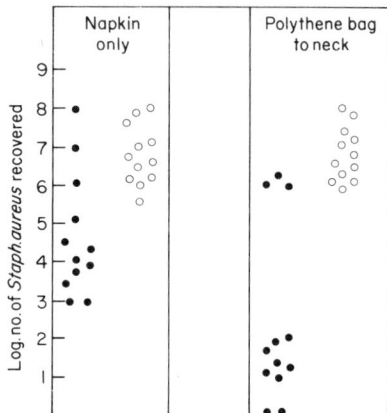

Fig. 8. Dispersal of *Staphylococcus aureus* by babies. Dispersal was estimated in the same way as for Fig. 7; ○, symptomless carriers; ●, babies with clinical pyoderma. After Eichenwald *et al.* (1960).

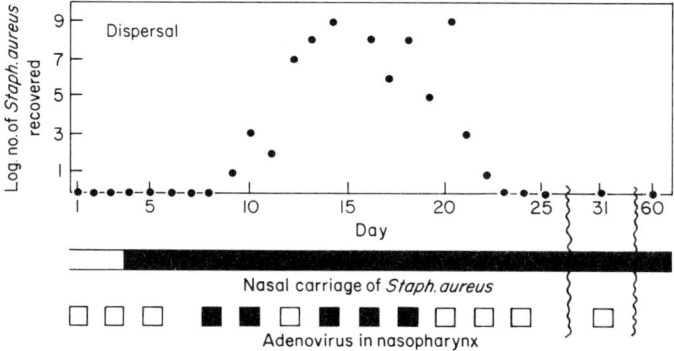

Fig. 9. Typical pattern of viral infection and dispersal of *Staphylococcus aureus*. Dispersal was estimated in the same way as for Figs 7 & 8. After Eichenwald *et al.* (1960).

and 5 non-disseminating carriers from each of 3 outbreaks showed complete correlation between carriage of respiratory virus and staphylococcal dissemination in 2 of these. No virus was detected in the third. These studies appear to demonstrate conclusively that in the presence of an upper respiratory virus considerable dispersal of staphylococci in very small particles may occur direct from the upper respiratory tract. No similar experience has, however, been reported during the last 12 years so that the significance of these observations is doubtful. The strains of staphylococcus, phage type 80/81, current at that time were epidemiologically very distinctive and their virtual disappearance may be the reason for the apparent non-recurrence of the phenomenon.

Although there is no reason to suppose that dispersal from the upper respiratory tract is involved, and indeed there are reasons to the contrary, dispersal of *Staph. aureus* in very small airborne particles has recently been reported from patients with extensive heavily infected burns (Hambreus, 1973).

(d) *Respiratory viruses*

If the secretions of the upper respiratory tract contain virus then the dispersal of these secretions will disseminate virus in essentially similar patterns to the dispersal of saprophytic or pathogenic bacteria.

Airborne transmission of infection seems most likely to be of epidemiological significance in connection with respiratory virus diseases from the common cold to influenza. The technical difficulties of quantitative studies of viral dispersal from infected individuals are, however, considerable and only a limited number of studies have been reported. The difficulties are increased by the relatively low level of viral infection of the secretion. The number of tissue culture infecting doses (TCID) rarely rises above 10^4/ml which is several powers of 10 below the density of bacterial growth which may be found. As a consequence, the number of virus-carrying particles dispersed is low, and very large air samples must be taken to obtain positive results. Artenstein and his colleagues (1967, 1968) have examined recruits with adenovirus infections. Individual patients in a small room of 1400 ft^3 (40 m^3) capacity were asked to talk and cough, and air samples were taken. Although these samples were as as large as 60-70 m^3, the majority were negative. On average only 1 TCID was found in between 15 and 30 m^3 of air and the median particle size was 10-15 μm. This would suggest a rate of dispersal of no more than *c.* 10 infective particles/h. They found no correlation between the clinical severity of disease and dispersion of infective particles nor did the presence or absence of viral infection appear to affect dispersal of meningococci from carriers of this organism. Although other workers (Couch *et al.*, 1966) have found a direct correlation between the quantity of virus present in respiratory secretion and dispersal in experimental Coxsackie A21 infections, this could not be shown in these studies. Individual variability in the amount of secretion dispersed was presumably so great as to mask the effects of the variation in numbers of viral particles in the secretion.

Recent work has established with a high degree of probability that foot and mouth disease in cattle, pigs and sheep may be wind borne over distances of tens of kilometres (Hugh Jones & Wright, 1970). This arises from the extremely large numbers of virus particles dispersed from the upper respiratory tract of infected animals. There is copious salivary flow and the density of infection in the secretion may exceed 10^8/ml (Hyslop, 1965). Pigs are the most prolific dispersers and may put up to nearly 10^5 particles carrying virus/h, with *c.* 70% >6 μm diam (Sellers & Parker, 1969; Donaldson *et al.*, 1970). It may be noted that both the density of virus in the secretion and the rate of dispersal into the air are c. 10^4 times those found by Artenstein et al. (1968) for human adenovirus infection.

An interesting sequel to this work is found in the gross viral contamination produced in the upper respiratory tract of those dealing with infected animals. The level of virus in their noses reached $>10^3$ TCID/ml and it was then possible to carry out studies on the redispersal of this from the human upper respiratory tract (Sellers, Donaldson & Herniman, 1970). Substantially less virus was found in the throat and saliva than in the nose (Fig. 10). Talking and

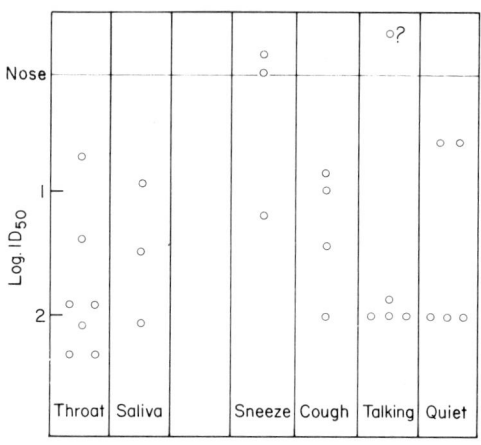

Fig. 10. Airborne foot and mouth virus collected and dispersed by man. After working in the room with the infected animal(s), swabs were taken from nose, throat and mouth, and the subject then performed the activity specified with his head in a large polythene bag from which samples were taken for virus assay. After Sellers et al. (1970).

coughing did not disperse into the test bag more than the rather high background level associated with the test procedures but sneezing, as would be expected, dispersed very much more. No human infections resulted from these massive doses but the virus could be passed on to other persons, and presumably to animals, by dispersal resulting from quiet conversation.

5. Conclusion

The secretions of the upper respiratory tract are regularly infected with saprophytic organisms e.g. the salivary streptococci. During normal activities, including talking, small particles of these secretions are expelled into the surrounding air. Salivary streptococci are therefore regularly to be found in the air of occupied places and afford an indication of the degree of pollution from human respiratory sources. Although it is the saliva of the mouth which is principally dispersed in this way, this is also contaminated by secretion from other parts of the respiratory tract e.g. the throat and lungs or the nose. Some activities such as coughing or snorting or some forms of sneezing may disperse directly secretions from the infected throat and nose.

The respiratory viruses of man are adapted to this situation. They are tolerant of drying to such an extent that they remain viable at least for the few tens of minutes needed to reach another site. Some viruses also appear to produce symptoms such as prolific secretion, coughing or sneezing which may greatly favour their dispersal.

Although the numbers of infected particles dispersed seem to be few, for many diseases, from 1-10/h for pulmonary tuberculosis or adenovirus, to name 2 infections which have been studied in detail, and these are diluted into a very large volume of air, this is not inconsistent with effective transmission of disease by this route. As Gordon (1904) pointed out, the human respiratory tract acts as a continuous air sampler which takes in and filters $c.$ 20 m^3 of air each day. Since the infecting dose of many of these diseases is very low, perhaps a single tubercle bacillus or <10 TCID of the respiratory viruses, a significant risk of infection results from as few airborne particles as one in many hundreds of cubic metres of air.

Although the data available are both scanty and of low quantitative accuracy the estimates shown graphically in Fig. 11 may be useful as an illustration of these points. A discussion of the dispersal of micro-organisms into the air and its epidemiological significance is given by Lidwell (1970).

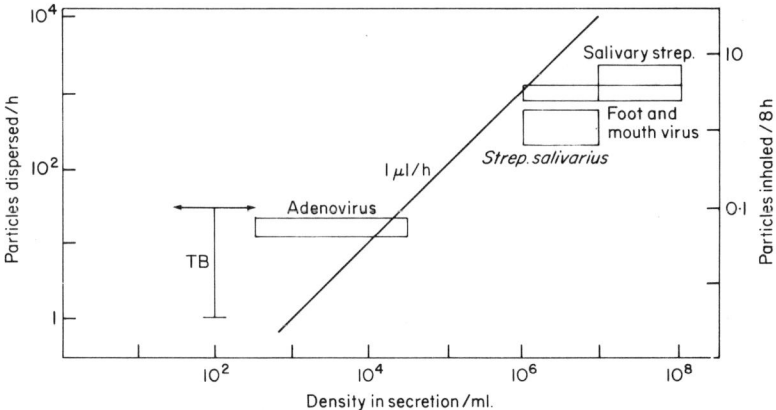

Fig. 11. Dispersal of micro-organisms from the respiratory tract. The numbers of particles inhaled in 8 h have been deduced for a sedentary person spending that time with an infected individual in a room of $c.$ 150 m^3 capacity ventilated at 5 air changes/h. The full line shows the result of the dispersal as particles 10-15 μm diameter of 1 μl of infected secretion/h. This reaches an upper limit at 10^4 particles/h.

6. Summary

There are 4 areas in the respiratory tract where micro-organisms may multiply and from which they may be dispersed as airborne droplets i.e. the nose, throat, oral cavity and lung. Each of these provides a different habitat with a different flora. There are however only 2 orifices, the mouth and the nostrils, considered as one, through which dispersal may take place. Frequent and substantial aerial dispersal arises from the mouth, in talking and sneezing, and involves primarily the saliva. This may be contaminated with secretion transferred from other areas. Coughing is a relatively inefficient mechanism for the production of small particles which can become directly airborne but may transfer material to the mouth or the environment. Simple breathing does not normally lead to any dispersal. The size of infected dispersed particles depends on the mechanics of spraying and the contaminated level of the secretion.

Since salivary streptococci are regular constituents of the mouth flora, their presence in the air is an indication of respiratory pollution. However, the large volumes of air inhaled, and the impracticability of controlling the extent of contamination, deprive this of its potential utility as an index of the risk of transmission of infection. Studies of the dispersal of particular pathogenic species e.g. *Mycobacterium tuberculosis* and respiratory tract viruses, have shown that, although the numbers of airborne particles dispersed which carry the particular micro-organisms may be very small, they could be sufficient to account for the observed rate of transmission of infection.

7. Acknowledgements

Figs 1, 2 and 3 are taken from: Take-off of bacteria and viruses. Symposium No. XVII of the Society for General Microbiology (Lidwell, 1967). Fig. 4 is taken from: Counts of airborne bacteria as indices of air hygiene (Reid *et al.*, *Journal of Hygiene, Camb.*, 1956).

8. References

AIR HYGIENE COMMITTEE. (1954). Air disinfection with ultra violet irradiation. *Spec. Rep. Ser. Med. Res. Coun.* No. 283. London: H.M.S.O.

ARTENSTEIN, M. S., MILLER, W. S., RUST, J. H. & LAMSON, T. H. (1967). Large-volume air sampling of human respiratory disease pathogens. *Am. J. Epidem.* **85,** 479.

ARTENSTEIN, M. S., MILLER, W. S., LAMSON, T. H. & BRANDT, B. L. (1968). Large-volume air sampling for meningococci and adenoviruses. *Am. J. Epidem.* **87,** 567.

BATES, H. J., POTTS, W. E. & LEWIS, M. (1965). Epidemiology of primary tuberculosis in an Industrial school. *New Eng. J. Med. Ap.* **8,** 714.

BOURDILLON, R. B. & LIDWELL, O. M. (1941). Sneezing and the spread of infection. *Lancet.* Sept. 27, p. 365.

BUCKLAND, F. E. & TYRELL, D. A. V. (1964). Experiments on the spread of colds. Laboratory studies on the dispersal of nasal secretion. *J. Hyg., Camb.* **62**, 365.
COUCH, R. B., CATE, T. R., DOUGLAS, R. G., GERONE, P. J. & KNIGHT, V. (1966). Effect of the route of inoculation on experimental viral disease in volunteers and evidence for airborne transmission. *Bact. Rev.* **30**, 517.
DAVIES, R. R. & NOBLE, W. C. (1962). Dispersal of bacteria on desquamated skin. *Lancet* ii, 1295.
DONALDSON, A. I., HERNIMAN, K. A. J., PARKER, J. & SELLERS, R. F. (1970). Further investigations on the airborne excretion of foot and mouth disease virus. *J. Hyg., Camb.* **68**, 557.
DUGUID, J. P. (1945). The numbers and sites of origin of the droplets expelled during expiratory activities. *Edinb. Med. J.* **52**, 385.
DUGUID, J. P. (1946). The size and duration of air carriage of respiratory droplets and droplet nuclei. *J. Hyg., Camb.* **44**, 471.
EICHENWALD, H. F., KOTSEVALOV, O. & FASSO, L. A. (1960). The 'cloud-baby': an example of bacterial-viral interaction. *Am. J. Dis. Child.* **100**, 161.
FLUGGE, C. (1897). Uber Luftinfektion. *Zeit. f. Hyg. u. Infektionskrankh.* **25**, 179.
FLUGGE, C., LATSCHENKO, HEYMANN, B., STICHER, R. & BENINDE, M. (1899). Die Verbreitung der Phthise durch staubförmiges Sputum und durch beim Husten verspritzte Tröpfchen etc. *Zeit. f. Hyg. u. Infektionskrankh.* **30**, 107, 125, 139, 163 and 193.
GORDON, M. H. (1904). Report on a bacterial test for estimating pollution of air. In *Ann. Rep. of Medical Officer of the Local Government Board for the year 1902-03.* London: H.M.S.O.
HAMBREUS, A. (1973). Dispersal and transfer of *Staphylococcus aureus* in an isolation ward for burned patients. *J. Hyg., Camb.* **71**, 787.
HAMBURGER, M. & ROBERTSON, O. H. (1948). Expulsion of group A haemolytic streptococci in droplets and droplet nuclei by sneezing, coughing and talking. *Am. J. Med.* **4**, 690.
HUGH JONES, M. E. & WRIGHT, P. B. (1970). Studies on the 1967-8 foot and mouth disease epidemic. *J. Hyg., Camb.* **68**, 253.
HYSLOP, N. St. G. (1965). Secretion of foot and mouth disease virus and antibody in the saliva of infected and immunised cattle. *J. Comp. Path.* **75**, 111.
JENNISON, M. W. (1942). Atomizing of mouth and nose secretions into the air as revealed by high-speed photography. In *Aerobiology*. Ed. F. R. Moulton. *Am. Ass. Adv. Sci.* Pub. No. 17. Washington, U.S.A.
KINGSTON, D., LIDWELL, O. M. & WILLIAMS, R. E. O. (1962). Epidemiology of the common cold III. *J. Hyg., Camb.* **60**, 341.
KOENIGER, H. (1900). Untersuchung über die Frage der Tröpfcheninfektion. *Zeit. f. Hyg. u. Infektionskrankh.* **34**, 119.
LIDWELL, O. M. (1963). Methods of investigations and analysis of results. In *Infection in Hospitals*. Eds R. E. O. Williams & R. A. Shooter, Oxford: Blackwell.
LIDWELL, O. M. (1967). Take-off of bacteria and viruses. In *Airborne Microbes*. Symposium No. 17. *Soc. gen. Microbiol.*
LIDWELL, O. M. (1970). Mikroorganismer-Levende stof i luften. In *Termisk og Atmosfaerisk Indeklima.* Copenhagen: Polyteknisk Forlag.
LIDWELL, O. M. & WILLIAMS, R. E. O. (1961*a*). Epidemiology of the common cold I. *J. Hyg., Camb.* **59**, 309.
LIDWELL, O. M. & WILLIAMS, R. E. O. (1961*b*). Epidemiology of the common cold II. *J. Hyg., Camb.* **59**, 321.
MEDICAL WORLD NEWS, (1965). New Twist on TB spread. **6**, No. 46 p. 34.
PERKINS, J. E., BAHLKE, A. M. & SILVERMAN, H. F. (1947). Effect of ultra-violet irradiation of classrooms on the spread of measles in large rural central schools. *Am. J. Pub. Hlth.* **37**, 529.
REID, D. D. LIDWELL, O. M. & WILLIAMS, R. E. O. (1956). Counts of airborne bacteria as indices of air hygiene. *J. Hyg., Camb.* **54**, 524-532.

REPORT. (1948). Studies in air hygiene. Medical Research Council Special Report Series No. 262. London: H.M.S.O. p. 231.
RILEY, R. L. & O'GRADY, F. (1961). Airborne Infection. New York: Macmillan. Ch. 5.
RILEY, R. J., MILLS, C. C., O'GRADY, F., SULTAN, L. V., WITTSTAAT, F. & SHIOPORI, D. N. (1962). Infectiousness of air from a tuberculosis ward. *Am. Rev. Resp. Dis.* **85**, 511.
SELLERS, R. F. & PARKER, J. (1969). Airborne excretion of foot and mouth disease virus. *J. Hyg., Camb.* **67**, 671.
SELLERS, R. F., DONALDSON, A. J. & HERNIMAN, K. A. J. (1970). Inhalation, persistence and dispersal of foot and mouth disease virus by man. *J. Hyg., Camb.* **68**, 565.
SHOOTER, R. A., SMITH, M. A. & HUNTER, C. J. N. (1959). A study of surgical masks. *Brit. J. Surg.* **47**, 246.
SULTAN, L., NYKA, W., MILLS, C., O'GRADY, F., WELLS, W. & RILEY, R. L. (1960). Tuberculosis disseminators. *Am. Rev. Resp. Dis.* **82**, 358.
WELLS, W. F. (1934). Airborne infection: droplets and droplet nuclei. *Am. J. Hyg.* **20**, 611.
WELLS, W. F., WELLS, M. W. & WILBER, T. S. (1942). The environmental control of epidemic contagions I. An epidemiologic study of radiant disinfection of air in day schools. *Am. J. Hyg.* **35**, 97.
WILLIAMS, R. E. O. & HIRCH, A. (1950). The detection of streptococci in air. *J. Hyg., Camb.* **48**, 504.
WILLIAMS, R. E. O., HIRCH, A. & COWAN, S. T. (1953). Aerococcus a new bacterial genus. *J. gen. Microbiol.* **8**, 475.
WILLIAMS, R. E. O., LIDWELL, O. M. & HIRCH, A. (1956). The bacterial flora of the air of occupied rooms. *J. Hyg., Camb.* **54**, 512-523.

Microflora of the Vagina During Pregnancy

ROSALINDE HURLEY, VALERIE C. STANLEY, BARBARA G. S. LEASK
AND J. DE LOUVOIS

*Queen Charlotte's Hospital for Women and
the Institute of Obstetrics and Gynaecology,
University of London, Goldhawk Road, London W6 0XG, England*

CONTENTS

1. Introduction 155
2. Anatomy of the female external genitalia and the vagina 156
3. The internal genitalia 156
4. Physiology of the vagina 158
5. Relationship of flora to vaginal pH and glycogen 158
6. Experimental procedures 159
   (a) General ecological survey of the vaginal flora of pregnant women . . . 159
   (b) Studies on yeast flora 161
7. Results 161
   (a) General ecological survey 161
   (b) Yeast flora 164
   (c) Clinicopathological correlations 165
8. Discussion 172
9. Acknowledgements 182
10. References 182

1. Introduction

THE MICROFLORA of the vagina has never been comprehensively studied, although many papers deal with the isolation of specific microbes, particularly with those that are associated with pathological conditions. Anatomical and physiological factors are known to influence the vaginal flora, but scant attention has been paid to geographical and ethnological determinants. Such factors are, of course, less accessible to study because of the taboos that prevail with respect to the organs of generation both in highly developed and in primitive civilisations.

Before the advent of the antibiotic era, vaginal examination during late pregnancy was very strongly, and rightly, discouraged save as a matter of obstetrical necessity, since the relationship between vaginal manipulation and serious infection had been established (Colebrook, 1936). The undesirability of repeated vaginal examination, especially late in pregnancy, militated against sequential studies of the vaginal flora throughout gestation. Such examinations are performed more frequently than formerly, and, if properly conducted, will not harm the patient. Thus, we believe that it is now safe to undertake studies of vaginal flora during pregnancy and, indeed, such studies must be made if we are to ascertain whether such belief is justified.

It is the purpose of this paper to present the preliminary findings of an ecological study of the vagina during pregnancy, together with more detailed observations on the yeast flora, the majority of which have been published elsewhere (see below). The studies do not include observations on *Chlamydia* spp. or viruses.

2. Anatomy of the Female External Genitalia and the Vagina

The external organs of generation are known as the vulva. Overlying the *symphysis pubis* is a fibrofatty pad, the *mons veneris*, covered by short, wavy hair, that first appears at puberty. Uniting anteriorly, at the *mons veneris*, lie 2 antero-posterior cutaneous folds, the *labia majora*. Their outer surfaces are covered with hair but their inner surfaces are composed of smooth shiny soft pinkish skin, similar to mucosa in appearance. The inner surfaces of the *labia majora* are provided with sebaceous glands and sweat glands. The *labia majora* lie on either side of the vagina, and unite posteriorly. Lying medial to the *labia majora* are 2 thick skin folds, devoid of hair, but containing a few sebaceous and sweat glands. These are the *labia minora*, that divide anteriorly to form the prepuce of the clitoris, and fuse posteriorly at the fourchette, the fold of skin where the posterior vaginal wall and the perineum unite. They are well supplied with blood vessels, and pinkish red in colour. The clitoris is the homologue of the male penis. The urethral meatus lies 2.5 cm infero-posterior to the clitoris. The female urethra is surrounded by numerous racemose glands which open by small orifices in the urethral wall; just within the meatus is a large pair called Skene's ducts or the peri-urethral glands.

The entrance to the vagina is called the introitus, partially closed in virgins by the hymen, which may be imperforate, cribriform, or may possess a single pin-hole aperture; normally, it has a free margin anteriorly. Bartholin's glands are a pair of oval, compound racemose glands, *c.* 1.25 cm in diam, lying laterally to the lower end of the vagina. Their ducts open externally and laterally to the base of the hymen. They are the homologues of Cowper's glands (para-urethral glands) in the male, and pour out a thin clear, alkaline mucus secretion at coitus. The perineum is a triangular shaped wedge of fat, muscle and connective tissue, covered in skin, separating the lower 3.5 cm of the posterior vaginal wall from the anal skin and rectum.

3. The Internal Genitalia

The vagina is the passage that leads from the uterus to the exterior. Its length is 7.5 cm along its anterior wall and 9 cm along its posterior wall. Its anterior and posterior walls are in apposition, except at the upper end, where they are held apart by the intravaginal portion of the cervix. The part of the vaginal cavity that surrounds the cervix is called the fornix — the posterior fornix is of importance, since its outer wall is directly related to the peritoneum (pouch of Douglas), which covers it for 0.5-2.5 cm. The vaginal wall is rugose and contains no glands. Although its inner coat is called 'mucous', like the skin, it is made of stratified epithelium, as is the lower third of the endocervical canal, and is 'mucous' only in so far as lubricated by the secretions of the cervix. Nearly one half of the cylindrical neck of the womb lies within the vagina, and is also clothed with stratified epithelium. The endocervical canal is lined with ciliated columnar epithelium, in its upper two-thirds, and contains deep glandular follicles which secrete clear, viscid, alkaline mucus. The uterus lies at a right angle to the vagina; and is normally anteverted, hence the greater length of the posterior wall. (Johnston & Whillis, 1947; McGregor, 1950).

The histology of the genital tract is well understood, particularly that of its lining membranes. The Müllerian ducts of the embryo, lined by columnar epithelium, fuse to form the Fallopian tubes, uterus, cervix and vagina of the adult. The epithelium of the Fallopian tubes becomes converted to ciliated columnar epithelium, that of the body of the uterus to low columnar epithelium, that of the cervix to non-ciliated high columnar epithelium, and that of the vagina to squamous epithelium. The change in vaginal epithelium is not due entirely to conversion of the original columnar epithelium, but to its replacement by squamous epithelium growing from below upwards, from the region of the urogenital membrane (Cruickshank & Sharman. 1934). In 36% of the newborn, the intravaginal portion of the cervix is covered by a single layer of cylindrical epithelium, with frequent persistence of cervical glands (Fischel, 1880). This anatomical anomaly is termed congenital erosion of the cervix.

Careful studies have been made on the histological appearances of the vaginal mucosa in foetuses, infants, and in prepubertal, postpubertal and sexually mature women. These studies can be summarized briefly. During intrauterine life, after the development of the epithelial covering of the vaginal lumen, the mucosa consists of 20 to 30 layers of large, vacuolated cells with an active basal layer, in every way resembling that seen during the sexually active years. After birth, the epithelium becomes progressively thinner, the cells smaller, the basal layer inactive; inactive epithelium is established within a month, and persists until puberty, when the initiation of hormonal activity awakens the dormant epithelium. The mucosa becomes thicker, pinker and velvety to the touch; 20-30 layers are built up rapidly, and it assumes the florid appearance of the mature vagina, thus to remain until the cessation of ovarian function. Since all the vagina save the distal portion is of the same embryonic origin, it is under the same physiological influences as the rest of the genital tract, and undergoes cyclical activity (Davis & Pearl, 1938).

4. Physiology of the Vagina

The presence of glycogen in the vaginal mucosa is controlled by the secretion of oestrogens and progesterone and has long been deemed important in maintaining vaginal acidity in the sexually mature woman (Kronig & Menge, 1897; Davis & Pearl, 1938). The glycogen content is dependent on the character of the epithelium. When oestrogenic hormone is active, the squamous epithelium is active, and glycogen is deposited. This occurs in the newborn, as the result of maternally derived hormone, and in the sexually mature woman. The vagina is acid, pH 4-5, and the microbial flora is rich in Döderlein's bacilli (Döderlein, 1892), showing what is called a grade 1 flora (Gordon, Hughes & Barr, 1966). During childhood, and after the menopause, oestrogenic activity is absent, glycogen is not deposited, and vaginal pH value is 6-7. The microbial flora is varied but cocci are said to predominate. Still other changes occur during pregnancy. As early as the 12th week, the epithelium begins to thicken, reaching c. 500 μm near term. There is active proliferation of the basal cells, and numerous mitotic figures evidence the intense activity. The whole thickness of the vacuolated cell layers is laden with glycogen, even those cells that are in process of desquamation. The amount of titratable lactic acid increases from 0.4-0.9% during pregnancy (Miura, 1928). After parturition the mucosa regresses rapidly, taking on the character of postmenopausal epithelium. The microbial flora and pH value cannot be studied readily at this time, due to the profuse, bloody, lochial discharges resulting from involution of the uterus. The mucosa is restored to normal at c. 6-8 weeks *post partum,* often before the reappearance of the menses.

The vaginal flora is probably also determined by the cyclical activity of the vagina of the non-pregnant, sexually mature woman (Murray, pers. comm.; Neary *et al.*, pers. comm.).

5. Relationship of Flora to Vaginal pH and Glycogen

There seems general agreement that lactobacilli play some role in the conversion of glycogen to lactic acid, and that they are responsible for the high degree of acidity, ranging in the mature female from pH 4.8-5.7 (Davis & Pearl, 1938). Kienlin (1926), however, showed that in the newborn, lactobacilli appear after the vagina becomes acid. Rogosa & Sharpe (1960) and Stewart-Tull (1964) suggested that lactobacilli did not initiate the splitting of glycogen, this being done by tissue enzymes, or enzymes of other bacteria. Wylie & Henderson (1969) showed that some strains of lactobacilli isolated from the human vagina do produce lactic acid from human glycogen, although not from that derived from oysters, the usual commercial source (Table 1). The massive impregnation of the vaginal mucosa with glycogen that occurs in pregnancy, and in diabetes mellitus, is thought to be related to the increase in the rate of isolation of *Candida* spp. (Salignon-Michelier, 1950).

Table 1
Speciation of 'wild strains' of lactobacilli from vagina (antenatally)

Lactobacillus sp.	Rogosa & Sharpe (1960)	Wylie & Henderson (1969)
Total	21	42
L. acidophilus*	14	11
L. fermentum	4	12
L. brevis	0	2
L. casei	2	6
L. leichmannii*	0	5
L. salivarius*	0	5
L. lactis	0	1
L. cellobiosus	1	0

* Able to produce acid from human vaginal glycogen.

6. Experimental Procedures

(a) *General ecological survey of the vaginal flora of pregnant women*

Specimens were obtained from 280 patients who booked at the antenatal clinics of Queen Charlotte's Hospital for Women but who were otherwise unselected. The sites sampled, the method of sampling, and the media inoculated in the clinics are shown in Table 2. The specimens were taken by one of 2 house surgeons who also noted all anatomical or physiological aberrations of the lower genital tract, according to defined criteria (Carrol, Hurley & Stanley, 1973). The specimen from the posterior fornix of the vagina was taken using a plastic foam sponge swab (3 x 1 x 1 cm) inserted in sponge holding forceps, after passage of a Cusco's speculum lubricated with KY jelly (Fig. 1).

Eleven basic media, incubated under different environmental conditions were used (Table 3), and were inoculated from the Stuart's transport medium.

Table 2
*Flora of lower genital tract in 280 unselected pregnant women**

Site sampled	Method of sampling	Medium used
Urethra	Charcoal swab	Chocolate agar & Thayer & Martin
Cervix	Charcoal swab	Chocolate agar & Thayer & Martin
Rectum	Charcoal swab	Thayer & Martin
Posterior fornix of vagina	Sponge swab	Stuart's Transport Medium

* Unpublished observations of de Louvois, Hurley & Stanley.

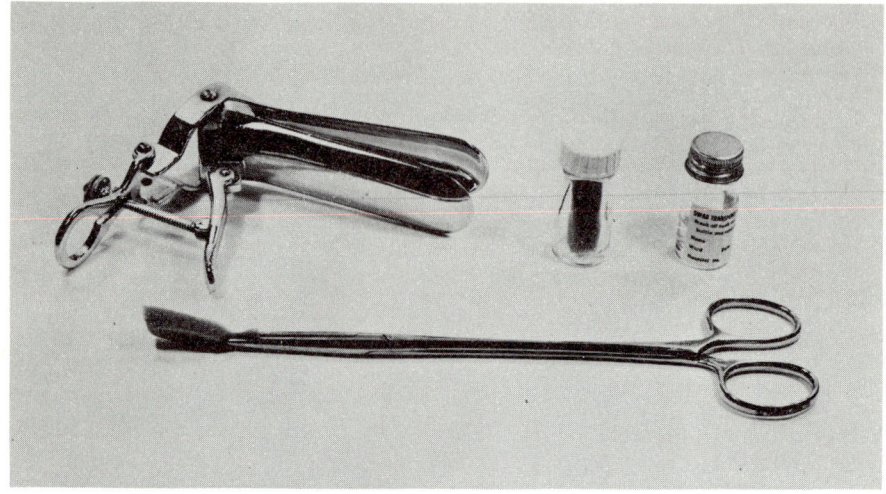

Fig. 1. Kit used to collect specimens for the posterior fornix of the vagina.

Table 3
Media used to study the microflora of lower genital tract in 280 unselected pregnant women*

Horse blood agar, incubated aerobically, and anaerobically for 48 h (Gaspak)
Heated horse blood agar, incubated in CO_2, for 48 h at 37° (Gaspak)
Crystal violet horse blood agar, incubated anaerobically for 48 h at 37°
Tellurite blood agar, incubated aerobically for 48 h at 37°
Nalidixic acid blood agar, incubated aerobically for 96 h at 37°
MacConkey agar, incubated aerobically for 48 h at 37°
Rogosa agar, incubated aerobically for 48 h at 30°
Sabouraud's agar, incubated aerobically for 48 h at 30°
Mycoplasma agar, incubated 10% CO_2 in N_2 for 8 days at 37°
Trichomonas medium, incubated for 4 days at 30°
Thayer & Martin's medium for 2 days at 37° in CO_2 (candle jar)

---

* Unpublished observations of de Louvois, Hurley & Stanley.

Further details of the media used and their preparation will be presented elsewhere. Plates were incubated for 48 h except those for *Mycoplasma* and *Listeria* spp. which were incubated for 8 days and 96 h, respectively.

No attempt was made to quantitate microflora, other than to record growth of particular microbes as heavy or scanty. All colony types were first Gram stained then subcultured to appropriate media for further examination. Speciation was not attempted for all groups. Staphylococci (Wilson & Miles, 1964; Cowan & Steel, 1965) were identified as *Staphylococcus aureus* if coagulase positive and deoxyribonuclease positive, and as *Staph. epidermidis* if coagulase negative and deoxyribonuclease negative. Micrococci were identified according to Cowan & Steel (1965). Streptococci were identified using the basic criteria of Cowan & Steel (1965), and Wilson & Miles (1964), and were classified according to their colonial appearance on blood agar (haemolysis), to their atmospheric requirements, to their ability to ferment lactose, to hydrolyse aesculin rapidly, to grow on MacConkey medium, and by serological typing for Lancefield groups, using 2 methods of typing. *Neisseria* spp. were speciated according to Cowan & Steel's (1965) criteria, and their sugar fermentation patterns were ascertained by the method of Flynn & Watkins (1972). The group designated as Gram variable cocco-bacilli did not fulfil the criteria of identification of any other group, and did not require X or V factors for growth. Lactobacilli were not speciated, and were defined as Gram positive, non-sporing rods, usually growing aerobically on Rogosa's agar at pH 5.4, but not growing on potassium tellurite medium nor on crystal violet blood agar. Corynebacteria were not speciated, but grew aerobically on potassium tellurite agar, and conformed in colonial and Gram stained appearance with the characteristics of the genus as described by Cowan & Steel (1965) and Wilson & Miles (1964). Enterobacteriaceae and *Pseudomonas* spp. were speciated, using the API test system*. *Bacteroides* spp. were not speciated, but were placed in the genus if they were Gram negative, non-sporing, non-motile anaerobes, growing on appropriate media. *Mycoplasma* spp. were speciated, using Oxoid media and BBL taxodiscs. *Trichomonas vaginalis* was identified by its morphology and characteristic motility in wet film. Yeasts were identified according to the criteria of Lodder & Kreger-van Rij (1952) and Lodder (1970), using sodium acetate agar for sporulation studies (Merritt & Hurley, 1972).

(b) *Studies on yeast flora*

The materials and methods used in these studies have been published elsewhere (Hurley & Morris, 1964; Hurley & Stanley 1972; Carroll, Hurley & Stanley, 1973; Hurley *et al.*, 1973). Table 4 shows the basis for the assignment to genus; Table 5 shows the basis for speciation of the pathogenic *Candida* spp., and Table 6 shows the more detailed criteria of identification used in our laboratories. The studies drawn upon include analyses of the yeast flora in 6629 specimens sent for diagnosis of vaginitis, over a 5-year period, together with 2 prospective surveys made on 1031 and 1085 pregnant women.

7. Results

(a) *General ecological survey*

Table 7 shows the distribution of micro-organisms in 280 unselected pregnant women, expressed in descending order of frequency. Lactobacilli, corynebacteria

* API Laboratory Products Ltd., Philpot House, Rayleigh, Essex.

Table 4
Identification of yeasts* from the vagina

Type of growth	Genera of yeasts
Pseudomycelium on corn meal agar	Candida spp.
Pseudomycelium absent or rudimentary	Torulopsis spp.
	Saccharomyces spp.
	Cryptococcus spp.
Sporulation on sodium acetate agar	Saccharomyces spp.
	Hansenula spp.
	Kluyveromyces spp.
	Pichia spp.

* Most grow well on blood agar; *Torulopsis glabrata* and *Saccharomyces* spp. do not.

Table 5
Identification of Candida* species†

Candida spp.	Occurrence of germ tube	Presence of chlamydospores	Assimilation of sucrose
albicans	+	+	+
stellatoidea	+	+ or −	−
tropicalis			
krusei			
pseudotropicalis	−	−	Not stated
parapsilosis (giant cells)			
guilliermondii			

* All produce pseudomycelium on corn meal agar. None of those listed assimilates nitrate.
† Speciated by auxanograms and zymograms.

and *Staph. epidermidis* were present usually as heavy growths, in most of the patients examined. Streptococci of various sorts were encountered frequently and the majority of the β-haemolytic streptococci were assigned to Lancefield Group B. No Group A streptococci were isolated. Such of the faecal streptococci as were grouped were assigned to Lancefield Group D. Groups C, F and G were also encountered. The percentage of *Escherichia coli* and of yeasts in the posterior fornix was high (19.3 and 24.3%, respectively). No *Listeria* or *Clostridium* spp. were isolated. Of the *Neisseria* spp., 3 were *N. pharyngis* and 1 was *N. catarrhalis*. *Neisseria gonorrhoeae* was not isolated. *Proteus mirabilis* was the only species of *Proteus* isolated. T-strain mycoplasmas were not sought but mycoplasmas were speciated; only *M. hominis* was isolated.

Table 6
Identification of Candida species

Species	Assimilation of						Germ Tubes	Chlamydospores
	Glucose	Galactose	Maltose	Sucrose	Lactose	Raffinose		
C. albicans	+	+	+	+	–	NT	+	+
C. stellatoidea	+	+	+	–	–	NT	+	+
C. tropicalis	+	+	+	+	–	NT	–	–
C. krusei	+	–	–	–	–	NT	–	–
C. pseudotropicalis	+	+	–	+	+	NT	–	–
C. parapsilosis	+	+	+	+	–	NT	–	–
C. guilliermondii	+	+	+	+	–	+	–	–
(T. glabrata)	+	–	–	–	–	NT	–	(–)

Species	Fermentation of					
	Glucose	Galactose	Maltose	Sucrose	Lactose	Raffinose
C. albicans	+	+(w)	+	–	–	–
C. stellatoidea	+	–	+	–	–	–
C. tropicalis	+	+	+	+	–	–
C. krusei	+**	–	–	–	–	–
C. pseudotropicalis	+	+	–	+	+	+/(w)
C. parapsilosis	+	+w/–	–	–	–	–
C. guilliermondii	+	+/(w)	–	+/(w)	–	+/(w)
(T. glabrata)	+	–	–	–	–	(–)

** Surface pellicle

Table 7
Flora of lower genital tract (HVS) in 280 unselected pregnant women *

Organism	No. of patients from whom isolated	% Incidence
Corynebacteria	234	83.4
Lactobacilli	229	81.8
Staphylococcus epidermidis	185	66.1
Faecal streptococci	116	41.4
Micrococci	103	36.8
Yeasts	68	24.3
Microaerophilic and anaerobic streptococci	61	21.7
Escherichia coli	54	19.3
Mycoplasma hominis	31	11.0
β haemolytic streptococci	26	9.2
Gram variable cocco-bacilli	19	6.8
Proteus spp.	17	6.1
Bacteroides spp.	15	5.4
Staphylococcus aureus	13	4.6
Non-haemolytic streptococci	12	4.3
Trichomonas vaginalis	7	2.5
Neisseria spp.	4	1.4
Klebsiella aerogenes	2	0.7
Pseudomonas aeruginosa	1	0.4

* Unpublished observations of de Louvois, Hurley & Stanley.

The study is still in progress. It has not reached the stage of final analysis, and no data are yet available on 'grouping' of microbes, relationship of flora to gestational stage, or relationship to disordered anatomy or physiology of the lower genital tract.

(b) *Yeast flora*

Table 8 shows the identification of 70 yeast isolates from 68 patients. The identification of the yeast tentatively named *Candida norvegensis* awaits confirmation. *Candida albicans, C. krusei* and *Torulopsis glabrata* are pathogenic yeasts; the others are almost certainly commensals or transient organisms. Some are powerful fermenters.

Yeasts were isolated from 1538 of 6629 vaginal swabs sent for diagnosis of vaginitis, and their distribution by species is shown in Table 9. *Candida albicans* predominated, comprising 93-95% of all isolates and occurred alone or in the presence of other fungi; next most commonly isolated was *Torulopsis glabrata* (3-5% of isolates). The percentage distribution of fungi remained remarkably constant over the 5-year period.

Table 8
Flora of lower genital tract (HVS) in 280 unselected pregnant women*

Identification of 70 isolates from 68 patients

Organism	No. found	Organism	No. found
Candida albicans	47	*Saccharomycopsis lipolytica*†	1
Torulopsis glabrata	11	*Torulopsis castellii*	1
C. albicans and *T. glabrata*	2	*Saccharomyces cerevisiae*	1
Rhodotorula rubra	2	*Saccharomyces uvarum*	1
Candida krusei	1	*Candida norvegensis*?	1

* Yeast identifications by B. G. S. Leask, according to Lodder & Kreger-van Rij (1952); Lodder (1970).
† Yarrow (1972).

Table 10 shows the distribution of yeast species in 2 groups of 1031, and 1085, women studied prospectively. Tables 11 and 12 compare the distribution of fungi isolated from specimens selected on diagnostic criteria, with the distribution obtaining in 1031 unselected patients studied prospectively. The latter study was conducted in 2 parts, and both in the winter (Table 11) and in the spring (Table 12) the percentage distribution differed in the selected and unselected groups. *Candida albicans* predominated in both but occurred relatively less frequently in the unselected group. The percentage of species other than *C. albicans* was higher in unselected patients. Table 13 compares the incidence of *Trichomonas vaginalis* and all species of yeasts isolated from unselected patients. The overall incidence of *T. vaginalis* was just under 6%, whereas that of yeasts was just under 18%. Although candida predominated in both seasons, trichomonas infestation was encountered relatively more frequently in winter, and candida infections in spring. The trichomonas-yeast ratio overall was 1 : 3, and the incidence of simultaneous infection with yeasts and trichomonads was 0.8%.

(c) *Clinicopathological correlations*

Clinical data were available for prospective studies on the incidence and likely significance of *C. albicans* and the other yeasts in the vaginas of pregnant women made on 1085 patients (de Fonseka, 1972), and on a further group of 303 patients (Carroll *et al.*, 1973). All were unselected pregnant women booking at the antenatal clinics of the same hospital. In the cases studied by de Fonseka (1972), clinical observations were made by diverse observers, of varying degrees of diligence, who completed a questionnaire, returning it to the laboratory with

Table 9

Identity of yeasts in yeast-containing vaginal swabs, 1966-1970 (series 1)

Yeast	Total no. of vaginal swabs containing yeasts in				
	1966	1967	1968	1969	1970
All yeasts	270	262	276	293	437
Total containing *Candida albicans*	257 (95)*	245 (94)	261 (95)	272 (93)	411 (94)
C. albicans only	253	239	255	262	399
C. albicans and *C. tropicalis*	0	1	0	0	0
C. albicans and *C. pseudotropicalis*	0	0	0	0	1
C. albicans and *C. krusei*	0	0	0	0	1
C. albicans and *Saccharomyces cerevisiae*	0	0	1	0	0
C. albicans and *Torulopsis glabrata*	4	5	5	10	10
C. stellatoidea	0	1	5	3	2
C. tropicalis	0	2	0	1	1
C. pseudotropicalis	0	0	0	0	1
C. parapsilosis	0	0	0	1	2
C. krusei	1	0	0	0	0
T. glabrata	9 (3)	10 (4)	9 (3)	15 (5)	16 (4)
T. inconspicua	1	0	0	1	0
T. holmii	0	1	0	0	0
S. cerevisiae	2	2	1	0	5
Rhodotorula glutinis and *Cryptococcus diffluens*	0	1	0	0	0

* No. in brackets is % of total.
By permission of the Editor of the *Journal of Obstetrics and Gynaecology of the British Commonwealth*.

Table 10

Identity and percentage distribution of yeast species in high vaginal swabs taken from two groups of pregnant women studied prospectively

Yeast	No. of swabs (= no. of patients) containing yeasts in			
	Series 2 (1031 patients)	% distribution of yeasts	Series 3 (1085 patients)	% distribution of yeasts
All yeasts	182 (17.7%)		*105 (9.7%)	
Candida albicans	139	76	87	83
Torulopsis glabrata	27	15	11	11
C. albicans and T. glabrata	1	<1	—	—
Saccharomyces cerevisiae	6	3	3	<3
C. stellatoidea	3	2	1	<1
C. tropicalis	2	1	2	<2
C. parapsilosis	2	1	—	—
T. holmii	1	<1	—	—
T. inconspicua	1	<1	—	—
Rhodotorula spp.	—	—	1	<1

* Patients with vulvovaginitis
By permission of the Editor of the *Journal of Obstetrics and Gynaecology of the British Commonwealth*.

Table 11

Percentage distribution of yeast species isolated from diagnostic specimens (series 1) and from 511 consecutive patients attending an antenatal booking clinic between November and January (series 2)

Species	Swabs sent for diagnosis containing yeast from patients in Series 1			Swabs from 511 patients in prospective Series 2
	Nov. 1966 to Jan. 1967	Nov. 1967 to Jan. 1968	Nov. 1968 to Jan. 1969	Nov. 1967 to Jan. 1968
All yeasts	86	77	84	77
Candida albicans (only)	78 (91%)	69 (90%)	74 (88%)	57 (74%)
C. albicans and Torulopsis glabrata	1	0	3	0
C. albicans (total)	79 (92%)	69 (90%)	77 (92%)	57 (74%)
T. glabrata	5 (6%)	6 (8%)	6 (7%)	14 (18%)
C. stellatoidea	0	1	1	2
C. tropicalis	0	0	0	1
C. parapsilosis	0 (2%)	0 (3%)	0 (1%)	1 (8%)
C. krusei	1	0	0	0
T. holmii	1	0	0	1
Saccharomyces cerevisiae	0	1	0	1

By permission of the Editor of the *Journal of Obstetrics and Gynaecology of the British Commonwealth*.

Table 12

Percentage distribution of yeast species isolated from diagnostic specimens (series 1) and from 520 consecutive patients attending a booking clinic between March and May (series 2)

Species	Swabs sent for diagnosis from patients in Series 1			Swabs from 520 patients in prospective Series 2
	March to May 1967	March to May 1968	March to May 1969	March to May 1968
All yeasts	59	54	63	105
Candida albicans (only)	56 (95%)	49 (91%)	60 (95%)	82 (78%)
C. albicans and Torulopsis glabrata	0	2	1	1
C. albicans (total)	56 (95%)	51 (94%)	61 (97%)	83 (79%)
T. glabrata	1 (2%)	2 (4%)	2 (3%)	13 (12%)
C. stellatoidea	0	1	0	1
C. tropicalis	1 (3%)	0 (2%)	0 (0%)	1 (9%)
C. parapsilosis	0	0	0	1
Rhodotorula glutinis and Cryptococcus diffluens	1	0	0	0
T. inconspicua	0	0	0	1
Saccharomyces cerevisiae	0	0	0	5

By permission of the Editor of the *Journal of Obstetrics and Gynaecology of the British Commonwealth*.

Table 13

Incidence of Trichomonas vaginalis and yeasts in pregnant women (series 2)

	Number of patients (Nov. 1967) to Jan. 1968	Number of patients (March to May 1968)	Total number of patients	Incidence		
				% of patients seen in winter	% of patients seen in spring	% of all patients
Yeasts only present	71	103	174	14	20	17
T. vaginalis only present	30	17	47	6	3	5
T. vaginalis and yeasts	6	2	8	1	0.4	0.8
Total no. of patients examined	511	520	1031			

By permission of the Editor of the *Journal of Obstetrics and Gynaecology of the British Commonwealth.*

the specimens requested; the patients who formed the subject of the second study were all examined carefully by one obstetrician (CJC) who recorded all observations and collected the specimens. Table 14 illustrates the importance of standardization of clinical observation to eliminate subjective error. It also emphasizes the importance of genuine, rather than notional, clinical co-operation in studies of this kind. The inferences drawn from the 2 sets of observations are quite different. The first study suggests that *C. albicans* is as likely as not to be associated with morbidity, and that in about half the cases it is present as a commensal, exciting neither symptoms nor signs. The second study shows that its presence is almost invariably associated with vaginal morbidity. That the laboratory's performance does not vary significantly can be seen from the incidence of isolation of *C. albicans* in the 2 series (17 and 16.5%, respectively) and from study of the figures shown in the preceding Tables.

Table 14

Clinical observer variation on two comparable groups studied in the same hospital

	Group A 1085 patients (Hurley et al., 1973)	Group B 303 patients (This report)
Patients with vaginal morbidity	39	86
C. albicans isolated	17	16.5
C. albicans associated with vaginal morbidity	8	16.2
C. albicans not associated with vaginal morbidity	9	0.3

The figures in the two columns are percentages.
By permission of the Editor of the *Journal of Obstetrics and Gynaecology of the British Commonwealth*.

Working from the more accurate data of Carroll, Stanley & Hurley (1973), Table 15 shows the nature of morbidity of the lower genital tract. Discharge, evident to the clinical observer, is very common during pregnancy; as is cervicitis. Other signs of frank morbidity are frequent and many women complain of irritation. Table 16 shows that 84% of isolates of *C. albicans* were associated with vaginitis; 14% were associated with other signs of morbidity; and that only once in 303 patients was the fungus isolated from a healthy lower genital tract.

Finally, from this study, and from studies made on the distribution and significance of precipitating antibodies to antigens of *C. albicans* in sera from pregnant women (Stanley, Hurley & Carroll, 1972; Stanley & Hurley, in press), we are able, having established the clinicopathological relationship of *C. albicans* to disease of the lower genital tract, to formulate the criteria upon which the diagnosis of candida vaginitis may be soundly based (Table 17).

Table 15

Morbidity of the lower genital tract in 303 unselected pregnant women

	Number	%
Complaint of irritation	43	14
Discharge	215	71
Vulvitis	28	9
Vaginitis	43	14
Plaques, or cheesy debris	16	5
Cervicitis	92	30
Cervical erosion	42	14
None of the above	44	14

By permission of the Editor of the *Journal of Obstetrics and Gynaecology of the British Commonwealth.*

Table 16

Isolation of C. albicans *from 50 of 303 pregnant women (HVS)*

Signs or symptoms of morbidity of lower genital tract	C. albicans isolated in association
Thrush plaques*	11 ⎫
Vaginitis	25 ⎬ 84%
Treated vaginitis (response to nystatin)	6 ⎭
Discharge only	5
Cervicitis	2
None	1 (2%)

* or 'cheesy' mucosa.

Table 17

Clinicopathological criteria for diagnosis of Candida vaginitis*

1. Presence of plaques of 'cheesy' mucosa, irrespective of isolation of *C. albicans*
2. *Signs* of vaginitis†, accompanied by isolation of *C. albicans*
3. Isolation of *C. albicans* (84% correlation with 1 and 2 above)
4. Presence of candida precipitins (64% correlation with 1 and 2 above)

† Mucosa reddened, swollen or granular
* Vulvitis does not occur independently of vaginitis

8. Discussion

Many difficulties beset studies of microbial ecology in man, other than those posed by anatomy, physiology and pathology. The site sampled is clearly relevant. No one would regard the whole of the gastro-intestinal tract, or the urinary tract, as a unit for ecological studies. The vagina may be deemed a single structure from the viewpoint of gross anatomy, but developmentally and physiologically it is far from uniform throughout its extent. Histologically, too, in many women the vaginal portion of the *cervix uteri* is aberrant. There is no reason to suppose that the flora of the posterior fornix is identical in relative distribution with that of the introitus, or other parts of the vagina, and the incidence of particular microbes sampled from various sites within it may show variation, the significance of which is difficult to assess. This point is illustrated by the studies of Carroll, Hurley & Stanley (1973) and the study reported here in preliminary form. In the former, the percentage of *C. albicans* isolated from the middle third of the lateral vaginal wall was 16.5%; in the current study, *C. albicans* was isolated from the posterior fornix in 17.6%. Sites other than the vagina may be chosen in searches for specific microbes, especially those that may have pathogenic potential or significance and that have been relatively little studied. McCormack, Rankin & Lee (1972) commented that the genital mycoplasmas, *Mycoplasma hominis* and T-strains, colonize the genital tract mucosa, including the external cervical os, vagina, vestibule and distal urethra. Archer (1968) had suggested that sampling of the urine yielded as much information on the distribution of T-strain mycoplasmas as did sampling of the genital tract, although Braun *et al.* (1970) believed that urine yielded fewer T-strains. Dunlop *et al.* (1969) noted that vaginal culture was better than cervical culture for T-strains. They compared samples taken from the urethra, vagina, cervix and posterior fornix and obtained more isolates of T-strains and *Mycoplasma hominis* from the vaginal cultures. McCormack, Rankin & Lee (1972) found that combination of urethral and vaginal cultures gave the highest yield. All these observations attest to the importance of the site or sites sampled in ecological studies made on the female genital tract. Archer's (1968) results, summarized in Table 18, suggest that sexual, as well as physiological factors may influence the incidence of these microbes. Shephard (1954) had noted that

Table 18

T-mycoplasma *strains in urine (from Archer, 1968)*

Type of patient	Number studied	% isolation of T-strains
Pregnant Women	100	58
Women attending		
Infertility Clinic	94	51
Geriatric patients	98	29
Nuns in an		
enclosed order	105	8

T-mycoplasmas, as well as other mycoplasmas, are more common in sexually active women whether venereally infected or not.

The methods of demonstration of particular microbes also pose problems. The difficulties may be illustrated by allusion to the far from uniform methodology used to demonstrate 2 of the most important pathogens of the female genital tract, *Trichomonas vaginalis* and *Neisseria gonorrhoeae*. Hughes, Gordon & Barr (1966) found that examination of smears stained by the Papanicolaou method yielded more positive identifications of *T. vaginalis* than examination of 'wet' swabs, or culture. They used charcoal impregnated swabs, but the 'wet' procedure is not precisely detailed. Table 19 illustrates some of their findings, and their relationship to the clinical sign of 'discharge'. The

Table 19

Frequency of detection of Trichomonas vaginalis *by laboratory methods in each of the three clinical subgroups*

	Number of patients	Cytology T. vaginalis positive	Culture T. vaginalis positive	Wet film T. vaginalis positive
Negative discharge	118	56 (47.4%)	41 (34.7%)	16 (13.6%)
Non-trichomonal discharge	127	71 (56.0%)	52 (40.9%)	19 (14.9%)
Trichomonal discharge	60	41 (68.3%)	35 (58.3%)	22 (36.6%)

By permission of the Editor of the *Journal of Obstetrics and Gynaecology of the British Commonwealth*.

various methods of demonstration of *T. vaginalis* have been tested in parallel frequently in surveys made at Queen Charlotte's Hospital for Women. The discrepancy between culture and examination of swabs immersed in warm saline is slight, of the order of <0.5%, in favour of an improved yield from culture. The Papanicolaou smear is virtually useless as a routine method, save under the scrutiny of an expert cytologist. However, it should be emphasized that the 'wet' method that suits us so well would not necessarily suit other diagnostic laboratories, since its success depends on rapid transfer of the specimen. The difficulties of isolating *N. gonorrhoeae* from clinical specimens are notorious (Flynn & Watkins, 1972) and there are many variations in techniques of collection, culture and biochemical identification.

Quite separate from the methodological difficulties, but inherent within them to a large extent, lies the problem of taxonomy. This is by no means resolved with respect to the vaginal flora. Possibly the most illustrative example is the organism variously called *Haemophilus vaginalis* (Gardner & Dukes, 1955) or *Corynebacterium vaginale* (Zinneman & Turner, 1962; 1963). Where the taxonomy is doubtful, it is impossible to determine whether the named microbe is commensal or pathogenic in the vagina. Lapage (1961) dealt with this problem in an exacting study, based on examination of several hundred routine gonococcal diagnosis plates, and 116 patients on whom there was accurate clinical assessment. Table 20 shows the characteristics of the organism as described by other authors (Leopold, 1953; Gardner & Dukes, 1955; Wurch &

Table 20

Some of the characteristics of Haemophilis vaginalis *as described by other authors*

Characteristic	Leopold 1953	Gardner & Dukes 1955	Wurch & Lutz 1955	Amies & Jones 1955
Gram stain	−	−	− or variable	−
Pleomorphism	some	some	some	marked
Colony size	small	small	small	larger
Lysis	β	vary	Greening & β	−
Oxidase	−	−	NS	−
Nitrate	−	NS	−	+
O_2 need	Microaerophil	Microaerophil	NS	Aerobic
Improved CO_2	NS	+	−	−
Glucose	A	A	A	A
Lactose	−	−	−	−
Penicillin	NS	R	S	S or R
Streptomycin	NS	R	S or R	S
Chloramphenicol	NS	R (S)	S	S
Tetracyclines	NS	S	S	S
Bacitracin	NS	S	S	NS

NS = not stated in report
Lapage, 1961
Various other sugars tested with some variable results (all authors).
By permission of Dr. S. P. Lapage.

Lutz, 1955; Amies & Jones, 1957) and Table 21 shows its distribution, according to various authors (Leopold, 1953; Gardner & Dukes, 1955; Wurch & Lutz, 1955; Ray & Maugham, 1956; Gardner, Dampeer & Dukes, 1957; Amies & Jones, 1957; Brewer, Halpern & Thomas, 1957; Döll, 1958; Lutz & Wurch, 1954). Lapage (1961) suggested that fastidious small Gram negative rods from the vagina fall into 2 types: *Haemophilus influenzae* or *H. parainfluenzae* (Thompson, 1936; Jacobsen, Mason & Arnold, 1937; Reymann, 1943; Henriksen, 1947), the incidence of which is low, and another group, probably diphtheroids. He concluded that his work showed that there were many pitfalls in the diagnosis of *H. vaginalis* vaginitis 'if this is a clinical entity, and if *H. vaginalis* is a species'! In our study (Table 7) we have not attempted to differentiate Gram variable cocco-bacilli, other than to search for the genus *Haemophilus* (Cowan & Steel,

Table 21

Distribution of Haemophilus vaginalis *in female patients*

Author	No. of cases	Type of case	% *H. vaginalis* isolated*	Diagnostic criteria
Leopold (1953)	58	Cervicitis +/− erosion	27	C
Gardner & Dukes (1955)	291	Private obstet. & gynae.	48.5	
	43	Other conditions e.g. cervicitis	0	WM,S,C.
	78	Normal controls	0	
Gardner, Dampeer & Dukes (1957)	1211	Vaginitis	44	WM,S,C.
Wurch & Lutz (1955)	500	Leucorrhoea	22	
Lutz & Wurch (1958)	100	Normal pregnancy	20	S,C.
	126	Trichomonas cases	28.5	
Ray & Maugham (1956)	231	Clinic obstet. & gynae.	39	WM,S
	75	Pregnancy & vaginitis	40	
Amies & Jones (1957)	371	Cases of cervicitis	5.1	WM,S,C.
	829	Grade III cervical smears	41 positive smears	S only available
Brewer *et al.* (1957)	211	Leucorrhoea	42	WM,S+/−C
Döll (1958)	300	Vaginitis	0	WM,S,C.

* The percentages are difficult to derive in some of these reports, but have been worked out from the available data. WM, Wet Mount. S, Stained smear. C, Culture.
Reproduced from Lapage (1961) by permission of Dr. S. P. Lapage.

1965). No *Haemophilus* spp. were found in 280 women, and in our diagnostic practice we have noted its isolation from the genital tract but rarely (Hurley, 1970a).

There is little doubt that organisms, hitherto undescribed and apparently unnamed, inhabit the vagina; some may well be pathogens. We have collections of such microbes, together with deviant forms of common vaginal organisms, awaiting further study.

In the face of difficulties of so fundamental a nature, the question of quantitation of flora would seem untimely asked! There are possibilities, such as the agar masking technique of Selwyn (pers. comm.) but any attempt at quantitation would, we believe, be premature at this stage. We have attempted nothing more than a division into 'heavy' or 'scanty' growth on solid media, using crudely standardized inocula. We have, however, been diligent in picking off all colony types for further study, in some cases proceeding to generic, and in others to specific, identification.

Table 22

The frequency of occurrence of various organisms in the normal vaginal flora

Organism	Jennison (pers. comm.)	Edmunds (1959)	Gordon et al. (1966)
Haemophilus vaginalis	NR	38.1	9.8
Corynebacterium (diphtheroids) spp.	NR	25.9	39.0
Mycoplasma (PPLO) spp.	NR	18.4	15.3
Lactobacilli (Döderlein)	NR	NR	81.0
Coliform	23.0	16.2	12.2
Aerobic streptococci	36.0	19.0	24.2
Streptococcus faecalis	5.5	12.9	NR
Anaerobic organisms	NR	18.2	21.0
Staphylococcus aureus	4.0	2.8	NR
Candida spp.	4.0	7.3	9.7

NR: Not recorded
By permission of Professor Langley.

Widely conceived studies on vaginal flora are notable for their paucity, and we have been unable to find many published references. A short account is given by Langley (1973), (Table 22) and incorporates the data of Edmunds (1959); Jennison (pers. comm.) and Gordon, Hughes & Barr (1966). Some of the findings of the last named appear as Table 23 giving more details of the microbial flora sought in their study, and relating the percentage incidence to categories of discharge. A brief account of the general nature of the flora is given by Hurley (1970b). Table 7 illustrates that, as one might expect, Gram positive rods of the genera *Lactobacillus* and *Corynebacterium* predominated in our study and were isolated from a high percentage of women. The lactobacilli were predominantly anaerobic and other anaerobes were also demonstrated. Haenel, Feldheim & Müller-Beuthow (1958) showed the predominance of anaerobic lactobacilli on the skin as in other sites of the body of man, and Evans et al. (1950) attributed the preponderance of anaerobic flora to the fact that skin bacteria grow principally in sebaceous glands, where anaerobic conditions prevail. The vagina is lined by squamous epithelium, as is the skin, but it contains no glands. Clearly, it is still favourable for the growth of anaerobes. *Staphylococcus epidermidis* was present in a high proportion of the women examined, and this, too, is noted by Wilson & Miles (1964) to be much the commonest of the numerous species of Gram positive cocci that abound on the skin and are part of its resident flora. Streptococci of various types were found in a relatively high percentage of women. Those that were grouped were predominantly Lancefield Groups D and B. Streptococci do not form part of the resident flora of the skin although they may be present as transient denizens. In view of the high proportion isolated, it seems unlikely that the vaginal epithelium contains antistreptococcal substances resembling long chain fatty acids and soaps, as does the skin (Burtenshaw, 1938; 1942). This too, is consonant with the histological structure.

Table 23

Analysis of microbial flora of the vagina in each of three clinical categories

Microbial flora	Clinical category		
	Discharge absent (%)	Non-trichomonal discharge (%)	Trichomonal discharge (%)
Grade I	43.0	38.1	16.2
Grade II	33.4	33.1	33.8
Grade III	23.6	28.8	50.0
Mycoplasma spp.	15.3	20.0	32.6
Haemophilus vaginalis	9.8	12.8	28.0
Aerobic streptococci	24.2	24.7	27.7
Lactobacilli	81.0	72.0	52.0
Corynebacteria	39.0	28.9	32.8
Candida spp.	9.7	10.2	6.8
Acinetobacter spp.	0.0	3.0	1.6
Neisseria spp.	0.0	0.8	0.0
Micrococci*	9.4	12.1	14.3
Enterobacteria†	12.2	9.0	15.6
Anaerobic flora‡	21.0	26.9	29.3

* Includes staphylococcus and micrococcus.
† Includes *Escherichia, Proteus-Providencia, Klebsiella-Aerobacter* groups.
‡ Includes anaerobic streptococci, *Veillonella, Bacteroides, Clostridium*. Gordon, Hughes & Barr (1966).

We were surprised by the percentage of women who harbour *Escherichia coli* and other pathogenic microbes in the vagina during pregnancy. The posterior fornix is inhabited by streptococci, yeasts, coliforms, mycoplasmas and certain anaerobes, all potentially harmful to mother and baby at and about the time of labour and delivery, even if they are not already associated with disease in the vagina or elsewhere, for example, with disease of the urinary tract.

Microbes may ascend into the amniotic sac, through ruptured membranes, and even through membranes that are apparently intact. After delivery, the biology of the vagina changes to resemble that of the postmenopausal woman, and cocci other than the gonococcus can and do invade the genital tract, formerly constituting a very grave hazard to life. Vaginal examination in late pregnancy, and other manipulations *per vaginam,* so abhorrent to our forbears before the antibiotic era, are becoming commonplace. At the same time less attention is focused on scrupulous methods of asepsis and antisepsis, sometimes with serious consequence (Jewett *et al.,* 1968).

The newborn are known to die of septicaemia, often caused by coliforms, and by bacteria that are relatively innocuous to the adult. Such infections may be acquired before or after abortion or delivery and are not always diagnosed during the all too short existence of the newborn. The true nature of the cause of perinatal death may be shown only at necropsy; the same consideration applies to microbial and viral meningitis occurring in the immediate neonatal period (Hurley, Norman & Pryse-Davies, 1969). The relevance of our ecological findings to serious disease in mother and baby, if any, will be analysed in another report.

The absence of gonococci in our series does not surprise us. We estimate the incidence of this pathogen at no more than 1 : 2500 of all our patients, and full diagnostic tests are encouraged in the antenatal period on those who are regarded as likely to harbour the organism (Hurley, 1970*c*).

For some years, studies on the pathogenesis of infections by *Candida* spp. have been conducted in our laboratories. We have paid particular attention to the role of *C. albicans* in the lower genital tract during gestation. Numerous studies have been made on the incidence of this fungus and other pathogenic candidas in various sites of the human body, and accounts are given by Winner & Hurley (1964) and Hurley (1967). The incidence of *C. albicans* in the vagina according to various authors is given by Winner & Hurley (1964). Its incidence relative to other yeast species in specimens submitted for diagnosis of vaginitis in pregnant women, over a 5-year period, together with its incidence in 4 groups of patients (1085, 1031, 303 and 280, respectively) studied prospectively at Queen Charlotte's Hospital is given in the figures accompanying this paper. Our findings are in close agreement with those of Oriel *et al.* (1972). The incidence of isolation of *C. albicans* from the vagina in our population is remarkably constant, and lies between 16 and 19%. Yeasts are isolated more frequently in spring than in winter; Herff (1895) reporting his observations on over 13,000 hospital admissions noted that candida infections were more common in warm than in cold weather. *Candida albicans* is always the predominating yeast species, with *Torulopsis glabrata* the next most common, and relative incidence of *C. albicans* compared with that of other yeast species is higher in specimens taken for diagnosis of vaginitis than in those taken from unselected women. This observation alone indicates that *C. albicans* is likely to be a vaginal pathogen. The pathogenicity of *C. albicans* is, of course, well known and fully documented but the results of many surveys, notably that of Mizuno (1961), suggested that it might be present in the vagina as a commensal, and that its presence did not necessarily indicate morbidity. Winner & Hurley (1964) state categorically that *C. albicans* is often found in the vagina without giving rise to any symptoms. However, the matter does not appear now, to at least one of these authors, as it appears to have appeared then!

Very careful clinicopathological studies are required to assess the role of a microbe in disease of superficial sites, particularly in the vagina, where instruments are required for examination. It is the relative inaccessibility of the vagina to examination that has led, we believe, to an erroneous interpretation of the role of the fungus. Mycotic vulvovaginitis was virtually rediscovered by Plass, Hesseltine & Borts (1931), who stated that the infection, although it had been known for 90 years under a variety of names, was then generally regarded as uncommon. The considerable writings that had appeared following Wilkinson's (1849) report of yeasts associated with profuse vaginal discharge had abated by the turn of the century, subsiding on a wave of scepticism based on doubt of the pathogenicity of yeasts. Medical microbiologists believed that yeasts could be present in the vagina without causing harm, and were part of the normal flora, although Castellani & Taylor (1925) had written "in the normal vagina monilias and other fungi appear to be constantly absent", adding "in the normal vaginal secretion monilia are present, if present at all, in extremely scanty numbers, being found neither in smears nor in cultures". Plass *et al.* (1931) remarked that as yeasts were so widely distributed in nature, it was surprising that they were not universally present in vaginal secretions. Their attention was drawn to mycotic vulvovaginitis by 2 pregnant patients, with vulvovaginitis of so severe a degree that their lives were miserable, since they were constantly annoyed by day and could not secure restful sleep by night. Cultures were negative for *Neisseria gonorrhoeae* but smears consistently showed the blastospores and pseudomycelium of a dimorphic fungus.

The most dramatic and striking feature of vulvovaginitis is the intense itching and soreness (pruritus) that may accompany it. This symptom was used by Plass and his colleagues (1931) as the index symptom for the cases they wished to study; their work related solely to patients with or without vulvovaginal irritation. After many years, the plain fact that pruritus had literally been selected as the index of vaginitis – a disease of the vaginal mucosa – became obscured, even to these workers, and Hesseltine (1940) stated that, clinically, mycotic vulvovaginitis was characterized by an invariable and constantly present symptom, pruritus. All had forgotten that the disease under study was vaginitis, not the symptom, pruritus, which is of diverse aetiology (*British Medical Journal*, 1973).

Carroll, Hurley & Stanley (1973) showed that only direct examination of the vagina reveals the presence or absence of vaginitis, which may not be deduced from a history of pruritus or discharge. Vulvitis does not occur independently of vaginitis. Their clinicopathological studies showed that isolation of *C. albicans* coincided with vaginitis in 84% of cases; and that in 14% it correlated with other signs or symptoms of morbidity (Table 16). They concluded that *C. albicans* is not part of the normal flora of the healthy vagina during pregnancy, that its presence indicates morbidity, and that it should be eliminated promptly by use of appropriate antimycotic agents, lest it lead to chronic vulvovaginitis, stubborn, and refractory to treatment.

For the pathologist, clinicopathological correlation is the objective of ecological studies, since it allows him to lay down diagnostic criteria for diseases caused by microbes that may well form part of the endogenous flora in body sites other than that under study (Table 17). With respect to clinicopathological studies on vaginal flora, the frequency of discharge is a factor of major importance, for, during pregnancy, the vagina is the source of increased physiological secretion, made up of extensive desquamation of glycogen rich cells from the much thickened epithelium, together with cervical mucus, and the microbial flora. The vagina may also be the source of pathological discharges, related to and caused by inimical factors, not all of which are pathogenic microbes. Carroll, Hurley & Stanley (1973) observed that 71% of all pregnant women had discharge evident to the clinical observer and usually sufficient to soil underclothing. When a symptom or sign is so frequent, it is easy for pathologists and obstetricians to claim that its very frequency argues in favour of a physiological origin. However, if the discharge is excessive, deviant, accompanied by discomfort for the patient and signs of disordered anatomy of the tract, it is likely to be pathological. We are becoming increasingly aware that much of the secretion that in the past may have been dismissed as physiological is, in fact, pathological, and may be associated with microbes such as mycoplasmas, *Chlamydia, Bacteroides* spp., or viruses, to which, hitherto, pathologists have paid but scant attention. During pregnancy, only 2 microbes, *C. albicans* and *Trichomonas vaginalis* are generally regarded as causes of microbial vaginitis. Other candidas and *Torulopsis glabrata* are likely to be pathogens in this site, as they are in others (Hurley & Morris, 1964; Carroll, Hurley & Stanley, 1973).

The quality and degree of close clinical observation of the vagina to detect aberrations from its normal anatomy and physiology is plainly crucial to interpretation of the role played by particular microbes. Should we decide, on defined clinicopathological criteria, that certain microbes are pathogenic in the vagina, we shall regard them as abnormal in that situation and do our best to eliminate them.

Regrettably, we are still in the position of trying to ascertain which microbes are pathogenic, so that it is impossible to give an account of the normal flora.

The study, a preliminary account of which is reported in Table 7, is still in progress, and will be extended to include T-strain mycoplasmas, *Chlamydia* spp. and viruses. Fuller reports will be published in due course, and clinicopathological correlations will be attempted.

Too little attention has been paid to the viral and chlamydial flora of the genital tract although it is known that *Chlamydia trachomatis* and many viruses, including herpes simplex, herpes zoster, variola, vaccinia, cytomegalovirus, and adenoviruses may all be isolated therefrom. While medical microbiologists are studying the pathogenesis of intra-uterine viral infection, we should do well to examine the viral flora of the genital tract in greater detail, since the importance of the ascending route of infection, and the dangers to the foetus of passing through the birth canal, have by no means been fully explored.

9. Acknowledgements

We are grateful to our colleagues, Dr. B. E. Andrews and Dr. Mair Thomas, for their advice on the isolation and identification of mycoplasmas; to Dr. M. T. Parker for grouping many of the streptococci; to Dr. D. Yarrow and Dr. Helen Buckley for their help in confirming the speciation of the rarer yeast isolates; to Mr. J. B. Jones and Mr. J. E. B. Foulkes who examined the patients who are the subject of Table 7, and to the nursing staff of Queen Charlotte's Hospital for their help and co-operation.

Much of this work was supported by grants from the Medical Research Council, The Department of Health and Social Security, and the Board of Governors of Queen Charlotte's Hospital for Women.

10. References

AIMES, C. R. & JONES, S. A. (1957). A description of *Haemophilus vaginalis* and its L forms. *Can. J. Microbiol.* **3**, 579.

ARCHER, J. F. (1968). 'T' strain mycoplasma in the female genital tract. *Brit. J. vener. Dis.* **44**, 232.

BRAUN, P., KLEIN, J. O., LEE, Y. H. & KASS, E. H. (1970). Methodological investigations and prevalence of genital mycoplasmas in pregnancy. *J. infect. Dis.* **121**, 391.

BREWER, J. I., HALPERN, B. & THOMAS, B. S. (1957). *Haemophilus vaginalis* vaginitis. *Amer. J. Obstet. Gynec.* **74**, 834.

BRITISH MEDICAL JOURNAL. (1973). Editorial **1**, 628.

BURTENSHAW, J. M. L. (1938). The mortality of haemolytic streptococcus on skin and on other surfaces. *J. Hyg., Camb.* **38**, 575.

BURTENSHAW, J. M. L. (1942). The mechanism of self-disinfection of the human skin and its appendages. *J. Hyg., Camb.* **42**, 184.
CARROLL, C. J., STANLEY, V. C. & HURLEY, R. (1973). Criteria for diagnosis of *Candida* vulvovaginitis in pregnant women. *J. Obstet. Gynec. Brit. Cwlth.* **80**, 258.
CASTELLANI, A. & TAYLOR, F. E. (1925). Vaginal monilias and vaginal moniliases. *J. Obstet. Gynaec. Brit. Emp.* **32**, 69.
COLEBROOK, L. (1936). The prevention of puerperal sepsis. *J. Obstet. Gynaec. Brit. Emp.* **43**, 691.
COWAN, S. T. & STEEL, K. J. (1965). *Manual for the Identification of Medical Bacteria.* Cambridge University Press.
CRUICKSHANK, R. & SHARMAN, A. (1934). The biology of the vagina in the human subject. *J. Obstet. Gynaec. Brit. Emp.* **41**, 369.
DAVIS, M. E. & PEARL, S. A. (1938). Biology of the human vagina in pregnancy. *Amer. J. Obstet. Gynec.* **35**, 77.
DE FONSEKA, C. I. (1972). The distribution of *Candida* species and their antibodies in pregnant women and the newborn; a clinical and experimental study. Ph.D. thesis. University of London.
DÖDERLEIN, A. S. G. (1892). *Das Scheidensekret und seine Bedeutung für das Puerperalfieber.* Leipzig: O. Durr.
DÖLL, W. (1958). Vorkommen von *'Haemophilus vaginalis'* bei 'unspezifischer Vaginitis'? *Zentbl. Bakt. ParasitKde* Abt. I. Orig. **171**, 372.
DUNLOP, E. M. C., HARE, M. J., JONES, B. R. & TAYLOR-ROBINSON, D. (1969). Mycoplasmas and 'non-specific' genital infection. *Br. J. vener. Dis.* **45**, 274.
EDMUNDS, P. N. (1959). *Haemophilus vaginalis* – Its association with puerperal pyrexia and leucorrhoea. *J. Obstet. Gynaec. Brit. Emp.* **66**, 917.
EVANS, C. A., SMITH, W. M., JOHNSTON, E. A. & GIBLETT, E. R. (1950). Bacterial flora of the normal human skin. *J. invest. Derm.* **15**, 305.
FISCHEL, W. (1880). Ein Beitrag zur Histologie der Erosionen der Portico Vaginalis Uteri. *Arch. f. Gynäkol.* **15**, 76.
FLYNN, J. & WATKINS, SHEENA A. (1972). A serum free medium for testing fermentation reactions in *Neisseria gonorrhoeae. J. clin. Path.* **25**, 525.
GARDNER, H. L., DAMPEER, T. K. & DUKES, C. D. (1957). The prevalence of vaginitis; a study in incidence. *Amer. J. Obstet. Gynec.* **73**, 1080.
GARDNER, H. L. & DUKES, C. D. (1955). *Haemophilus vaginalis* vaginitis. *Amer. J. Obstet. Gynec.* **69**, 962.
GORDON, A. M., HUGHES, H. E. & BARR, G. T. D. (1966). Bacterial flora in abnormalities of the female genital tract. *J. clin. Path.* **19**, 429.
HAENEL, H., FELDHEIM, G. & MÜLLER-BEUTHOW, W. (1958). Zur mikrobiologischen Ökologie des Menschen. *Zentbl. Bakt. ParasitKde* Abt. I. Orig. **172**, 73.
HENRIKSEN, S. D. (1947). Gram-negative diplo-bacilli from the genito-urinary tract. *Acta path. et microbiol. Scand.* **24**, 184.
HERFF, O. von (1895). Ueber Scheidenmykosen (*Colpitis mycotica acuta*). *Samml. Klin. Vortr.* **137**, 493.
HESSELTINE, H. C. (1940). Vulval and vaginal mycosis and trichomoniasis. *Amer. J. Obstet. Gynec.* **40**, 641.
HUGHES, H. E., GORDON, A. M. & BARR, G. J. D. (1966). A clinical and laboratory study of trichomoniasis of the female genital tract. *J. Obstet. Gynaec. Brit. Cwlth.* **73**, 821.
HURLEY, R. (1967). The pathogenic *Candida* species: A review. *Rev. Med. vet. Mycol.* **6**, 159.
HURLEY, R. (1970*a*). Haemophilus endometritis in women fitted with Lippes loop. *Brit. Med. J.* **1**, 566.
HURLEY, R. (1970*b*). Section XI, Microbiology, In *Scientific Foundations of Obstetrics and Gynaecology.* Eds E. E. Philipp, J. Barnes & M. Newton. London: Heinemann.
HURLEY, R. (1970*c*). Infant gonorrhoea resulting from untreated mother. In Medical Forum, *Modern Medicine,* **9**, 871.

HURLEY, R., LEASK, B. G. S., FAKTOR, J. A. & DE FONSEKA, C. I. (1973). Incidence and distribution of yeast species and of *Trichomonas vaginalis* in the vagina of pregnant women. *J. Obstet. Gynaec. Brit. Cwlth.* **80**, 252.
HURLEY, R. & MORRIS, E. D. (1964). The pathogenecity of *Candida* species in the human vagina. *J. Obstet. Gynaec. Brit. Cwlth.* **71**, 692.
HURLEY, R., NORMAN, A. P. & PRYSE-DAVIES, J. (1969). Massive pulmonary haemorrhage in the newborn associated with Coxsackie B virus infection. *Brit. Med. J.* **3**, 636.
HURLEY, R. & STANLEY, V. C. (1972). Candida vaginitis in pregnant women. In *The Diagnosis and Chemotherapy of Urogenital Bacterial, Protozoal and Mycotic Infections.* Ed. F. Gasparri. Firenze: Edizioni Mediche P. Periti.
JACOBSEN, F., MASON, H. C. & ARNOLD, L. (1937). Laboratory diagnosis in chronic gonorrhoea of the female. *J. Lab. Clin. Med.* **23**, 729.
JEWETT, J. F., REID, D. E., SAFON, L. E. & EASTERDAY, C. L. (1968). Continuing risk of puerperal sepsis. *J. Amer. Med. Ass.* **206**, 344.
JOHNSTON, T. B. & WHILLIS, J. (1947). *Gray's Anatomy,* 29th Ed. London: Longmans, Green & Co.
KIENLIN, H. (1926). Die Reaktion des Vaginalsekrets Neugeborener. *Zentbl. Gynäk.* **50**, 644.
KRONIG & MENGE (1897). (Cited by Davis & Pearl, 1938).
LANGLEY, F. A. (1973). In *Postgraduate Obstetrical and Gynaecological Pathology.* Eds H. Fox & F. A. Langley. Oxford: Pergamon Press.
LAPAGE, S. P. (1961). *Haemophilus vaginalis* and its role in vaginitis. *Acta path. microbiol. Scand.* **52**, 34.
LEOPOLD, S. (1953). Heretofore undescribed organism isolated from genitourinary system. *U.S. Armed F. Med. J.* **4**, 263.
LODDER, J. (1970). (Ed.) *The Yeasts: A Taxonomic Study.* 2nd Ed. Amsterdam: North-Holland Publishing Company.
LODDER, J. & KREGER-VAN RIJ. N. J. W. (1952). *The Yeasts: A Taxonomic Study.* Amsterdam: North-Holland Publishing Company.
LUTZ, A. & WURCH, Th. (1954). Recherches sur la sensibilité aux antibiotiques d'éléments de la flore vaginale au cours de la gestation. *Bull. Féd. Socs. Gynéc. Obstét. Lang. fr.* **6**, 115.
McCORMACK, W. M., RANKIN, J. S. & LEE, Y. H. (1972). Localization of genital mycoplasmas in women. *Amer. J. Obstet. Gynec.* **112**, 920.
McGREGOR, A. L. (1950). *Synopsis of Surgical Anatomy,* 7th ed., Bristol: John Wright & Sons.
MERRITT, A. E. & HURLEY, R. (1972). Evaluation of sporulation media for yeasts obtained from pathological material. *J. med. Microbiol.* **5**, 21.
MIURA, H. (1928). *Kyoto Ikadaigaku Zasshi (Mitt. med. Akad. Kioto),* **2**, Heft 1.
MIZUNO, S. (1961). *Vulvovaginal candidiasis: survey of vulvovaginal candidiasis in Japan.* Tokyo: Research Committee of Candidiasis, Education Ministry of Japan, 19.
ORIEL, J. D., PARTRIDGE, B. M., DENNY, M. J. & COLEMAN, J. C. (1972). Genital yeast infections. *Brit. med. J.* **2**, 761.
PLASS, E. D., HESSELTINE, H. C. & BORTS, I. H. (1931). Monilia vulvovaginitis. *Amer. J. Obstet. Gynec.* **21**, 320.
RAY, J. L. & MAUGHAM, S. M. (1956). *Haemophilus vaginalis* as etiological agent in vaginitis. *West. J. Surg.* **64**, 581.
REYMANN, F. (1943). Cultivation of gonococci as diagnostic method in gonorrhoea in women. 1. Morphological and cultural aspects of gonococci. *Acta derm.-venerol.* **24**, 130.
ROGOSA, M. & SHARPE, M. E. (1960). Species differentiation of human vaginal lactobacilli. *J. gen. Microbiol.* **23**, 197.
SALIGNON-MICHELIER, A. (1950). Contribution à l'étude des moniliases vaginales chez la femme enceinte. Thèse de Pharm. Marseille.

SHEPHARD, M. C. (1954). The recovery of pleuropneumonia-like organisms from negro men with and without non-gonococcal urethritis. *Amer. J. Syph.* **38**, 113.

STANLEY, VALERIE C., HURLEY, R. & CARROLL, C. J. (1972). Distribution and significance of Candida precipitins in sera from pregnant women. *J. med. Microbiol.* **5**, 313.

STEWART-TULL, D. E. S. (1964). Evidence that vaginal lactobacilli do not ferment glycogen. *Amer. J. Obstet. Gynec.* **88**, 676.

THOMPSON, L. (1936). Other oxydase positive bacteria found in cultures made for *Neisseria gonorrhoeae. J. Bact.* **31**, 82.

WILKINSON, J. S. (1849). Some remarks upon the development of epiphytes with the description of new vegetable formation found in connexion with the human uterus. *Lancet* **2**, 448.

WILSON, G. S. & MILES, A. A. (1964). Topley and Wilson's Principles of Bacteriology and Immunity 5th ed. London: Edward Arnold.

WINNER, H. I. & HURLEY, R. (1964). *Candida albicans.* London: J. & A. Churchill Ltd.

WURCH, Th. & LUTZ, A. (1955). Étude du contenu vaginal dans 500 cas de leucorrhées Cytologie – Microbiologie. *Revue française Gyn. Obstét.* **50**, 289.

WYLIE, G. & HENDERSON, A. (1969). Identity and glycogen-fermenting ability of lactobacilli isolated from the vagina of pregnant women. *J. med. Microbiol.* **21**, 363.

YARROW, D. (1972). Four new combinations in Yeasts. *Antonie van Leeuwenhoek* **38**, 357.

ZINNEMAN, K. & TURNER, G. C. (1962). Taxonomy of *Haemophilus vaginalis. Nature, Lond.* **195**, 203.

ZINNEMAN, K. & TURNER, G. C. (1963). The taxonomic position of *Haemophilus vaginalis (Corynebacterium vaginale). J. Path. Bact.* **85**, 213.

Some Factors Associated with Geographical Variations in the Intestinal Microflora

B. S. DRASAR

Department of Bacteriology, St. Mary's Hospital Medical School, London W2 1PG, England

CONTENTS

1. Introduction 187
2. Mechanisms controlling the microflora 187
   (a) The small intestine 187
   (b) The large intestine 188
3. Geographical variations in the microflora 190
   (a) The small intestine 190
   (b) The large intestine (faeces) 191
4. Factors influencing mechanisms controlling the microflora . . . 192
   (a) Race 192
   (b) Tropical residence 194
   (c) Diet 194
5. Conclusion 194
6. References 195

1. Introduction

THE COMPOSITION, distribution and activity of the intestinal flora depends on the composition of the intestinal contents in various parts of the gut and on the antibacterial systems of the host. Both the physiology of the host and the diet are important; in some parts of the intestine interactions between bacteria undoubtedly play an important role (Table 1).

2. Mechanisms Controlling the Microflora

(a) *The small intestine*

The duodenum and jejunum of normal people in the U.S.A. and W. Europe contain few bacteria ($<10^3$/ml of intestinal contents). Streptococci and lactobacilli are the bacteria most usually isolated. In the ileum a richer and more permanent flora is found (Legler & Zeitler, 1962).

Gastric acid controls the entry of bacteria into the small intestine; the distribution of bacteria within the small intestine is determined by the movement of intestinal contents (Dixon, 1960). The type of diet and frequency of feeding affects the secretion of gastric acid and the rate of gastric emptying thus indirectly influences the distribution of intestinal bacteria (Drasar *et al.*, 1969).

Table 1

Some factors influencing the intestinal flora

Factors influencing the intestinal flora
(1) *Host physiology*
Intestinal secretions
Intestinal mucosa
Immune mechanisms
(2) *Environmental factors*
Bacterial contamination
Diet
Antibacterial drugs
(3) *Bacterial interactions*

Bile salts (Percy-Robb & Collee, 1972), IgA antibody (Shearman *et al.*, 1972) and lysozyme (Florey, 1930) are among the factors that select those bacteria able to survive in the intestine. Interactions between bacteria are probably not important in view of the sparse population.

(b) *The large intestine*

The faecal flora is assumed to refer to that of the large intestine. A small number of parallel studies suggest that the flora of the large intestine is similar to that of faeces though the intestinal contents are more dilute than are those of the rectum (Seeliger & Werner, 1963).

The colon can be thought of as a fermenter (Boni, 1967) but it should be remembered that the composition of the medium depends upon the activity of the small intestine and is thus unknown (Fig. 1).

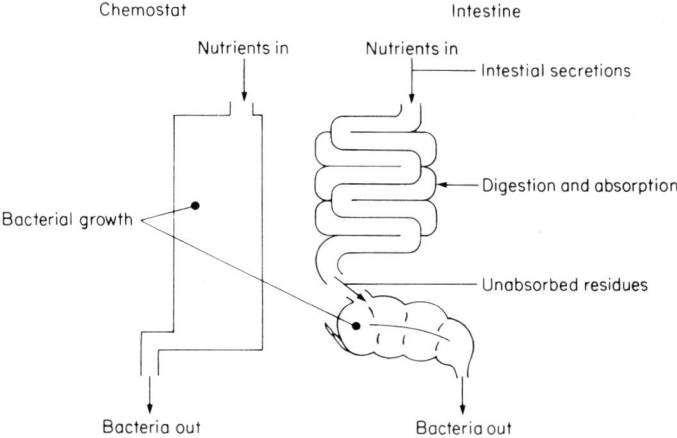

Fig. 1. Comparison of the intestine with a continuous culture fermenter.

Diet has long been considered to be one of the most important factors in controlling the faecal flora (Dugeon, 1926). However, attempts to change this flora in adult man by dietary manipulation have, in general, produced equivocal results (Drasar et al., 1973): only the studies of Hoffman (1964) made on one subject, suggest consistent changes.

Contributions to the nutrients available to the flora are from 2 major sources, the diet and the body (Table 2). Digestion and absorption from the intestine is an efficient process and the major dietary components available to the flora must be non-absorbable residues such as plant steroids, fibre and food additives. The efficiency of absorption probably accounts for the lack of effect of most dietary changes on the flora. Metabolic studies in which the food additive cyclamate was added to the diet demonstrated cyclamate-dependent changes in the metabolism by the flora (Drasar et al., 1972). Presumably this compound was not absorbed, and passed directly into the colon to exert ecological pressure on the flora (Fig. 2). Not all compounds act directly, some components of the diet may be absorbed and excreted in the bile, compounds such as stilboestrol may be enterohepatically circulated (Smith, 1973). The reabsorption and circulation of such non-nutrients may depend upon the distribution within the small intestine of bacteria able to hydrolyse the conjugates formed by the liver.

Table 2

Sources of nutrients in the intestine

Sources of nutrients	Substances available	Notes
Intestinal residue	Pancreatic enzymes Bile salts Neutral steroids Intestinal mucus Secretory IgA Cells	Autodigestion occurs 300-500 mg/day lost in faeces 500-1000 mg/day lost in faeces 50-200 g of mucosa shed/day

Dietary residue	Unavailable 'Fibre'
	  carbohydrate
	Natural anutrients
	  e.g. plant steroids
	Food additives
	  e.g. cyclamate
		4-8 g/day of crude fibre enters
		  the colon
		
		up to 4 g/day intake

Biliary excretions	Endogenous
	  e.g. bile pigments
	Exogenous
	  e.g. coumarin
	     stilboestrol	

Note: the figures are based on N. American and European data.

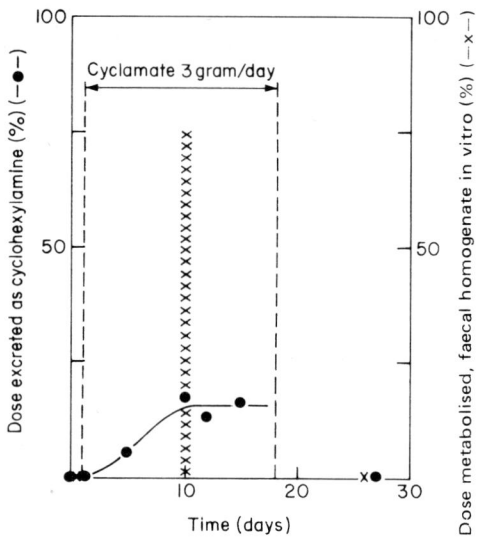

Fig. 2. The metabolism of cyclamate by a human subject during a 30 day period. Cyclamate was administered from day 2 to day 18. x, % conversion of C^{14} cyclamate to cyclohexylamine by faecal homogenate; ●, % of dose administered excreted in the urine as cyclohexylamine.

The body contributes intestinal cells, digestive residues and biliary excretion products. Auto-digestion and absorption recycles some of the intestinal residue. The amount of bile salt reaching the colon can be diminished by decreasing the intake of fat in the diet (Hill, 1971).

Bacterial interactions are probably very important in the colon but their mechanism and exact influence are not well understood. Colicines may be important in the control of *Escherichia coli* (Branche et al., 1963) but factors influencing the dominant species, *Bacteroides fragilis*, wait to be investigated.

3. Geographical Variations in the Microflora

(a) *The small intestine*

Apparently healthy people living in S. India have very many more bacteria in the small intestine than do people in N. America and W. Europe (Fig. 3). Jejunal contents from S. Indian subjects yield approximately one hundred times as many bacteria as samples from English subjects, furthermore, enterobacteria, bacteroides and other faecal organisms occur more commonly in jejunal specimens from S. Indians (Bhat et al., 1972). A similar large number of bacteria can be demonstrated in the small intestine of children in Guatemala (Mata et al., 1972).

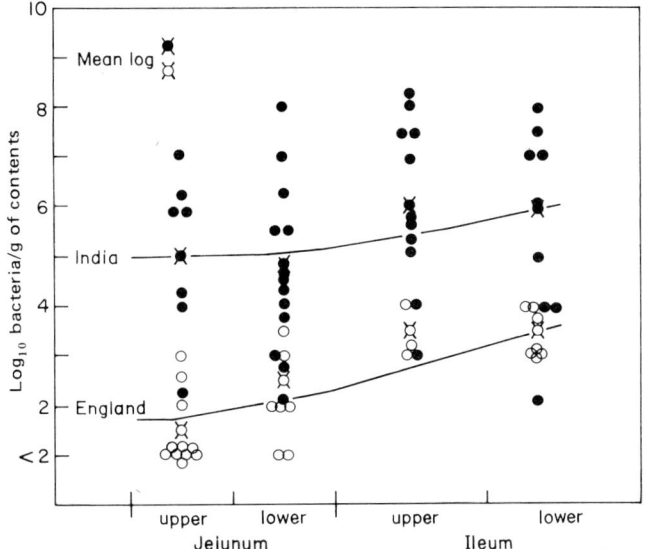

Fig. 3. Distribution of bacteria in small intestine. From the data of Drasar et al. (1969) and Bhat et al. (1972). ●, Indian samples; ○, English samples.

(b) *The large intestine (faeces)*

The comparison of the bacterial flora of faeces from subjects in different countries, consuming different diets and living in very different socio-economic circumstances, was undertaken as part of a larger study on the aetiology of cancer of the colon (Hill et al., 1971; Drasar & Hill, 1972). The groups of people studied differed in many ways (Table 3); the Indians, Japanese and Ugandans were poor villagers in simple rural communities whereas the Americans and British lived in modern cities.

The results of these investigations are summarized in Table 4. Non-sporing anaerobic bacteria formed the largest bacterial group in all the samples examined. Only in the Japanese samples did the number of facultative bacteria approach that of the anaerobic. *Bacteroides fragilis, Bifidobacterium adolecentis* and *Eubacterium aerofaciens* were the bacterial species most often isolated (Peach et al., 1972). Bacteroides occured in the greatest numbers in specimens from the developed countries, whereas eubacteria were prominent in those from India and Japan. Enterococci were most numerous in specimens from Japan and Uganda. *Streptococcus faecalis* was the dominant enterococcus from English, Scots and Americans but *Strep. faecium* predominated among Indian, Japanese and Ugandan isolates. *Escherichia coli* was the most common enterobacterium isolated; however, many other groups were represented, especially among Indian, Japanese and Ugandan isolates. *Clostridium perfringens* and *Cl. bifermentans*

Table 3

The human populations studied

Country of origin	Populations studied	Diet
England	Staff of St. Mary's Hospital Medical School	Mixed Western
Scotland	Staff of Edinburgh University Medical School	Mixed Western
U.S.A.	Staff of the Center for Diseases Control, Atlanta	Mixed Western
India*	Villagers : Controls for sprue study from near Vellore, South India	Rice, no meat, little fat
Uganda	Villagers : Parents of children attending a malnutrition clinic, Kyandondo Region	Matoke (boiled mashed bananas)
Japan	Villagers : Farmers and housewives near Yonaga	Rice, fruits, vegetables, little meat and fat

* This population was also studied by Bhat *et al.* (1972).

occurred in all groups; *Cl. paraputrificum* was common in English, Scots and American specimens but rare from other sources. In this context it is of interest to note that, among the clostridia, 90% of the English isolates could be assigned to a species whereas only 65% of the Ugandan isolates could be so classified. This difference probably reflects the sources of the strains used in the establishment of the classification. One might speculate that many as yet unclassified organisms occur in other groups, their presence masked by the dominant species.

4. Factors Influencing Mechanisms Controlling the Microflora

(a) *Race*

The nutrients available to the bacteria in the intestine depend, at least in part, on the ability of the intestine to digest and absorb nutrients. While the expression of digestive function may be influenced by the environment some variations are probably of genetic origin.

There are striking differences in the incidence of lactase deficiency among adults of various ethnic groups. Thus, the condition occurs only in *c.* 6% of the English and Americans of European origin. However, among indigenous Ugandans, and Americans of African origin, 70% or more of people are deficient in lactase (Neale, 1971).

Faecal specimens from Americans of both European and African extraction, living in Atlanta, were examined (Table 5). The socio-economic status and diet of the 2 groups were comparable. Only the numbers of lactobacilli differed significantly between the groups. It is tempting to attribute the greater numbers

Table 4

Faecal microflora of different human populations

Diet	Country	Mean $\log_{10}$ no. of bacteria/g of faeces					
		Enterobacteria	Enterococci	Lactobacilli	Clostridia	Bacteroides	Bifidobacteria	Eubacteria*
Largely carbohydrate	India	7.9	7.3	7.6	5.7	9.2	9.6	9.5
	Japan	9.4	8.1	7.4	5.6	9.4	9.7	9.6
	Uganda	8.0	7.0	7.2	5.1	8.2	9.4	9.3
Mixed Western	England	7.9	5.8	6.5	5.7	9.8	9.9	9.3
	Scotland	7.6	5.3	7.7	5.6	9.8	9.9	9.3
	U.S.A.	7.4	5.9	6.5	5.4	9.7	9.9	9.3

* Numbers estimated on the basis of isolates identified.

Table 5

Faecal microflora of Americans

Subjects investigated	Mean & (range) $\log_{10}$ no. of bacteria/g of faeces					
	Enterobacteria	Enterococci	Lactobacilli	Clostridia	Bacteroides	Gram +, non-sporing anaerobes
African extraction (12 subjects)*	7.3 (6.2-9.0)	5.0 (4.1-8.9)	8.0 (4.5-9.0)	5.1 (2.6-6.2)	9.8 (9.1-10.0)	10.0 (8.4-10.3)
European extraction (22 subjects)*	7.4 (4.9-9.7)	5.9 (3.0-8.7)	6.5 (3.6-9.3)	5.4 (2.3-6.0)	9.7 (8.2-10.8)	10.0 (8.1-10.6)

* One sample from each subject examined.

of these organisms isolated from subjects of African origin to lactase deficiency in the host, but this conclusion awaits substantiation.

(b) *Tropical residence*

The small-intestinal mucosa of normal people, both indigenous and expatriate, in tropical regions differ both in structure, as judged from biopsy material, and function, as judged by D-xylose absorption, from that of normal people in temperate regions (Lindenbaum, 1968). On transfer to a temperate climate the mucosa of expatriates reverts to normal, and that of tropical people become 'normal' as judged by W. European and N. American standards (Gerson *et al.*, 1971; Lindenbaum *et al.*, 1971).

(c) *Diet*

Diet influences many body functions, (Table 6): these effects of diet are most obvious in malnutrition. The decreased production of intestinal cells (e.g. Deo *et al.*, 1965) and secretions decreases the amount of nutrients available to the flora from the body. However, the reduced efficiency of digestion resultant upon these changes may result in an increase in the nutrients available from dietary sources. Furthermore, the decreased secretion of bile salts may also influence bacterial growth in the small intestine.

Table 6

Some body systems impaired by malnutrition

Intestinal structure and function	Mucosal cell replacement impaired Digestive secretions diminished e.g. tryspin bile salts
Immune systems	Antibody synthesis impaired Phagocytic activity reduced

Immune systems are also influenced by malnutrition (Scrimshaw *et al.*, 1968) and a decrease of secretory IgA might occur. In patients with hypo-γ-globulinaemia an increase in the number of anaerobes in the small intestine has been observed (Parkin *et al.*, 1972); although this is partially explicable in terms of the achlorhydria accompanying the disease, some is probably due to the immunoglobulin deficiency.

5. Conclusion

The differences in distribution of bacteria in the small intestine may be explicable in terms of the influence of diet and tropical residence on intestinal function, together with an effect of diet on IgA secretion. Some indirect influence of diet on distribution was postulated by Gorbach et al. (1970) to explain the variations of the small-intestinal flora in the Indians studied by them.

Similarly, the differences in faecal flora may result from the influence of the variations in fat intake upon the bile salts entering the colon. The differences in faecal bile salt concentration (Hill et al., 1971) would support this. However, dietary influence on bile salt synthesis and intestinal function, together with climatic influences, cannot be finally discarded. Variations in intestinal structure and function are at least as important as diet in controlling the intestinal flora.

6. References

BHAT, P., SHANTAKUMARI, S., RAJAN, D., MATHAN, V. I., KAPADIA, C. R., SWARNABIA, C. & BAKER, S. J. (1972). Bacterial flora of the gastro-intestinal tract in South Indian control subjects and patients with tropical sprue. *Gastroenterology* **62**, 11.
BONI, P. (1967). Growth dynamics of the intestinal flora considered as a continuous-flow culture : a thermodynamic and hydrodynamic qualitative approach. *Riv. Ist. sieroter. ital.* **42**, 261.
BRANCHE, W. C., YOUNG, V. U., ROBINET, H. G. & MASSEY, E. D. (1963). Effect of colicine production in *Escherichia coli* in the normal human intestine. *Proc. Soc. exp. Biol. Med.* **114**, 198.
DEO, M. G., SOOD, S. K. & RAMLINGASWAMI, V. (1965). Experimental protein deficiency. *Archs. Path.* **80**, 14.
DIXON, J. M. S. (1960). The fate of bacteria in the small intestine *J. Path. Bact.* **79**, 131.
DRASAR, B. S., SHINER, M. & McLEOD, G. M. (1969). Studies on the intestinal flora. 1 : The bacterial flora of the gastrointestinal tract in healthy and achlorhydric persons. *Gastroenterology,* **56**, 71.
DRASAR, B. S. & HILL, M. J. (1972). Intestinal bacteria and cancer. *Amer. J. clin. Nutr.* **25**, 1399.
DRASAR, B. S., RENWICK, A. G. & WILLIAMS, R. T. (1972). The role of the gut flora in the metabolism of cyclamate. *Biochem. J.* **129**, 881.
DRASAR, B. S., CROWTHER, J. S., GODDARD, P., HAWKSWORTH, G., HILL, J. M., PEACH, S., WILLIAMS, R. E. O. & RENWICK, A. (1973). The relation between diet and the gut microflora in man. *Proc. Nutr. Soc.* **32**, 49.
DUGEON, L. S. (1926). Study of the intestinal flora under normal and abnormal conditions. *J. Hyg., Camb.* **25**, 119.
FLOREY, H. W. (1930). The relative amounts of lysozyme present in the tissues of some animals. *Brit. J. exp. Path.* **11**, 251.
GERSON, C. D., KENT, T. H., SAHA, J. R., SIDDIGI, N. & LINDENBAUM, J. (1971). Recovery of small intestinal structure and function after residence in the Tropics. II : Studies of Indians and Pakistanis living in New York City. *Ann. intern. Med.* **75**, 41.
GORBACH, S. L., BANWELL, J. G., JACOBS, B., CHATTERJEE, B. D., MITRA, R., SEN, N. N. & MAZUMDER, D. M. G. (1970). Tropical sprue and malnutrition in West Bengal. *Amer. J. clin. Nutr.* **23**, 1545.
HILL, M. J. (1971). The effect of some factors on the faecal concentration of acid steroids, neutral steroids and urobilins. *J. Path.* **104**, 239.
HILL, M. J., CROWTHER, J. S., DRASAR, B. S., HAWKSWORTH, G., ARIES, V. & WILLIAMS, R. E. O. (1971). Bacteria and the aetiology of cancer of the large bowel. *Lancet* **1**, 95.
HOFFMAN, K. (1964). Untersuchungen über die Zusammensetzung der Stuhlflora während eines langdauernden Ernährungsversuchs mit kolenhydratreicher, mit Fettreicher und mit Eiweissreicher. *Zentbl. Bakt. ParasitKde.* Abt. I, Orig. **192**, 500.

LEGLER, F. & ZEITLER, G. (1962). Die Dünn und Dickdarm Flora bei einigen gastrointestinalen Krankheitsbildern. *Dt. med. Wschr.* **87,** 695.

LINDENBAUM, J. (1968). Small intestine dysfunction in Pakistanis and Americans resident in Pakistan. *Amer. J. clin. Nutr.* **21,** 1023.

LINDENBAUM, J., GERSON, C. D. & KENT, T. H. (1971). Recovery of small intestinal structure and function after residence in the Tropics. I. *Ann. intern. Med.* **74,** 218.

MATA, L. J., MEJICANOS, M. L. & JIMÉNEZ, F. (1972). Studies on the indigenous gastrointestinal flora of Guatemalan children. *Amer. J. clin. Nutr.* **25,** 1391.

NEALE, G. (1971). Disaccharidase deficiencies. *J. clin. Path.* **24,** Suppl. **5,** 22.

PARKIN, D. M., McCLELLAND, D. B. L., O'MOORE, R. R., PERCY-ROBB, I. W., GRANT, I. W. B. & SHEARMAN, J. C. (1972). Intestinal bacterial flora and bile salt studies in hypogammaglobulin-anemia. *Gut.* **13,** 182.

PEACH, S. L., FERNANDEZ, F., JOHNSON, K. & DRASAR, B. S. (1972). The classification of some non-sporing strict anaerobes. *J. med. Microbiol.* **5,** P XIV.

PERCY-ROBB, I. W. & COLLEE, J. G. (1972). Bile acids : A pH dependent antibacterial system in the gut. *Brit. med. J.* **3,** 813.

SCRIMSHAW, N. S., TAYLOR, C. E. & GORDON, J. E. (1968). *Interactions of Nutrition and Infections.* Geneva: W.H.O.

SEELIGER, H. & WERNER, H. (1963). Recherches qualitatives et quantitatives sur la flore intestinale de l'homme. *Annls Int. Pasteur, Paris* **105,** 911.

SHEARMAN, D. J. C., PARKIN, D. M. & McCLELLAND, D. B. L. (1972). The demonstration and function of antibodies in the gastrointestinal tract. *Gut* **13,** 483.

SMITH, R. L. (1973). *The Excretory Function of Bile.* London: Chapman & Hall.

The Bacterial Flora of the Upper Gastrointestinal Tract in Children both in Health and Disease

CHARLOTTE M. ANDERSON, D. N. CHALLACOMBE
AND JUDITH M. RICHARDSON

The University of Birmingham Institute of Child Health,
The Nuffield Building, Francis Road, Birmingham B16 8ET

CONTENTS

1. Introduction 197
2. The normal intestinal microflora in infants and children 198
3. Alterations in small-intestinal microflora 199
4. Relationship of *E. coli* in the upper small intestine to diarrhoeal symptoms . . 201
5. References 202

1. Introduction

UNTIL the last two decades, study of the normal microbial flora of the gut has been limited to identification of organisms present in the mouth and throat and in the faeces. For the study of the flora of the living bowel, luminal contents must be obtained by intubation or needling of the bowel during operation. The results obtained by intubation, particularly of the small intestine, may be difficult to interpret for technical reasons such as: uncertainty of the level from which specimens are taken; difficulty of ensuring travel of the tube down to lower levels in reasonable time; carrying of contamination from above during the procedure and the possible alteration of local conditions by the presence of the tube, a 'foreign body'. For accurate documentation there must also be detailed identification of all organisms present.

The wide diversity of techniques for obtaining specimens, as well as variation of microbiological culture methods, makes it difficult to compare the results of various workers in this field. For instance, the choice of selective and non-selective culture media must be wide, first, so as not to inhibit the growth of selected species and, secondly, so as not to allow more prevalent strains to overgrow those present in small numbers. Anaerobic organisms also have fastidious requirements (Drasar, 1967) and it is only recently that new techniques, such as those described by this author, have come into common use, thus rendering earlier studies on the anaerobic flora, particularly of the small intestine, difficult to interpret.

During the last decade there have been many studies of the normal and abnormal flora of the adult gut (Donaldson, 1964, 1970; Floch, Gorbach & Luckey, 1970; Gorbach, 1971) but few in children.

2. The Normal Intestinal Microflora in Infants and Children

The alimentary tract at birth is either sterile or contains only a few micro-organisms (Wilson & Miles, 1955) but within a few days of birth a profuse flora is established in the mouth and large intestine, the mouth flora consisting predominantly of Gram positive organisms normally considered non-pathogenic, and the colonic flora predominantly of Gram negative organisms and anaerobes such as *Lactobacillus* and *Bacteroides* spp.

Studies of the small intestinal microflora in westernized communities are few Davison, 1925; Barbero *et al.*, 1952; Anderson & Langford, 1958). The latter authors, working in association with Cregan & Hayward (1953) and using similar methods of assessing bacterial growth but studying their subjects at operation and by intubation techniques, confirmed their findings for adults that the upper small intestine in health was usually sterile or contained very few organisms. There were more organisms in the ileum and *Escherichia coli* was sometimes cultured from there but not from higher up the small intestine. Although careful anerobic cultures were made, the exact methods of Drasar (1967) were not used. However, more recent studies of Challacombe *et al.* (1974 *a*), using these anaerobic cultures were made, the exact methods of Drasar (1967) were not used. small intestine of healthy infants. These latter authors studied their patients by an intubation technique introduced by Rhea & Kilby (1970), using a long polythene feeding catheter (Argyl 5FG, 91cm in length) to the distal end of which a gold bead was attached so that the tip could be weight directed. Challacombe *et al.* agree substantially with the earlier findings of Anderson & Langford (1958) although they assessed qualitative growth numerically. In 13 infants without gastroenterological symptoms or signs, 7 had sterile duodenal contents, the remaining 6 growing organisms as shown in Table 1. By contrast, the gastric juice was sterile in only 1 infant, 15 types of organism being found in the others, and in 6, *E. coli* was present but failed to appear in the subsequent

Table 1

Bacterial flora in duodenal juice of 3 groups of children

	Control patients	Patients with	
		Chronic diarrhoea	Post-intestinal surgery
No. of patients	13	7	7
Samples sterile	7	0	1
Types of organisms	8	14	14
Mean total bacterial count ($\log_{10}$/ml)	6	7.3	6.4
Coliforms ($\log_{10}$/ml)	0	5	6

duodenal specimens. The organisms isolated from the duodenum in the 6 infants whose juice was not sterile were also present in the nose, throat and gastric juice of each individual patient, suggesting contamination from higher levels, as has been shown to occur following meals and could well occur during intubation. However, it is notable that *E. coli* disappeared, suggesting that there may be some coliform inhibiting mechanism in the normal upper small intestine.

Challacombe *et al.* (1974 *a*) consider it difficult to assign an upper limit to the number of bacteria which can be present in the duodenum before the count is considered abnormal. Kalser *et al.* (1966) suggested 10^3/ml but some infants studied by Challacombe *et al.* (1974 *a, b*) had somewhat higher counts. Obviously the limit fixed must depend on techniques of obtaining specimens. This was conclusively demonstrated by the latter workers who showed a progressive rise in total bacterial counts the longer the tube was left *in situ* in the duodenum and also a qualitative change, the number of *E. coli* increasing markedly. During earlier studies, Anderson & Langford (pers. comm.) had also observed this phenomenon, pointing to the difficulty of interpretation of studies by different authors when exact details of procedures and their timing may not be given.

In summary, the upper small intestine of the normal child, as in the adult, in a westernized community is usually sterile, or contains a sparse Gram positive flora, rarely $> 10^4$. There is almost certainly a mechanism by which the growth of Gram negative organisms, such as *E. coli* and other 'faecal type' micro-organisms, is prevented until the ileum is reached. Whether this mechanism has a mechanical basis depending on normal intestinal motility patterns; whether secreted immune globulins or the microbe-inhibiting properties of bile play a part; or whether pH and oxygen reduction potentials are important, is not yet clear.

3. Alterations in Small-Intestinal Microflora

Overgrowth of organisms in the small intestine can occur readily as indicated by the changes referred to from leaving an intubation tube *in situ* for a number of hours (Challacombe *et al.*, 1974 *a*). Obvious sources of contamination of the small intestine with a high concentration of organisms occur predominantly in adults, such as from an entero-colic fistula, or an infected biliary system draining into the duodenum (Scott & Khan, 1968), or in pernicious anaemia when poor acid secretion allows a greater number of organisms to reach the small intestine, or from stagnation of bowel contents resulting from surgical procedures causing blind loops, strictures or disturbances in motility.

Accompanying acute gastroenteritis in adults (Cohen *et al.*, 1967; Gorbach *et al.*, 1971) and in children (Thompson, 1955), a general increase in small-intestinal flora has been observed, the latter author in particular drawing attention to the presence of enteropathogenic *E. coli* in the duodenum as well as in the stools during such an infection.

Studies of alterations in the bacterial flora of the upper small intestine in children in association with gastrointestinal disorders are few but Anderson & Langford (1958) found little difference from normal in infants with coeliac disease or cystic fibrosis and concluded from this that malabsorbed food products in the small intestine were not of themselves responsible for encouraging bacterial overgrowth. However, in pursuing the concept that normal intestinal motility was important in maintaining relative sterility in the small intestine, they studied children with intestinal obstruction leading to intestinal stasis (Bishop & Anderson, 1960) and their findings were of some interest, in relation to the source of colonization of the small intestine. In a study of 12 newborn and 13 older children, a profuse flora of faecal type could be demonstrated above the point of obstruction when this was below the duodeno-jejunal flexure. The small intestine below the site of *complete* obstruction was sterile in infants born with this condition, i.e. small-intestinal atresia, indicating the source of infection above an obstruction to be ingestion and not retrograde spread from the colon. With obstruction proximal to the duodeno-jejunal flexure, the small intestine and stomach tended to remain bacteriologically normal.

An abnormally profuse but non-specific bacterial microflora has been demonstrated in the upper small intestine of infants with protracted diarrhoea and carbohydrate intolerance, especially when this was complete, affecting all sugars even monosaccharides (Gracey, Burke & Anderson, 1969; Lifshitz, Coello-Ramirez & Gutierres-Topete, 1970; Coello-Ramirez, Lifshitz & Zuniga, 1972). Burke & Anderson (1966) showed bacterial contamination of the small intestine following neonatal gastrointestinal surgery. In not all of these studies were anaerobic organisms studied but coliforms were constantly present in the upper small intestine.

Challacombe *et al.* (1974 *b*) have recently studied 2 abnormal groups; first, 7 infants with chronic diarrhoea (i.e. $>$ 4 loose fluid-containing stools/day for $>$ 2 weeks), a heterogeneous group, all being considered to have chronic non-specific gastroenteritis but 2 showing some evidence of monosaccharide intolerance and 1 secondary lactose intolerance; secondly, 7 infants who were failing to thrive and had persistent diarrhoea following partial resections of the large or small intestine. Compared with their previously mentioned study of infants without gastrointestinal disease (Challacombe *et al.*, 1974 *a*), there was an increase in the mean of the total micro-organism count and an increased number of types of organisms in the 2 groups of diarrhoeal patients (Table 2). *Escherichia coli* was frequently isolated from the duodenum of diarrhoeal patients but not from those of the control group. In 4 patients whose faeces were studied, the particular *E. coli* serotypes isolated from mouth, stomach and

Table 2

Range of organisms isolated from stomach and duodenal contents of 13 control children

Range of organisms	Frequency of isolation	
	Stomach (15 types)	Duodenum (8 types)
Coagulase positive staphylococci	4	1
Coagulase negative staphylococci	12	5
Micrococci	2	0
Streptococcus viridans	10	2
Non-haemolytic streptococci	1	0
Enterococci	3	0
Pneumococci	2	0
Lactobacilli	1	0
Neisseria spp.	4	2
Diphtheroids	2	1
Haemophilus spp.	1	2
Proteus spp.	2	0
E. coli	5	0
Klebsiella spp.	3	1
Veillonella spp.	1	1

Gastric juice sterile in 1 sample; duodenal juice sterile in 7 samples.

duodenum were also found in the faeces, confirming colonization of the whole gut in such abnormal states.

4. Relationship of *E. coli* in the Upper Small Intestine to Diarrhoeal Symptoms

The significance of the relationship of the findings of Challacombe *et al.* (1974 *b*) of increased bacterial numbers and types, including *E. coli*, in the small intestine in infants with non-specific persistent diarrhoea, to the abnormal and fluid nature of the stools is not yet clear. However, what is clear is that the floral relationships of the small intestine are relatively easily disturbed and more attention must be directed to determining whether the presence of so-called non-enteropathogenic serotypes of *E. coli* or other Gram negative organisms in the duodenum and jejunum has any deleterious effect on the metabolic function of the small-intestinal mucosa which might result in persistent diarrhoea, or malabsorption. Coello-Ramirez *et al.* (1972) discuss the consequent metabolic possibilities and offer the suggestion that so-called non-enteropathogenic *E. coli* may in fact be excreting an enterotoxin which has an effect on the mucosa.

It may be necessary for us to take a new look at the method of typing *E. coli* strains for enteropathogenicity, and it well may be that some strains not considered pathogenic may have the ability to colonize the upper gut, and produce enterotoxin. Whether they will do this in the healthy gut is another question. The possibility that the primary illness in patients with persistent diarrhoea and proliferation of *E. coli* in the upper gut has another cause, e.g. viral infection, alteration of intestinal motility, or immunological reaction, remains to be elucidated. The effect of *E. coli* toxin on the upper small gut mucosa is to alter intestinal fluid and electrolyte movements thus stimulating diarrhoea.

5. References

ANDERSON, C. M. & LANGFORD, R. F. (1958). Bacterial content of small intestine of children in health, in coeliac disease and in fibrocystic disease of the pancreas. *Brit. med. J.* **1**, 803.
BARBERO, G. J., RUNGE, G., FISCHER, D., CRAWFORD, M. N., TORRES, F. E. & GYORGY, P. (1952). Investigation of the bacterial flora, pH and sugar content in the intestinal tract of infants. *J. Pediat.* **40**, 152.
BISHOP, R. F. & ANDERSON, C. M. (1960). The bacterial flora of the stomach and small intestine in children with intestinal obstruction. *Arch. Dis. Childh.* **35**, 487.
BURKE, V. & ANDERSON, C. M. (1966). Sugar intolerance as a cause of protracted diarrhoea following surgery of the gastrointestinal tract in neonates. *Aust. Paediat. J.* **2**, 219.
CHALLACOMBE, D. N., RICHARDSON, J. M. & ANDERSON, C. M. (1974 *a*). The bacterial microflora of the upper gastrointestinal tract in infancy. 1: Infants without diarrhoea. *Archs. Dis. Childh.* (in the press).
CHALLACOMBE, D. N., RICHARDSON, J. M., ROWE, B. & ANDERSON, C. M. (1974 *b*). The bacterial microflora of the upper gastrointestinal tract in infancy. II: Patients with protracted diarrhoea. *Archs. Dis. Childh.* (in the press).
COELLO-RAMIREZ, P., LIFSHITZ, F. & ZUNIGA, V. (1972). Enteric microflora and carbohydrate intolerance in infants with diarrhoea. *Paediatrics* **49**, 233.
COHEN, R., KALSER, M. H., ARTEAGA, I., YAWN, E., FRAZIER, D., LEITE, C. A., AHEARN, D. G. & ROTH, F. (1967). Microbial intestinal flora in acute diarrhoea disease. *J. Amer. Med. Ass.* **201**, 835.
CREGAN, J. & HAYWARD, N. J. (1953). The bacterial content of the healthy human small intestine. *Brit. med. J.* **1**, 1356.
DAVISON, W. C. (1925). The duodenal contents of infants in health and disease and following diarrhoea. *Am. J. Dis. Child.* **29**, 743.
DONALDSON, R. M. Jr. (1964). Normal bacterial populations of the intestine and their relation to intestinal function. *New Eng. J. Med.* **270**, 938, 994, 1050.
DONALDSON, R. M. Jr. (1970). Small bowel bacterial overgrowth. *Adv. internal Med.* **16**, 191.
DRASAR, B. S. (1967). Cultivation of anaerobic intestinal bacteria. *J. Path. Bact.* **94**, 417.
FLOCH, M. H., GORBACH, S. L. & LUCKEY, T. D. (1970). Symposium. The intestinal microflora. *Amer. J. Clin. Nutr.* **23**, 1425, 1545.
GORBACH, S. L. (1971). Intestinal microflora. *Gastroenterology* **60**, 1110.
GORBACH, S. L., BANWELL, J. G., CHATTERGEE, B. D., JACOBS, B. & SACK, R. B. (1971). Acute undifferentiated human diarrhoea in the tropics. 1: Alterations in intestinal microflora. *J. clin. Invest.* **50** , 881.
GRACEY, M., BURKE, V. & ANDERSON, C. M. (1969). Association of monosaccharide malabsorption with abnormal small-intestinal flora. *Lancet* **2**, 384.
KALSER, M. H., COHEN, R., ARTEAGA, I., YAWN, E., MAYORAL, L., HOFFERT, W. R. & FRAZIER, D. (1966). Normal viral and bacterial flora of the human small and large intestine. *New Engl. J. Med.* **274**, 500.

LIFSHITZ, F., COELLO-RAMIREZ, P. & GUTIERRES-TOPETE, G. (1970). Monosaccharide intolerance and hypoglycemia in infants with diarrhoea. 1: Clinical course of 23 infants. *J. Pediat.* **77,** 595.

RHEA, J. W. & KILBY, J. O. (1970). A nasojejunal tube for infant feeding. *Paediatrics* **46,** 36.

SCOTT, A. J. & KHAN, G. A. (1968). Partial biliary obstruction with cholangitis producing a blind loop syndrome. *Gut,* **9,** 187.

THOMPSON, S. (1955). The role of certain varieties of *Bacterium coli* in gastroenteritis of babies. *J. Hyg., Camb.* **53,** 357.

WILSON, G. S. & MILES, A. A. (1955). In *Principles of Bacteriology and Immunology,* (Topley & Wilson, eds), 4th ed. London: Edward Arnold.

Clostridium perfringens (Cl. welchii) in the Human Gastro-intestinal Tract

J. G. COLLEE

University Medical School, Edinburgh EH8 9AG, Scotland

CONTENTS

1. Introduction 205
2. Occurrence in human faeces 206
3. Distribution in the human gut 207
4. The ingestion of *Cl. perfringens* 208
5. Growth control mechanisms in the gastro-intestinal tract 208
6. Host protective factors 209
7. Endogenous pathogenic potential in the gut 209
8. Skin contamination with faecal clostridia 210
9. Specific intestinal infections 210
10. *Clostridium perfringens* food poisoning 211
11. The borderland of cases and carriers 212
12. Toxigenicity and sporulation 214
13. Challenges 215
14. Summary 215
15. Acknowledgement 216
16. References 216

1. Introduction

THE WIDESPREAD OCCURRENCE of *Clostridium perfringens (Cl. welchii)* in nature has been ably reviewed by Willis (1969) who commented on the organism's normal presence in soil and sewage and in the human and animal intestinal tract where its vegetative and spore forms are abundant. The primary natural habitat is the gut of animals and man but the possibility that *Cl. perfringens* multiplies significantly under certain conditions in soil merits consideration. Willis (1969) summarized evidence on the distribution of the various types (A–E) of the organism and noted that type A strains are by far the most common under normal (non-infected) conditions. It is well recognized that haemolytic and non-haemolytic strains occurring within the type A group are described according to the effects of their colonies on horse-blood agar and that a proportion of non-haemolytic strains producing markedly heat-resistant spores have been typically but not exclusively associated with *Cl. perfringens* food poisoning (see below). Haemolytic type A strains – the so-called 'classical' strains – are more common and more numerous in the normal healthy human gut than are non-haemolytic strains, and markedly heat-resistant strains are usually in a minority.

2. Occurrence in Human Faeces

Clostridium perfringens is a most unlikely candidate for commensalism in the human gut and the importance of its commensal role is in doubt. The wide range of toxins and aggressins that it produces includes some with necrotizing, cytolytic or lethal properties and, under suitable conditions *in vitro* and *in vivo*, it can outgrow many other bacteria and destroy various tissues. Nevertheless, organisms with this great potential for harm can often be recovered from the faeces of healthy human subjects in moderately large numbers of c. 10^4/g wet weight. Of the clostridia that may be isolated from the human gut from time to time, *Cl. perfringens* is the commonest and the most numerous; however, its occurrence is recorded quite irregularly. This partly reflects variations in procedures used by different workers and the different primary aims of the various investigations. For example, the selection of dilutions of the material to be plated and the practical range of selective media chosen for a particular study markedly influence the likely range of isolates. If *Cl. perfringens* is recognized primarily on the basis of its colony morphology and haemolytic effect on horse-blood agar, non-haemolytic strains may be missed, whereas their phospholipase activity or their sulphite-reducing activity might declare their presence on other suitable media.

Despite these considerations, it is clear that there is a true and marked host-to-host variation in the numbers of *Cl. perfringens* in the human gut and that there are wide fluctuations in the numbers excreted in the faeces of any individual subject.

Smith & Crabb (1961) reported faecal counts of *Cl. perfringens* ranging from 0–$10^{5.9}$/g for 10 normal adults, with a median count of $10^{3.2}$/g. In studies with babies, Smith & Crabb (1961) showed that the numbers of *Cl. perfringens* in faeces fluctuated widely. Two of 5 babies excreted large numbers of *Cl. perfringens* within a few days of birth but the numbers declined rapidly and high counts did not persist beyond the first week. In one of the 5 babies, *Cl. perfringens* was again excreted in considerable numbers at 28 weeks, when the baby was eating an appreciable amount of solid food. Smith & Crabb (1961) noted that, on occasion and inexplicably, a sudden very transient increase in the excretion of *Cl. perfringens* occurred. Floch, Gershengoren & Freedman (1968) recovered clostridia from faeces of 3 of 8 healthy subjects in counts ranging from $10^{6.3}$–$10^{8.7}$. Mata, Carrillo & Villatoro (1969) investigated the faecal flora of 19 breast-fed children, 12 weanlings and 12 adults. Seven of the 12 samples from adults yielded clostridia but the technical procedures used would fail to recover organisms that were not present in significant numbers in a million-fold dilution. Only 3 or 4 of the 31 children had detectable numbers of clostridia in their faeces, but when clostridia were found they were present in large numbers. Cooke (1967) found $10^{5.8}$–$10^{6.8}$ of unspecified clostridia/g in the faeces of 3 of

20 normal persons and clostridia were not demonstrable (<1 in 0.003g) in faeces sampled from the others. Hill et al. (1971) and Drasar & Crowther (pers. comm.) reported a mean count of $10^{4.2 \pm 1.8}$ for phospholipase-positive clostridia and $10^{5.7 \pm 1.1}$ for phospholipase-negative clostridia/g in faecal samples from 68 English subjects and similar counts were obtained with samples from 16 Scottish subjects.

Although there are some parallels in observations made on the occurrence of Cl. perfringens in the alimentary tract of man and animals, it is important to consider different hosts separately. Smith & Crabb noted that some animal species might be identified on the basis of the distinctive composition of their faecal flora (Smith, 1965) and this has important implications in relation to the choice of the research model for a specific investigation; the present paper deals specifically with observations on the human gut flora.

Despite the variable reports obtained in the quantitative studies referred to above, Cl. perfringens has long been regarded authoritatively as an index of faecal contamination (Willis, 1956) and some of its characters make it better than Escherichia coli for this purpose. The statements that type A strains of Cl. perfringens are constantly and uniformly present in human faeces (Smith, 1955; Wilson & Miles, 1955) were confirmed – with due allowance for the ambiguities of constancy and uniformity – by the work of Collee, Knowlden & Hobbs (1961) who also demonstrated the frequent occurrence of non-haemolytic strains in faeces. These studies were greatly facilitated by the selective solid media of Lowbury & Lilly (1955) and of Willis & Hobbs (1959).

3. Distribution in the Human Gut

Clostridium perfringens occurs in significant numbers in the large gut but not in the normal stomach or small gut of man. For example, Vince et al. (1972) did not isolate clostridia from the fasting gastric juice of any of 20 normal control subjects, or from the mid-small intestine of any of 11 normal people. Clostridia were isolated from the contents of the terminal ileum of 2 of 12 control subjects at counts in the range of $10^4 - 10^5$/ml. There are many other studies that support these observations in fasting subjects. The general occurrence of Cl. perfringens in the large gut is adduced from studies that predominantly relate to faecal specimens. In addition to the evidence given earlier in this paper, the findings of Sutton (1966 a) confirm and extend currently accepted views. Sutton's investigation of 250 faecal specimens from 100 primary schoolchildren, 50 patients with minor illnesses and 100 aboriginal persons gave median faecal counts of $(3.5-4.5) \times 10^4$ Cl. perfringens/g. The range of excretion studied in one person over a 3 week period was $1.5 \times 10^3 - 2.05 \times 10^5$.

4. The Ingestion of *Cl. perfringens*

Sporing forms of *Cl. perfringens* of varying heat resistance and marked chemical resistance regularly contaminate our cooked and uncooked foods (Nakamura & Schulze, 1970) and, being frequently ingested, regularly traverse the gastro-intestinal tract. Hobbs *et al.* (1953) isolated heat-resistant strains from various raw meats, and Sylvester & Green (1961) investigated the effects of cooking infected meat. McKillop (1959) isolated *Cl perfringens* from 72% of uncooked foods received at a hospital kitchen; 90% of samples of dust were positive and 24 of 46 samples of cooked chicken were positive for *Cl. perfringens*. Willis (1957) recovered haemolytic and non-haemolytic strains from a city water supply. Strong, Canada & Griffiths (1963) examined various American foods for *Cl. perfringens;* the highest recoveries were from raw meat, poultry and fish with counts that ranged from 10–1180 cells/g. These authors cite the evidence of many other workers who have isolated *Cl. perfringens* from foods.

Accordingly, we may ingest considerable numbers of *Cl. perfringens* in our diet from time to time, but viable counts for cooked food or for foods eaten raw at table are still required with differential counts for vegetative cells and spores before we can make a quantitative assessment of our daily intake of clostridia.

5. Growth Control Mechanisms in the Gastro-intestinal Tract

If *Cl. perfringens* is not generally allowed to multiply in the gut, it must either be ingested in very large numbers from time to time in the diet or it must occasionally escape from our 'normal' control to produce the high faecal counts that are regularly reported. An alternative but less attractive argument is that the organism regularly multiplies in the gut and that its numbers are subsequently reduced to a variable extent by various factors that operate *in vivo*.

The factors that control the numbers of *Cl. perfringens* in the gut and hold the production or actions of its toxins in check are not understood. In experiments in which *Cl. perfringens* was grown in fluid cultures with various *Bacteroides* spp., the growth rates of each of these species *in vitro* did not appear to be suppressed by the other organism (Collee & Slater, unpublished results). It is doubtful whether such findings bear any relationship to a complex intestinal ecology in which bacterial division rates are much slower and complex factors operate (Gorbach, 1971). The acidity of some sections of the gastrointestinal tract is said to inhibit bacterial growth, but it is clear that more complex antimicrobial systems are involved. Bile acids secreted in the bile in the form of conjugates can be deconjugated by anaerobic bacteria including *Bacteroides* species and *Cl. perfringens*. Percy-Robb & Collee (1972) showed that unconjugated bile acids have marked bactericidal and bacteriostatic effects that depend on pH and operate effectively against *Bacteroides* species and *Cl. perfringens*. It is proposed that unconjugated bile acids may be involved in a homeostatic mechanism that inhibits the growth of certain bacteria including *Cl. perfringens* in the small gut. The known association between the occurrence of *Cl. perfringens* type A in the gut of chickens and the susceptibility of these birds to the growth-promoting effect of penicillin (Lev & Forbes, 1959) suggests that the organism somehow suppresses the growth rate of some natural hosts. Studies with germ-free animals and various intestinal bacteria support this concept which has been recently reconsidered by Donaldson (1970) in relation to malabsorption and bacterial overgrowth in the small bowel of man. Anaerobic bacteria are certainly associated with such clinical conditions, but *Cl. perfringens* is not now specifically associated with any recognized chronic malabsorption syndrome in man.

6. Host Protective Factors

Factors that protect host epithelial cells from clostridial attack may include peristaltic activity and secretion of mucus. If toxins are produced in significant amount in the gut, there are various ways in which they may be prevented from operating against host tissues. The observations of Howie, Duncan & Mackie (1953) on the growth of *Cl. perfringens* in the stomach remnant of patients after partial gastrectomy are of interest. These investigators found that 12 of 15 patients examined during the first week after partial gastrectomy showed numerous *Cl. perfringens* in films and cultures of the contents of the stomach remnant. It may be that the post-operative management of these patients combined with the surgical interference to upset the normal control mechanisms, including gastric acid secretion, that inhibit the growth of *Cl. perfringens* in the stomach. These findings led to an investigation of neutralizing substances occurring in the human gut and found to inactivate the α-toxin (phospholipase) of *Cl. perfringens* (Goudie & Duncan, 1956). Factors considered included local antitoxin, lipoprotein complexes and proteolytic enzymes (Goudie, 1959).

7. Endogenous Pathogenic Potential in the Gut

Apart from its debated commensal role, *Cl. perfringens* may be associated with various pathological conditions in the human gut. The organism is sometimes (but not commonly) involved in infections of the biliary tract, or in appendicitis, or in gas gangrene occurring as a complication of mesenteric thrombosis, intestinal strangulation, or accidental or operative abdominal injury. In the past, similar features observed in patients with shock associated with abdominal emergencies such as intestinal obstruction and shock associated with gas gangrene encouraged the view that *Cl. perfringens* was partly responsible for both of these situations. Much of the research was done with dogs and it may be unwise to draw parallels too closely to clinical observations in man. The results of the careful work of Nagy & Weipers (1968) have now shown, even in studies with dogs, that no clear relationship exists between death in experimental intestinal obstruction and the occurrence of any single bacterial species or specific bacterial group. The lethal effects seem to be associated with the invasion of the peritoneal cavity by multiplying bacteria and not directly with a clostridial toxin. This is very much in line with the views of Bullen and his colleagues (Bullen, 1970) concerning a possible mechanism of pathogenicity of *Cl. perfringens*, but a probably different concept is extended by Nagy & Weipers (1968) to a wide range of organisms that may be associated with peritonitis.

The pre-operative administration of antibiotics to patients destined for bowel surgery is hotly debated by those who maintain on the one hand that 'bowel sterilization' is impossible and, on the other, by those that seek at least to lessen the load of likely pathogens in the planned operative area. Neomycin is a common constituent of the antibacterial preparations used for this purpose. The potential pathogens currently regarded as most likely to cause trouble as a result of spillage of bowel contents at operation are various cocci and coliform organisms. *Clostridium perfringens* is involved infrequently and this is fortunate because it is resistant to neomycin. It should be noted that *Bacteroides* spp. are also classically resistant to the aminoglycoside antibiotics and the common occurrence of post-operative bacteroides infections merits more attention than clostridial infection in this regard.

8. Skin Contamination with Faecal Clostridia

Skin contamination with faecal strains of *Cl. perfringens* is of particular concern to the surgeon. Spores can be recovered readily from the skin of healthy ambulant subjects (Collee & Watt, 1971), and Ayliffe & Lowbury (1969) recovered considerable numbers of *Cl. perfringens* from the skin of the buttocks and thighs of hospital patients. The marked resistance of *Cl. perfringens* spores to chemical antiseptics that can be used on the skin (Sykes, 1970) allows these organisms to survive normal pre-operative skin-cleaning procedures. Operations on the hip or thigh of middle-aged or elderly patients with a degree of vascular impairment are therefore associated with a recognized risk of clostridial contamination; subsequent germination of *Cl. perfringens* spores in the devitalized tissues of such patients occasionally occurs and post-operative gas gangrene results (Parker, 1969). This serious complication can be avoided by the pre-operative administration of an antibiotic to these patients. Penicillin is the antibiotic of choice.

9. Specific Intestinal Infections

Clostridium perfringens may be associated specifically, on occasion with necrotizing infections of the jejunum or colon.

Enteritis necroticans is a severe necrotizing jejunitis first reported by Zeissler & Rassfeld-Sternberg (1949) in Germany and now classically associated with the studies of Murrell and his colleagues (Murrell *et al.*, 1966) on 'pig-bel', a similar disease affecting natives of New Guinea after they have over-indulged in a traditional feast of pork. Indirect evidence incriminates *Cl. perfringens* type C with this syndrome. This subject is reviewed comprehensively and succinctly by Willis (1969).

Type C strains of *Cl. perfringens* cause a well-recognized form of enteritis in piglets, and Arbuckle (1972) recently observed that these bacteria attach to the intestinal epithelium of the area of gut involved. Jones & Rutter (1972) drew analogies between that observation and the known adhesion of strains of pathogenic *Escherichia coli* possessing the K88 antigen to the mucosa of the small intestine in piglets.

The possibility that the haemagglutinin of *Cl. perfringens* might represent a mechanism of adhesion for enteropathogenic type A strains was explored by Collee (1961, 1965) who found that freshly isolated strains did not produce this factor and that the *Cl. perfringens* haemagglutinin was not related to the fimbriae of various coliform organisms. The systems considered by Arbuckle (1972) and by Jones & Rutter (1972) are quite distinct from structural or soluble haemagglutinins. It may be worth while to ask again whether *Cl. perfringens* can adhere to some site in the human gut when it multiplies and causes trouble.

10. *Clostridium perfringens* Food Poisoning

The recognition of the *Cl. perfringens* food-poisoning syndrome in the decade that followed World War II and the considerable incidence of this condition in Britain during the 1960's and 1970's illustrates the organism's ability to take advantage of our changing habits in bulk food processing and communal eating while our standards of hygiene have not been adequately upgraded. Our raw meats are faecally contaminated and may contain spores of typical food-poisoning strains of *Cl. perfringens*. These strains occur widely and have been demonstrated by Hobbs *et al.* (1953) in market samples of meat, in human and animal faeces, in dust and on insects.

When meat is cooked in bulk, heat gain is slow (Sylvester & Green, 1961) and subsequent cooling may be slow. The heat drives off dissolved oxygen and the meat maintains the anaerobic environment which falls slowly through a temperature range that allows rapid bacterial multiplication (Norval & Collee, 1957; Barnes, Despaul & Ingram, 1963). The heat-shocked spores are activated and germinate readily under these conditions (Roberts, 1968). *Clostridium perfringens* can multiply rapidly at temperatures in the 37–45° range (Collee, Knowlden & Hobbs, 1961). A cooked meat broth culture is therefore ingested by the unsuspecting victim and, as a bulk-cooked food is typically involved, many people may be at risk. It appears that *Cl. perfringens* passes the acid barrier of the stomach in such a meal and begins to sporulate at some point in the gut. An enterotoxic factor is elaborated during this process and the enterotoxin is liberated when sporulation is complete and lysis releases the free spore. Our knowledge of the sequence of events (Fig. 1) owes much to the studies of Dische & Elek (1957), Hauschild, Hilschimer & Thatcher (1967), Hauschild & Thatcher (1968), Duncan & Strong (1969, 1971), Hauschild, Niilo & Dorward (1971), Niilo (1971), Stark & Duncan (1971), Strong, Duncan & Perna (1971), Duncan, Strong & Sebald (1972) and Duncan (1973).

The characteristic syndrome of abdominal cramping pain and diarrhoea usually occurs c. 9–14 h after ingestion of the infected food, and symptoms usually subside within a day or so. Vomiting and pyrexia are typically absent. The effective challenge dose seems to be of the order of 10^8-10^9 organisms and the numbers of clostridia in the faeces of those involved in an outbreak range from 10^5-10^8/g with many spores present. A transient carrier state ensues for some weeks at least. In some outbreaks it is clear that only a relatively small proportion of those at risk seek medical help and it is likely that there is a considerable variation in the size of the challenge doses consumed and in the degree of host susceptibility. The significant factor that precipitates the *Cl. perfringens* food poisoning syndrome in man seems to be ingestion of an adequately high challenge dose of viable organisms. The nature of the food in which the dose is administered may be important; meat foods are predominantly involved, but milk dishes and leguminous seeds have also been associated with outbreaks.

Although the enteropathogenic potential of *Cl. perfringens* has been associated with a typical food poisoning subgroup of type A strains with characteristically heat-resistant spores, it is now clear that classical type A strains and haemolytic or non-haemolytic strains with spores of low or intermediate heat resistance can produce an enterotoxic effect under suitable conditions (Hall, *et al.*, 1963; Taylor & Coetzee, 1966; Sutton & Hobbs, 1968). This confirms the early observation of McKillop (1959). Moreover, the experiments of Hauschild *et al.* (1967) and Hauschild & Thatcher (1967) confirmed the enterotoxicity of cultures of 'heat sensitive' strains. Sutton & Hobbs (1968) drew attention to our tendency to regard the spores of classical strains as heat sensitive whereas they are only relatively so; these spores are certainly less markedly heat resistant than the spores of typical food poisoning strains, but they can survive some of our cooking procedures.

11. The Borderland of Cases and Carriers

Sutton (1966*b*) distinguished quantitatively between recent cases and carriers of food poisoning strains of *Cl. perfringens* and he related high carriage rates (15–25%) to communal feeding and to poor hygiene. It is disturbing that Dische

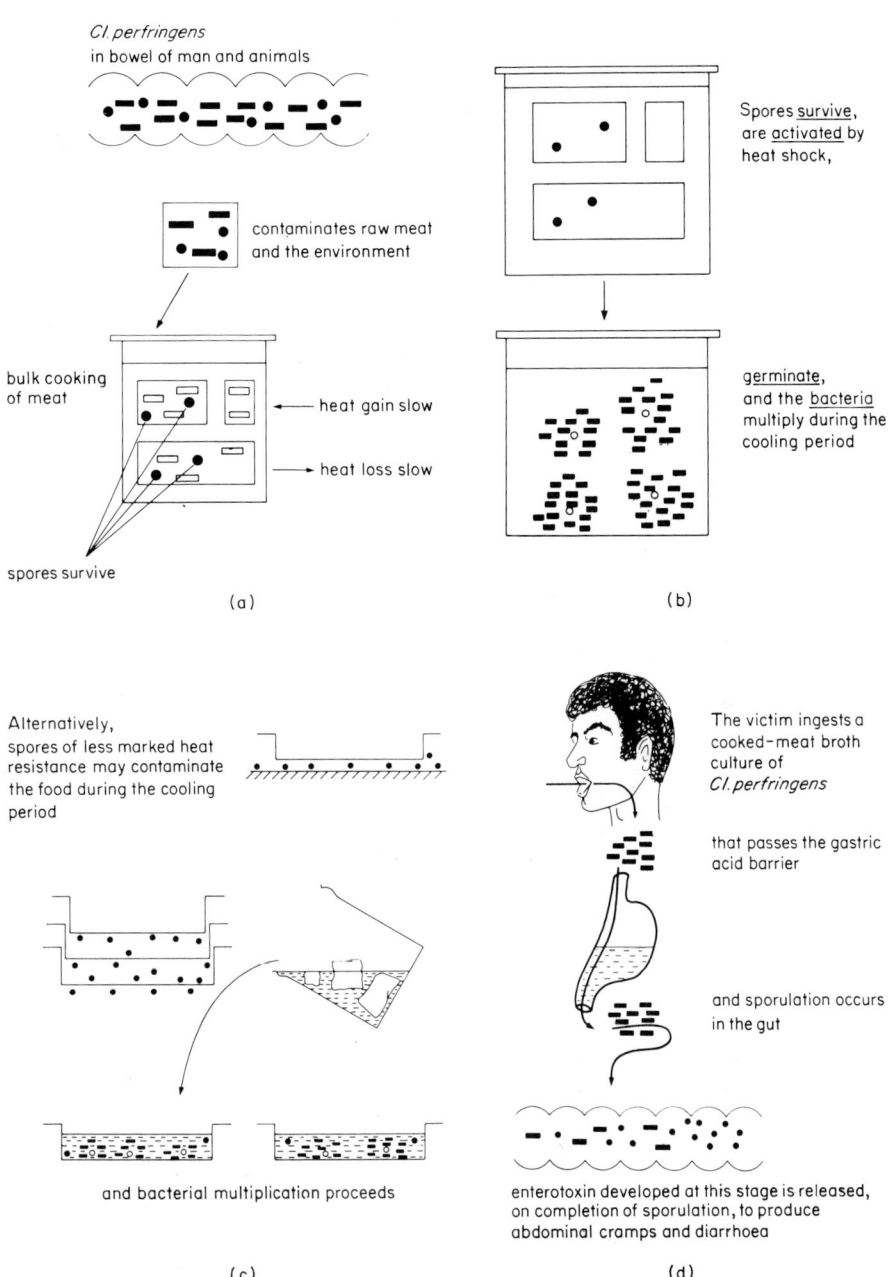

Fig. 1. Sequence of events that may lead to *Clostridium perfringens* food poisoning. (a) Contamination of meat that is then cooked in bulk. (b) Survival of heat-resistant spores, activation, germination and growth. (c) Alternative mode of contamination of outside of dish from working surface or floor. Stacking of dishes contaminates inside surfaces and the inoculum may grow in cooked food subsequently distributed into these dishes. (d) Ingested organisms sporulate at some point in the gut and release enterotoxin.

& Elek (1957) and Leeming, Pryce & Meynell (1961) found high carriage rates among hospital personnel and their families and in-patients. Sutton extended the high-carrier list to include boarding-school students and aboriginal natives, and Turner & Wong (1961), by choosing to study Chinese hospital patients from large family groups in Hong Kong, demonstrated the highest carrier rate (63%) recorded in the absence of a recognized outbreak of overt illness. These studies have contributed significantly to our knowledge but, as we now know that classical strains may also cause food poisoning, the extensive carriage of so-called typical food poisoning strains in a population group is only one index of an unsatisfactory situation that appears to be world-wide. Sutton & Hobbs (1968) noted that multiplication of strains of *Cl. perfringens* already present in the gut has apparently occurred simultaneously with the multiplication of a food poisoning strain in some patients. This important observation merits further study, but our procedures presently available for the identification of individual strains are inadequate.

Dubos *et al.* (1965) pointed out that the populations of some recognized commensal species in the gut increase rapidly after birth and, having become very numerous, thereafter fall sharply in number whereas other species persist at high and relatively constant numbers. These workers argue that a species possessing some degree of infectivity elicits a protective response that then eliminates the potential pathogen or at least limits its numbers under normal conditions. The somatic heterogeneity of *Cl. perfringens* strains would pose problems for such a system if a specific immunogenic mechanism is primarily involved, but it may be that the regular dietary ingestion of strains and the associated adjuvant effect of endotoxin from coliform organisms in the gut keeps the system primed (Thomas, McSween & White, 1973) against this versatile pathogen. The fact that the very high challenge doses involved in food poisoning appear to produce only a transient infection suggests that an efficient clearing mechanism is involved.

12. Toxigenicity and Sporulation

Yamagishi, Ishida & Nishida (1964) investigated strains of *Cl. perfringens* derived from soil and demonstrated an inverse relationship between the heat resistance of the spores and the toxigenicity of the vegetative forms of individual strains. These workers assayed their strains for α, θ and μ toxins, and Weiss & Strong (1967) confirmed their findings in relation to the production of α toxin. Strains that contaminate food may well be derived from soil and 'typical food poisoning strains' are certainly relatively weakly toxigenic and markedly heat resistant. However, they produce enterotoxin and, as this is directly related to sporulation (Duncan *et al.*, 1972) it follows that the inverse relationship of 'classical' toxigenicity cannot embrace enterotoxigenicity. Nor can it embrace the production of all exoenzymes. Two food poisoning strains investigated by Collee (1965) produced no neuraminidase, but the careful work of Moss, Schekter & Cherry (1967) demonstrated that neuraminidase could be produced by some typical food poisoning and classical strains. Recent work by Fraser & Collee (unpublished) confirms that although some British reference strains of typical food poisoning *Cl. perfringens* are neuraminidase negative, at least one is strongly neuraminidase positive and its spores are markedly heat resistant.

13. Challenges

Clostridium perfringens provides a model for important qualitative and quantitative bacteriological studies. Public Health concern with the organism primarily reflects our knowledge that it is an index of faecal contamination and that *Cl. perfringens* food poisoning is a direct indictment of the catering procedures with which it is associated. The regular occurrence of *Cl. perfringens* in the human gut and on faecally soiled skin is of surgical importance in relation to post-operative gas gangrene. The organism's regular occurrence as a *contaminant* in various exudates and its capacity for rapid growth in certain media can be most misleading in clinical bacteriology.

The present state of our knowledge of the enteropathogenicity of *Cl. perfringens* in man is intriguing and confusing, and the distinction between commensalism, transient colonization, and infection is not clear. The fundamental problems posed by the organism—the remarkable differences in degree of heat resistance of its spores, its apparent control of the growth of certain hosts, host control of the organism's commensal growth in man, the organism's capacity for extremely rapid growth and the different mechanisms of pathogenicity involved when it gets out of control in the gut or in devitalized tissue – remain largely unexplained. These are humbling observations and stimulating challenges.

14. Summary

Clostridium perfringens (Cl. welchii) forms a minor component of the faecal flora of healthy adults; type A strains occur quite often in numbers of $c.\ 10^4/g$ (wet weight) of faeces. The organism does not normally occur in significant numbers in the stomach or small intestine of man. Host factors that may control its colonization of the small gut, its restricted proliferation in the large gut and its aggressive potential in the gastro-intestinal tract are not well understood.

The enteropathogenic potential of *Cl perfringens* for man is clearly established and several syndromes are recognized. Enteritis necroticans and 'pig-bel' are relatively unusual and severe conditions associated with type C strains of *Cl. perfringens* in the human gut; these conditions resemble enterotoxaemic diseases produced by various types of *Cl. perfringens* in animals. Some cases of post-gastrectomy diarrhoea in man have been associated with abnormal proliferation of *Cl. perfringens* type A in the proximal bowel.

Clostridium perfringens sporulates effectively at some point in the intestine. Ingestion of a large dose of *Cl. perfringens* and sporulation of the organisms *in vivo* seem to be important factors in the production and release of the enterotoxic factor that is now regarded as the cause of the *Cl. perfringens* food poisoning syndrome. The enterotoxin seems not to be one of the already labelled toxins or aggressins of *Cl. perfringens*.

Clostridium perfringens is an index of faecal contamination. Its regular occurrence in the bowel and on faecally soiled skin is of surgical importance. In addition to the organism's unacceptably frequent occurrence as a specific cause of food poisoning, its frequent clinically undeclared presence in food merits general consideration in relation to current standards of hygiene.

15. Acknowledgement

The help of Dr Brian Watt in the preparation of this paper is gratefully acknowledged and I thank the Medical Research Council for a grant to support studies of anaerobes of medical importance.

16. References

ARBUCKLE, J. B. R. (1972). The attachment of *Clostridium welchii (Cl. perfringens)* type C to intestinal villi of pigs. *J. Path.* **106**, 65.

AYLIFFE, G. A. J. & LOWBURY, E. J. L. (1969). Sources of gas gangrene in hospital. *Brit. med. J.* **2**, 333.

BARNES, E. M., DESPAUL, J. E. & INGRAM, M. (1963). The behaviour of a food poisoning strain of *Clostridium welchii* in beef. *J. appl. Bact.* **26**, 415.

BULLEN, J. J. (1970). Role of toxins in host-parasite relationships. In *Microbial Toxins* Vol. 1, Eds S. J. Ajl, S. Kadis & T. C. Montie. New York & London: Academic Press.

COLLEE, J. G. (1961). The nature and properties of the haemagglutinin of *Clostridium welchii*. *J. Path. Bact.* **81**, 297.

COLLEE, J. G. (1965). The relationship of the haemagglutinin of *Clostridium welchii* to the neuraminidase and other soluble products of the organism. *J. Path. Bact.* **90**, 13.

COLLEE, J. G., KNOWLDEN, J. A. & HOBBS, B. C. (1961). Studies on the growth, sporulation and carriage of *Clostridium welchii* with special reference to food poisoning strains. *J. appl. Bact.* **24**, 326.

COLLEE, J. G. & WATT, B. (1971). Changing approaches to the sporing anaerobes in medical microbiology. In *Spore Research 1971*. Eds A. N. Barker, G. W. Gould, & J. Wolf. London & New York: Academic Press.

COOKE, E. M. (1967). A quantitative comparison of the faecal flora of patients with ulcerative colitis and that of normal persons. *J. Path. Bact.* **94**, 439.

DISCHE, F. E. & ELEK, S. D. (1957). Experimental food poisoning by *Clostridium welchii*. *Lancet* **2**, 71.

DONALDSON, R. M. (1970). Small bowel bacterial overgrowth. In *Advances in Internal Medicine*, **16**, 191.

DUBOS, R., SCHAEDLER, R. W., COSTELLO, R. & HOET, P. (1965). Indigenous, normal, and autochthonous flora of the gastrointestinal tract. *J. exp. Med.* **122**, 67.

DUNCAN, C. L. (1973). Time of enterotoxin formation and release during sporulation of *Clostridium perfringens* type A. *J. Bact.* **113**, 932.

DUNCAN, C. L. & STRONG, D. H. (1969). Ileal loop fluid accumulation and production of diarrhoea in rabbits by cell-free products of *Clostridium perfringens. J. Bact.* **100**, 86.

DUNCAN, C. L. & STRONG, D. H. (1971). *Clostridium perfringens* type A food poisoning. I. Response of the rabbit ileum as an indication of enteropathogenicity of strains of *Clostridium perfringens* in monkeys. *Infection and Immunity* **3**, 167.

DUNCAN, C. L., STRONG, D. H. & SEBALD, M. (1972). Sporulation and enterotoxin production by mutants of *Clostridium perfringens. J. Bact.* **110**, 378.

FLOCH, M. H., GERSHENGOREN, W. & FREEDMAN, L. R. (1968). Methods for the quantitative study of the aerobic and anaerobic intestinal bacterial flora of man. *Yale J. biol. Med.* **41**, 50.

GORBACH, S. L. (1971). Intestinal microflora. *Gastroenterology* **60**, 1110.

GOUDIE, J. G. (1959). The nature of a neutralizing substance for *Clostridium welchii* alpha-toxin in faeces. *J. Path. Bact.* **78**, 17.

GOUDIE, J. G. & DUNCAN, I. B. R. (1956). *Clostridium welchii* and neutralizing substances for *Clostridium welchii* alpha-toxin in faeces. *J. Path. Bact.* **72**, 381.

HALL, H. E., ANGELOTTI, R., LEWIS, K. H. & FOSTER, M. J. (1963). Characteristics of *Clostridium perfringens* strains associated with food and food-borne disease. *J. Bact.* **85**, 1094.

HAUSCHILD, A. H. W. & THATCHER, F. S. (1967). Experimental food poisoning with heat-susceptible *Clostridium perfringens*, type A. *J. Food Sci.* **32**, 467.

HAUSCHILD, A. H. W., NIILO, L. & DORWARD, W. J. (1967). Experimental enteritis with food poisoning and classical strains of *Clostridium perfringens* type A in lambs. *J. Infectious Dis.* **117**, 379.

HAUSCHILD, A. H. W., HILSCHIMER, R. & THATCHER, F. S. (1967). Acid resistance and infectivity of food-poisoning *Clostridium perfringens. Can. J. Microbiol.* **13**, 1041.

HAUSCHILD, A. H. W. & THATCHER, F. S. (1968). Experimental gas gangrene with food-poisoning *Clostridium perfringens* type A. *Can. J. Microbiol.* **14**, 705.

HAUSCHILD, A. H. W., NIILO, L. & DORWARD, W. J. (1971). The role of enterotoxin in *Clostridium perfringens* type A enteritis. *Can. J. Microbiol.* **17**, 987.

HILL, M. J., DRASAR, B. S., ARIES, V., CROWTHER, J. S., HAWKSWORTH, G. & WILLIAMS, R. E. O. (1971). Bacteria and aetiology of cancer of large bowel. *Lancet* **1**, 95.

HOBBS, B. C., SMITH, M. E., OAKLEY, C. L., WARRACK, G. H. & CRUICKSHANK, J. C. (1953). *Clostridium welchii* food poisoning. *J. Hyg., Camb.* **51**, 75.

HOWIE, J. W., DUNCAN, I. B. R. & MACKIE, L. M. (1953). The growth of *Clostridium welchii* in the stomach after partial gastrectomy. *Lancet* **2**, 1018.

JONES, G. W. & RUTTER, J. M. (1972). Role of the K88 antigen in the pathogenesis of neonatal diarrhoea caused by *Escherichia coli* in piglets. *Infection and Immunity* **6**, 918.

LEEMING, R. L., PRYCE, J. D. & MEYNELL, M. J. (1961). *Clostridium welchii* and food poisoning. *Brit. med. J.* **1**, 50.

LEV, M. & FORBES, M. (1959). Growth response to dietary penicillin of germ-free chicks and of chicks with a defined intestinal flora. *Brit. J. Nutrit.* **13**, 78.

LOWBURY, E. J. L. & LILLY, H. A. (1955). A selective plate medium for *Clostridium welchii. J. Path. Bact.* **70**, 105.

MATA, L. J., CARRILLO, C. & VILLATORO, E. (1969). Faecal microflora in healthy persons in a preindustrial region. *Appl. Microbiol.* **17**, 596.

McKILLOP, E. J. (1959). Bacterial contamination of hospital food with special reference to *Clostridium welchii* food poisoning. *J. Hyg., Camb.* **57**, 31.

MOSS, C. W., SCHEKTER, M. A. & CHERRY, W. B. (1967). Distribution of neuraminidase among food-poisoning strains of *Clostridium perfringens. Appl. Microbiol.* **15**, 718.

MURRELL, T. G. C., ROTH, L., EGERTON, J., SAMELS, J. & WALKER, P. D. (1966). Pig-bel: enteritis necroticans. A study in diagnosis and management. *Lancet* 1, 217.
NAGY, L. & WEIPERS, W. L. (1968). A study of bacteria and lethal factors in fluids from experimental intestinal obstruction in dogs. *J. Path. Bact.* 95, 199.
NAKAMURA, M. & SCHULZE, J. A. (1970). *Clostridium perfringens* food poisoning. *Ann. Rev. Microbiol.* 24, 359.
NIILO, L. (1971). Mechanism of action of the enteropathogenic factor of *Clostridium perfringens* type A. *Infection & Immunity* 3, 100.
NORVAL, J. & COLLEE, J. G. (1957). Observations on contamination of large steak pies with *Clostridium welchii*. *Scot. med. J.* 2, 427.
PARKER, M. T. (1969). Post-operative clostridial infections in Britain. *Brit. med. J.* 2, 671.
PERCY-ROBB, I. W. & COLLEE, J. G. (1972). Bile acids: a pH dependent antibacterial system in the gut? *Brit. med. J.* 2, 813.
ROBERTS, T. A. (1968). Heat and radiation resistance and activation of spores of *Clostridium welchii*. *J. appl. Bact.* 31, 133.
SMITH, H. W. (1965). Observations on the flora of the alimentary tract of animals and factors affecting its composition. *J. Path. Bact.* 89, 95.
SMITH, H. W. & CRABB, W. E. (1961). The faecal bacterial flora of animals and man: its development in the young. *J. Path. Bact.* 82, 53.
SMITH, L. D. (1955). *Introduction to the Pathogenic Anaerobes*. Chicago: University of Chicago Press.
STARK, R. L. & DUNCAN, C. L. (1971). Biological characteristics of *Clostridium perfringens* type A enterotoxin. *Infection & Immunity* 4, 89.
STRONG, D. H., CANADA, J. C. & GRIFFITHS, B. B. (1963). Incidence of *Clostridium perfringens* in American foods. *Appl. Microbiol.* 11, 42.
STRONG, D. H., DUNCAN, C. L. & PERNA, G. (1971). *Clostridium perfringens* type A food poisoning. II. Response of the rabbit ileum as an indication of enteropathogenicity of strains of *Clostridium perfringens* in human beings. *Infection & Immunity* 3, 171.
SUTTON, R. G. A. (1966 a). Enumeration of *Clostridium welchii* in the faeces of varying sections of the human population. *J. Hyg., Camb.* 64, 367.
SUTTON, R. G. A. (1966 b). Distribution of heat-resistant *Clostridium welchii* in a rural area of Australia. *J. Hyg., Camb.* 64, 65.
SUTTON, R. G. A. & HOBBS, B. C. (1968). Food poisoning caused by heat-sensitive *Clostridium welchii*. A report of five recent outbreaks. *J. Hyg., Camb.* 66, 135.
SYKES, G. (1970). The sporicidal properties of chemical disinfectants. *J. appl. Bact.* 33, 147.
SYLVESTER, P. K. & GREEN, J. (1961). The effect of different types of cooking on artificially infected meat. *Medical Officer* 105, 231.
TAYLOR, C. E. D. & COETZEE, E. F. C. (1966). Range of heat resistance of *Clostridium welchii* associated with suspected food poisoning. *Mon. Bull., Minist. Hlth.* 25, 142.
THOMAS, H. C., McSWEEN, R. N. M. & WHITE, R. G. (1973). Role of the liver in controlling the immunogenicity of commensal bacteria in the gut. *Lancet* 1, 1288.
TURNER, G. C. & WONG, M. M. (1961). Intestinal excretion of heat-resistant *Clostridium welchii* in Hong Kong. *J. Path. Bact.* 82, 529.
VINCE, A., DYER, N. H., O'GRADY, F. W. & DAWSON, A. M. (1972). Bacteriological studies in Crohn's disease. *J. med. Microbiol.* 5, 219.
WEISS, K. F. & STRONG, D. H. (1967). Some properties of heat-resistant and heat-sensitive strains of *Clostridium perfringens*. 93, 21.
WILLIS, A. T. (1956). Anaerobes as an index of faecal pollution in water. *J. appl. Bact.* 19, 105.
WILLIS, A. T. (1957). Observations on the anaerobes present in a city water supply, with special reference to *Clostridium welchii*. *J. appl. Bact.* 20, 53.
WILLIS, A. T. (1969). *Clostridia of Wound Infection*. London: Butterworths.
WILLIS, A. T. & HOBBS, G. (1959). Some new media for the isolation and identification of clostridia. *J. Path. Bact.* 77, 511.

WILSON, G. S. & MILES, A. A. (1955). (eds) *Topley & Wilson's Principles of Bacteriology and Immunology.* 4th ed. London: Arnold.

YAMAGISHI, T., ISHIDA, S. & NISHIDA, S. (1964). Isolation of toxigenic strains of *Clostridium perfringens* from soil. *J. Bact.* **88,** 646.

ZEISSLER, J. & RASSFELD-STERNBERG, L. (1949). Enteritis necroticans due to *Clostridium welchii* type F. *Brit. med. J.* **1,** 267.

Enteric and Salmonella Infection: the Carrier State

J. H. McCoy

*Public Health Laboratory, Hull Royal Infirmary,
Anlaby Road, Kingston-upon-Hull, England*

CONTENTS

1. Introduction 221
2. Development of the carrier state 223
3. The chronic carrier 223
   (a) Frequency of excretion 223
   (b) Numbers of bacilli excreted 223
4. Prevention of typhoid fever 225
5. Salmonella infection 225
6. References 227

1. Introduction

"THE FIRST authenticated case of cholera which occurred in London in the autumn of 1848, was that of John Harnold, a seaman of the steamship Elbe, newly arrived from Hamburg, where the disease was prevailing. He died in a lodging near Horsleydown, near the river. The next case was that of a man who came to lodge in the same room: a few hours afterwards cases occurred in Lower Fore Street, Lambeth, and in White Hart Court, Chelsea, amongst people who had no water for drinking or any other purpose, except what was obtained by dipping a pail into the Thames. Thus, the cholera poison from John Harnold appeared to be distributed like the seeds of a river-side plant, some of which germinate and grow up by the side of their parent, whilst others are conveyed some distance by the tide and take root on another part of the shore." (Snow, 1853).

By painstaking inquiry into individual infections and by analysis of deaths as related to the several water supplies of London during epidemics in 1831-2, 1848-9, and 1853-4, Snow (1855) advanced the concept that the cause of cholera was a parasitic micro-organism propagating only in the human intestine and disseminated by the ingestion of excreta.

In particular the cholera epidemic of 1853-4 afforded an opportunity to observe the effect of supplies of pure and contaminated water distributed in the same districts. In districts supplied only by sewage-contaminated Thames water, the deaths from cholera/100,000 living were 114; in districts supplied by both sewage-contaminated Thames water, and sewage-free Thames water, 60 deaths occurred from cholera; in districts supplied only by sewage-free Thames water, no deaths from cholera were recorded.

Snow did suggest that "at least one of the continued fevers – typhoid fever with ulceration of the small intestine – is also propagated in the same way as cholera" but it was William Budd (1873) who elaborated the epidemiology of typhoid fever, again as the result of investigation of the circumstances in which epidemics originated and were propagated.

Budd considered typhoid fever to be self propagating, the specific poison breeding and multiplying within the body of the infected person, and cast off chiefly in the discharges from the diseased intestine. Once cast off, the poison could communicate the fever to other persons by contaminating drinking water or by infecting the air. Budd also recognized that the specific poison might be conveyed to milk and to butter and that linen, wearing apparel, bedding and other porous fabrics contaminated by the discharges of the patient served also to transmit the infection, although "the proportion of cases that originate in these sources is, no doubt, comparatively small".

In the sanitary circumstances of his time when few potable water supplies were free from contamination with excreta, Budd emphasized that control of the enteric fevers could be accomplished by the destruction of the specific poison in the discharges of the sick and in soiled linen and bedding by treatment with chemicals. An acute observer, Budd noted however, that cases of typhoid fever were continually arising which "cannot be traced to contact with a known infection". Although he used analogies drawn from the behaviour of infectious diseases (smallpox and syphilis) known to be conveyed by specific agents, to exclude the spontaneous origin of typhoid fever, the reason for the occurrence of enteric infection in the absence of known infection did not become apparent until after the discovery of the typhoid bacillus in 1884.

Koch (1902) reaffirmed that in typhoid fever the patient or convalescent was the source of further infection in man and again stressed the necessity for efficient disinfection of excreta. At Koch's recommendation bacteriological stations were instituted in typhoid-ridden districts, notably in South West Germany. In addition the repeated bacteriological examination of the stools of patients convalescent from typhoid fever led to the discovery of the chronic carrier which provided the explanation for Budd's (1873) epidemiological observation that not all cases of typhoid fever could be traced to a known human infection.

"That persons apparently quite healthy could harbour typhoid bacilli after an attack of the fever, emit them continuously or periodically in the excreta and thus act as potential sources of infection was a new fact in the aetiology of enteric fever" (Ledingham & Arkwright, 1912), which led since that date to the virtual eradication of indigenous enteric fever from the British Isles.

2. Development of the Carrier State

In enteric fever, the bacilli enter the body by the alimentary tract, penetrate the wall of the intestine and are carried *via* the lymphatic channels, the mesenteric lymph glands and the thoracic duct, to the blood stream. Removal of the bacilli from the circulation is effected by the cells of the reticulo-endothelial system, particularly those of the liver and spleen, where a phase of active multiplication occurs. At the end of this incubation period bacilli again enter the blood stream and are diffused widely throughout the body (onset of infection). During the period of infection bacilli are discharged into the intestine *via* the common bile duct from the infected liver and/or gall bladder, and appear in the excreta.

Natural recovery from the infection is brought about by the production of antibodies which can be demonstrated in the blood from the first week of the infection and which continue to appear in increasing concentration during subsequent weeks. In a proportion of cases, however, foci of typhoid bacilli remain indefinitely in bone marrow, liver, biliary tract, spleen or urinary tract. Those present in the liver, biliary or urinary tracts are discharged in the excreta.

A clear cut distinction therefore exists between *temporary excretors* who excrete only during an attack of enteric fever and convalescence, and *chronic carriers* who (following infection) excrete during the duration of their lives. Chronic faecal carriers are more common than chronic urinary carriers. Women become chronic carriers more frequently than men.

3. The Chronic Carrier

(a) *Frequency of excretion*

Table 1 (modified from Table 5; *Report,* 1961) shows the frequency of excretion in 29 chronic typhoid carriers, 27 females and 2 males, from the majority of whom consecutive weekly samples of faeces were examined. Intermittency of excretion was the rule. No carrier gave 100% of positive specimens. At one extreme, 68 out of 69 samples from one carrier were positive (99%); at the other extreme, 1 out of 106 samples from another carrier (1%). This carrier was discovered accidentally: the result of administration of a purgative being mistaken for an acute attack of diarrhoea. For the duration of her life, while routine weekly samples of formed stools remained negative, the administration of a purgative always resulted in positive stools.

(b) *Numbers of bacilli excreted*

Variation in the frequency of excretion of *Salmonella typhi* in chronic carriers is most likely to be a function of the numbers of bacilli excreted. Table 2,

Table 1

Chronic typhoid carriers. Frequency of excretion

Number of carriers	Faecal samples examined	No. of samples from which *Salmonella typhi* was isolated	Frequency of excretion of *Salm. typhi* by individuals. (% of samples positive)
8	545	516	90*, 91, 93, 94, 96, 96, 97*, 99.
6	442	373	80, 83, 86, 86, 86, 87
2	116	90	77, 78
2	200	84	42, 43
1	34	11	32
2	49	11	22, 24
3	279	41	10*, 13, 16
5	231	7	1, 3*, 3*, 3*, 9

* Monthly examination of faecal samples. Others weekly.

Table 2

Chronic carriers. Numbers of bacilli excreted

Number of carriers	Known duration of carrier state (years)	No. of *Salm. typhi*/g of faeces ($\times 10^6$)									
		0.005	0.50	0.55	0.60	1.0	2.5	4.0	4.5	45.0
3	3		1	1						
9	12	4	1		1	1	1	1	1	1

modified from Thomson (1954), shows the range of numbers of bacilli excreted by 12 known chronic carriers of known duration. *Salmonella typhi* was not isolated from 4 carriers. In the others the numbers excreted ranged from 5×10^5 to 4.5×10^7 of *Salm. typhi*/g of faeces. It is known from *post mortem* examinations that whilst almost pure cultures of *Salm. typhi* are frequently obtained from the upper portion of the small intestine in the region of the entrance of the common bile duct, at lower levels the bacilli become less and less numerous and may be absent from the contents of the large intestine.

This disappearance has been ascribed to the bacteriostatic action of the normal, mainly anaerobic, flora of the large intestine, which can, however, be reduced or abolished by the action of streptomycin (Miller & Bohnhoff, 1968).

A similar effect has been demonstrated in experimental typhoid fever in man (Hornick *et al.*, 1970). Whereas clinical typhoid did not develop in 14 volunteers each given 1000 viable typhoid bacilli orally, one of 4 volunteers treated with streptomycin became ill.

4. Prevention of Typhoid Fever

Typhoid fever is limited strictly to man, transmitted usually indirectly through food and water contaminated with excretions of acute infections and chronic carriers. Prevention is therefore indirect and is effected largely by ensuring that potable water supplies are free from intestinal micro-organisms and that, so far as is possible, water used in milk and food production attains the same standard.

There is no effective method of treatment for the cure of the chronic carrier. However, the control of water supplies for drinking and for milk production has reduced the number of outbreaks of typhoid fever, whilst the disease in individual patients responds rapidly to chloramphenicol. Treatment by chloramphenicol has also reduced the number of patients who become chronic carriers.

Of 83 cases of typhoid fever investigated between 1941–4, 4 became chronic carriers (4.8%) (Cruickshank, 1947), but of 496 cases in the Aberdeen typhoid outbreak in 1964, treated with chloramphenicol, only 5 chronic carriers resulted (1%). (Russell, Sutherland & Walker, 1966).

5. Salmonella infection

In contrast with the enteric fevers of which man is the only source, chronic human carriers play little part in human salmonellosis of the food poisoning type, in which food animals form the reservoir of infection.

Salmonella food poisoning is characterized in the human by an acute illness with vomiting and diarrhoea after a short incubation period which rarely exceeds 48 h. Clinical recovery is rapid. In contrast to the enteric fevers invasion of the blood stream occurs rarely. The infection is thus confined to the intestine and in consequence the chronic carrier state is rare. Salmonellae are excreted in the faeces from the onset of illness and in convalescence. Whilst it is impossible to predict the duration of excretion in an individual, the duration of excretion has been investigated in outbreaks of infection following the consumption of an infected meal.

In general the picture which emerges is that the number of excretors remains more or less constant for *c.* 3 weeks after which the number of excretors declines at a constant rate of some 40 – 50%/week until only a few excretors remain, some of whom excrete for a comparatively long time but eventually cease excreting. The duration of excretion in children with *Salm. typhi-murium* infection of whom only a few received treatment with antibiotics (Lennox, Harvey & Thomson 1954) has been contrasted with the duration of excretion in another outbreak in which nearly all of the children infected received treatment with antibiotics to which the infecting strain of *Salm. typhi-murium* was sensitive *in vitro* (Dixon, 1965). Very few children ceased to excrete the organism in the first 4 weeks. Thereafter the fall in the number of excretors occurred more slowly than in untreated children (Table 3).

Table 3

Duration of excretion in Salm. typhi-murium *infection*

| | No. of subjects excreting *Salm. typhi-murium* in week ||||||||
	1	3	5	7	9	11	16	18
Untreated	64	55	20	3	2	1	1	0
Treated	67	63	50	30	20	10	1	0

Abridged from Dixon (1965).

There would seem to be no doubt but that the administration of antibiotics prolonged the time of excretion in the treated children as specimens in both outbreaks were examined by the same methods. This effect may be ascribed to the effect of antibiotics on the normal inhabitants of the intestine.

In the control of human salmonellosis the most that can be done is to exclude from food handling all persons known to be excreting until at least 3 negative samples have been obtained at intervals of at least a week. Control of the animal reservoir of salmonellae, which is the source of human salmonellosis, presents greater difficulties. Table 4 shows the changes in the 10 serotypes of salmonellae most frequently isolated from human infections since 1958. The Table is characterized by the disappearance of some serotypes as important sources of human infections and the emergence of others. These changes may result from

Table 4

Serotypes and rank of the 10 most common serotypes of salmonellae

Serotype	1958-63	1964-67	1968-72
typhi-murium	1	1	1
heidelberg	2	5	6
newport	3	9	
enteritidis	4	3	2
stanley	5		10
thompson	5		
saint paul	6		8
bredeney	7	7	9
menston	8		
brandenburg	9	2	
anatum	10	6	
panama		4	3
dublin		8	
indiana		10	7
agona			4
virchow			5

factors over which little control can be exercised, as for example, the growth of intensive rearing of animals for food, changes in animal husbandry and variations in the bacteriological quality of animal feed.

6. References

BUDD, W. (1873). *Typhoid Fever: its nature, mode of spreading and prevention.* London: Longmans Green.
CRUICKSHANK, J. C. (1947). Typhoid fever in Devon. The value of phage-typing in a rural area. *Mon. Bull. Min. Hlth. Lab. Serv.* **6**, 88.
DIXON, J. M. S. (1965). Effect of antibiotic treatment on duration of excretion of *Salmonella typhi-murium* by children. *Brit. Med. J.* **2**, 1343.
HORNICK, R. B., GREISMAN, S. E., WOODWARD, T. E., DU PONT, H. L., DAWKINS, A. T. & SNYDER, M. J. (1970). Typhoid Fever: Pathogenesis and Immunologic Control. *New Engl. J. Med.* **283**, 686.
KOCH, R. (1902). Die Bekämpfung des Typhus. Vertrag gehalten in der Sitzung des wissenschaftlichen Senats bei der Kaiser Wilhelms-Akademie am 28. Nov. 1902.
LEDINGHAM, J. C. G. & ARKWRIGHT, J. A. (1912). *The Carrier Problem in Infectious Diseases.* London: Edward Arnold.
LENNOX, M., HARVEY, R. W. S. & THOMSON, S. (1964). An outbreak of food poisoning due to *Salmonella typhi-murium* with observations on the duration of infection. *J. Hyg., Camb.* **52**, 312.
MILLER, C. P. & BOHNHOFF, M. (1968). Changes in the mouse's enteric microflora associated with enhanced susceptibility to *Salmonella* infection following streptomycin treatment. *J. inf. Dis.* **113**, 59.
REPORT. (1961). The detection of the typhoid carrier state. Report of a PHLS Working Party on the Bacteriological Examination of Waterworks Employees. *J. Hyg., Camb.* **59**, 231.

RUSSELL, E. M., SUTHERLAND, A. & WALKER, W. (1966). Ampicillin for persistent typhoid excretors including a clinical trial in convalescence. *Brit. Med. J.* **2**, 555.

SNOW, J. (1853). *On Continuous Molecular Changes, more particularly in their Relation to Epidemic Diseases.* London: Churchill.

SNOW, J. (1855). *On the mode of Communication of Cholera.* 2nd ed. London: J. & A. Churchill.

THOMPSON, S. (1954). The number of bacilli harboured by enteric carriers. *J. Hyg., Camb.* **52**, 67.

The Effect of Antimicrobial Agents on Human Faecal Flora: Studies with Cephalexin, Cyclacillin and Clindamycin

VERA L. SUTTER AND S. M. FINEGOLD

Wadsworth VA Center Hospital, Los Angeles and the Department of Medicine, UCLA School of Medicine, Los Angeles, California, U.S.A.

CONTENTS

1. Introduction 229
2. Experimental 230
   (a) Subject material 230
   (b) Collection and processing of faecal specimens 230
   (c) Bacteriological studies 230
3. Results 232
   (a) Cephalexin 232
   (b) Cyclacillin 232
   (c) Clindamycin 234
4. Discussion 236
5. Acknowledgements 239
6. References 239

1. Introduction

THE RELATIONSHIP between intestinal bacteria and their host remains to be more fully elucidated but a great deal of evidence is accumulating that this relationship is of extreme importance. Intestinal bacteria have been shown to play a role in normal physiological processes as well as pathophysiological processes. Disturbances in the normal distribution or balance of these bacteria are associated with malabsorption syndromes, hepatic coma and infections related to the bowel. It is desirable that antimicrobial agents used for therapy of systemic infection have little or no effect on normal faecal flora so that the above disturbances are minimized. Additionally, it is important that colonization of the bowel with antibiotic-resistant, potential pathogens such as the *Klebsiella-Enterobacter-Serratia* group, *Proteus* spp., *Pseudomonas* spp. and *Staphylococcus aureus* should be minimal because intestinal carriers of these organisms are known to serve as reservoirs of infection in the hospital.

The purpose of this report is to present data on the effects of the newer antibiotic agents, cephalexin, cyclacillin and clindamycin, on the aerobic and anaerobic flora in human faeces.

2. Experimental

(a) *Subject material*

The subjects studied were adult males hospitalized for a variety of conditions, but not involving the gastrointestinal tract, or were healthy adult volunteers from the hospital or laboratory staff. All were male except for one female in each group. Faecal specimens were collected prior to administration of the drug and again following several days of therapy. All drugs were administered orally.

Nine subjects were given cephalexin and specimens from 3 of these were analysed for both aerobic and anaerobic bacteria, while specimens from 6 were analysed for aerobic bacteria only. Nine subjects were treated with cyclacillin. Three subjects were studied for alteration in aerobic and anaerobic faecal bacteria and 6 were studied for aerobic bacterial alteration only. Ten subjects were given clindamycin and specimens from 5 subjects were studied both aerobically and anaerobically, while those from 5 were studied aerobically only. Dosages and period of treatment with the 3 drugs are given in Tables 2-7.

(b) *Collection and processing of faecal specimens*

Specimens were collected in unsterile plastic coated paper containers and were usually processed immediately upon receipt in the laboratory (½-1 h after collection). Occasionally, it was necessary to refrigerate a specimen for a few hours until processing could be done.

A portion of *c.* 1 g was taken from the centre of the specimen, weighed in a sterile tube which was being gassed with CO_2, then diluted with an appropriate amount of 0.05% yeast extract solution to make a 10^{-1} dilution. This tube contained sterile glass beads and was sealed with a butyl rubber stopper. The 10^{-1} dilution was thoroughly emulsified on a vortex mixer. Serial 10-fold dilutions were then made in the yeast extract solution.

(c) *Bacteriological studies*

Volumes of 0.1 ml of various dilutions were plated on a variety of non-selective and selective media as shown in Table 1. The anaerobic atmosphere was established by evacuating and replacing the atmosphere of either Brewer or Gaspak jars 5 times with N_2, then finally filling with a mixture of 10% CO_2, 10% H_2 and 80% N_2. Palladium-coated aluminia catalyst was used in the jars. Incubation was at 37° for different periods of time (Table 1).

After incubation, total colony counts were made from the aerobic and anaerobic blood agar plates and different colony types enumerated, purified and identified. Also, colonies on selective media were enumerated, purified and identified. In general, aerobic or facultative bacteria and *Candida* spp. or yeasts

Table 1

Media and methods employed for bacteriological studies

Medium	Source or reference	Purpose	Dilutions plated	Incubation
Blood agar (BA)	Sutter et al. (1972)	Total aerobic and anaerobic counts, predominant bacteria	$10^{-2}, -4, -6, -8$	Aerobic, 1-3 days / Anaerobic, 2-4 days
Kanamycin-vancomycin BA	Sutter et al. (1972)	*Bacteroides*		Anaerobic, 2-4 days
Kanamycin-vancomycin Laked BA	Sutter et al. (1972)	*Bacteroides melaninogenicus*	$10^{-1}, -2, -4, -6$	Anaerobic, 2-7 days
Modified FM agar	Ohtani (1970)	*Fusobacterium (Sphaerophorus)*		
Rifampin BA	Sutter et al. (1972)	*Clostridium, Eubacterium*	$10^{-2}, -4, -6, -8$	Anaerobic, 2-4 days
Neomycin BA	Sutter et al. (1972)	*Clostridium*, anaerobic cocci		
Egg yolk – neomycin-nagler	Sutter et al. (1972)	*Cl. perfringens*	$10^{-1}, -2, -4, -6$	Anaerobic, 1-3 days
Blood agar – heated dilutions	Sutter et al. (1972)	*Clostridium*	$10^{-2}, -4, -6$	Anaerobic, 2-4 days
		Bacillus	$10^{-2}, -3, -4$	Aerobic, 1-3 days
Eugonagar + maltose	Sutter et al. (1972)	*Bifidobacterium*	$10^{-2}, -4, -6, -8$	
LBS (Rogosa)	BBL	*Lactobacillus*	$10^{-1}, -2, -4, -6$	Anaerobic, 2-4 days
Modified *Veillonella* agar	Sutter et al. (1972)	Gram-negative cocci	$10^{-2}, -4, -6, -8$	
Desoxycholate agar	BBL	Enterobacteriaceae	$10^{-2}, -4, -6, -8$	Aerobic, 1-3 days
Cetrimide agar	Brown & Lowbury (1965)	Fluorescent pseudomonads	$10^{-1}, -2, -4$	Aerobic, 2-7 days
Polymyxin agar	Finegold & Sweeney (1961)	*Staphylococcus aureus*		Aerobic, 2-4 days
Bile-aesculin-azide agar	Pfizer PSE	Group D streptococci	$10^{-1}, -2, -4, -6$	Aerobic, 1-3 days
Mitis-salivarius agar	Difco	Other streptococci		Aerobic, 1-3 days
Molybdate agar	MacLaren & Armen (1958)	*Candida* and yeasts		Aerobic, 3-7 days

were identified according to criteria outlined by Bailey & Scott (1966) and anaerobic bacteria were identified by criteria outlined in the Anaerobic Bacteriology Manual (Sutter et al., 1972) or the Anaerobe Laboratory Manual (Holdeman & Moore, 1972).

3. Results

Quantitative and qualitative determinations of pre-treatment specimens were compared with those of specimens taken during therapy. Differences in counts of organisms of 4 log cycles or more were considered to be due to drug effect.

(a) *Cephalexin*

Most of the normal bacterial flora of subjects treated with cephalexin remained essentially the same during the course of therapy (Table 2, 3). The organisms which remained unchanged were *Escherichia coli*, Group D streptococci, *Bacteroides fragilis*. *Lactobacillus*, *Eubacterium* and *Clostridium* spp. Organisms which were usually lost were various *Streptococcus* spp. other than Group D and the anaerobic cocci. One patient, Subject 1, had no detectable *Bacteroides fragilis* in the pre-treatment specimen but regained these bacteria in counts of 10^{10} during therapy. This patient had agammaglobulinemia.

Opportunistic pathogens were increased or acquired in 6 patients and decreased or eliminated in 4. Bacteria belonging to the *Klebsiella-Enterobacter-Serratia* group were acquired by 2 subjects and remained in both of 2 subjects in whom they were detected prior to therapy. *Proteus-Providencia* organisms were acquired in one subject and eliminated in one while *Pseudomonas* was acquired in one patient and eliminated in another. Coagulase positive staphylococci were eliminated in the one patient in whom this organism was detected prior to therapy. *Candida* spp. or yeasts were acquired in 2 subjects and eliminated in one. They remained in all of 3 subjects in whom they were detected prior to therapy.

(b) *Cyclacillin*

Bacteriological results of the studies with subjects treated with cyclacillin are shown in Tables 4 & 5. With the exception of 2 subjects, No. 7 who lost *E. coli*, and No. 3 who lost group D streptococci, most organisms of the normal faecal flora remained unchanged. Streptococci other than Group D were eliminated in 3 subjects, remained unchanged in one and were gained by one.

Opportunistic pathogens were increased or acquired in 9 instances. Bacteria in the *Klebsiella-Enterobacter-Serratia* group were acquired by 5 individuals and increased in another. One subject (No. 5) acquired *Pseudomonas* in addition to

Table 2

Effect of cephalexin on aerobic and facultative flora

Colony count*/g of wet faeces per subject†

Organism	1 ‡Pre-tr	1 6th day	2 Pre-tr	2 7th day	3 Pre-tr	3 6th day	4 Pre-tr	4 6th day	5 Pre-tr	5 10th day	6 Pre-tr	6 12th day	7 Pre-tr	7 6th day	8 Pre-tr	8 14th day	9 Pre-tr	9 13th day
Total count	9	8	8	8	7	7	7	8	7	8	6	9	7	8	8	7	8	8
Escherichia coli	8	7	8	7	7	5	7	8	7	6	6	8	6	4	8	7	8	7
Klebsiella-Enterobacter-Serratia	—	—	6	5	5	3	—	5	—	—	—	—	—	—	—	—	—	7
Proteus-Providencia	3	7	6	4	—	—	—	—	5	3	—	—	7	7	—	—	—	—
Pseudomonas	4	—	—	—	—	—	6	—	—	—	—	—	5	8	—	—	—	—
Bacillus	8	—	—	—	5	3	—	—	5	—	7	—	5	—	—	—	—	—
Group D streptococci	8	9	8	8	6	6	—	8	7	4	5	8	4	6	8	6	8	8
Streptococci (not Group D)	7	—	6	7	6	—	—	—	—	3	5	—	7	—	—	—	—	—
Staphylococcus, coagulase positive	4	—	—	—	—	—	—	—	—	—	—	—	—	—	—	—	—	—
Candida-yeasts	3	6	—	7	—	—	7	4	—	—	—	—	3	—	—	4	3	4

* Results are expressed as the colony count (1/log no. of bacteria/g of wet faeces).
† Each subject received 2 g of cephalexin/day.
‡ Pre-tr, pre-treatment.

Table 3

Effect of cephalexin on anaerobic flora

	Colony count*/g of wet faeces per subject†					
	1		2		3	
Organism	Pre-treatment	6th day	Pre-treatment	7th day	Pre-treatment	6th day
Total count	10	10	10	10	10	10
Bacteroides fragilis	–	10	9	9	10	9
Bact. melaninogenicus	–	–	4	–	–	–
Acidaminococcus	–	–	7	–	–	–
Peptostreptococcus	9	7	9	–	–	–
Lactobacillus	9	7	9	6	–	–
Bifidobacterium	–	–	8	7	9	–
Eubacterium	10	10	9	9	9	7
Clostridium perfringens	7	7	4	–	–	–
Cl. ramosum	–	–	–	8	9	7
Other Clostridium spp.	8	9	6	9	9	7

* Results are expressed as the colony count (1/log no. of bacteria/g of wet faeces).
† Each subject received 2 g of cephalexin/day.

the *Klebsiella* organisms. *Proteus-Providencia* organisms and coagulase positive staphylococci were acquired by one subject each. *Candida* spp. or other yeasts were eliminated in 4 subjects and remained unchanged in 2.

(c) Clindamycin

Results of the studies on patients treated with clindamycin are shown in Tables 6 & 7. Striking changes occurred in the normal anaerobic flora. *Bacteroides fragilis* was eliminated in all 5 of the subjects whose specimens were studied anaerobically. Anaerobic cocci, lactobacilli and *Bifidobacterium* were also eliminated. The anaerobic flora which remained following clindamycin therapy was limited to *Eubacterium* and *Clostridium* spp. Species found in high numbers were *Eubacterium aerofaciens, E. lentum, E. ventriosum* and in 2 subjects, *Eubacterium* spp. which could not be identified further. The *Clostridium* spp. found in high numbers were *Cl. difficile, Cl. innocuum, Cl. oroticum* and *Cl. ramosum*.

Opportunistic pathogens were acquired in 15 instances and decreased or eliminated in 5. *Klebsiella-Enterobacter-Serratia* organisms were acquired in 6 subjects and increased in 2. *Proteus-Providencia* was also acquired by one of these subjects (No. 10) but lost in 2 of them (Nos 2 & 4). *Pseudomonas* was acquired in one (No. 7), increased in one (No. 2) and lost in one (No. 5). *Candida* spp. or other yeasts were gained by 4 subjects and lost by 2.

Table 4

Effect of cyclacillin on aerobic and facultative flora

Colony count*/g of wet faeces per subject‡

Organism	1 Pre-tr‡	1 8th day	2 Pre-tr	2 8th day	3 Pre-tr	3 7th day	4 Pre-tr	4 7th day	5 Pre-tr	5 7th day	6 Pre-tr	6 7th day	7 Pre-tr	7 7th day	8 Pre-tr	8 7th day	9 Pre-tr	9 7th day
Total count	6	8	9	9	9	8	6	7	8	8	9	9	9	9	8	9	7	8
Escherichia coli	5	5	9	9	7	8	6	7	8	7	6	3	7	—	8	9	6	7
Klebsiella-Enterobacter-Serratia	—	4	8	8	—	6	—	—	—	7	6	4	3	9	—	6	—	6
Proteus-Providencia	—	—	7	7	8	6	—	—	—	—	—	—	—	—	—	4	—	—
Pseudomonas	—	—	—	—	—	—	—	—	—	7	3	5	—	—	—	—	—	—
Bacillus	5	5	—	8	5	4	3	4	—	—	3	—	—	—	4	6	5	4
Group D streptococci	6	7	8	9	5	—	4	5	8	7	6	5	9	6	3	5	7	8
Streptococci (not Group D)	5	—	—	—	—	—	5	5	5	—	6	—	—	9	—	—	—	—
Staphylococcus, coagulase positive	—	—	—	—	—	—	—	—	—	—	—	—	—	—	—	6	—	—
Candida-yeasts	3	4	3	—	3	—	—	—	5	—	3	—	5	5	—	—	—	—

* Results are expressed as the colony count (1/log no. of bacteria/g of wet faeces).
† Each subject received 2 g of cyclacillin/day.
‡ Pre-tr, pre-treatment.

Table 5

Effect of cyclacillin on anaerobic flora

	Colony count*/g of wet faeces per subject					
	1		2		3	
Organism	Pre-treatment	8th day	Pre-treatment	8th day	Pre-treatment	7th day
Total count	10	10	10	10	10	10
Bacteroides fragilis	10	10	10	10	10	10
Bacteroides sp.	9	—	—	—	—	—
Fusobacterium	3	—	—	—	—	—
Veillonella	—	—	8	—	4	5
Acidaminococcus	10	8	9	9	—	—
Megasphaera	—	—	8	8	—	—
Peptococcus	—	—	—	—	9	9
Peptostreptococcus	—	—	8	9	—	8
Lactobacillus	7	5	6	6	—	9
Bifidobacterium	9	9	—	9	10	10
Eubacterium	6	8	8	—	9	9
Clostridium perfringens	—	3	4	—	3	—
Cl. ramosum	6	6	9	8	8	—
Other Clostridium spp.	6	7	9	9	9	9

* Results are expressed as the colony count (1/log no. of bacteria/g of wet faeces).
† Each subject received 2 g of cyclacillin/day.

4. Discussion

The results indicate that cephalexin caused relatively minor suppression of normal aerobic and anaerobic flora. Acquisition or retention of potential pathogens was sporadic. These results are comparable to those obtained with oral tetracyclines in a prior study (Finegold et al., 1967).

Cyclacillin also caused relatively minor suppression of normal faecal flora. Colonization of the subjects with potential pathogens was more frequent in the patients treated with cyclacillin than in those treated with cephalexin and was similar to other broad spectrum penicillins in this regard (Finegold et al., 1967).

Therapy with clindamycin resulted in major changes in normal anaerobic faecal flora and allowed frequent colonization with and proliferation of potential pathogens, particularly those in the *Klebsiella-Enterobacter-Serratia* group. However, the changes were not as drastic as those occurring in patients treated with lincomycin (Finegold et al., 1966). The loss of *Bacteroides fragilis*, lactobacilli and *Bifidobacterium* spp. and replacement with large numbers of *Clostridium* and *Eubacterium* spp. could represent a potentially hazardous situation in patients treated with clindamycin. A significant percentage of strains of the species of *Clostridium* found have recently been reported resistant to

Table 6

Effect of clindamycin on aerobic and facultative flora

Colony count*/g of wet faeces per subject†

Organism	1 Pre-tr‡	1 11th day	2 Pre-tr	2 12th day	3 Pre-tr	3 6th day	4 Pre-tr	4 8th day	5 Pre-tr	5 7th day	6 Pre-tr	6 10th day	7 Pre-tr	7 7th day	8 Pre-tr	8 8th day	9 Pre-tr	9 7th day	10 Pre-tr	10 7th day
Total count	9	9	7	8	6	8	8	10	7	9	7	9	8	9	6	9	8	9	8	9
Escherichia coli	8	8	7	7	6	8	6	8	7	8	5	8	8	7	6	4	8	8	8	8
Klebsiella-Enterobacter-Serratia	5	6	—	5	—	8	4	9	5	9	—	7	—	—	—	8	—	8	—	7
Proteus-Providencia	7	8	3	—	—	—	3	—	6	5	—	—	—	—	—	—	—	—	—	8
Pseudomonas	—	—	3	8	—	—	—	—	3	—	—	—	—	7	—	—	—	—	—	—
Bacillus	—	3	3	—	4	—	—	—	4	—	—	—	—	—	4	—	3	—	—	—
Group D streptococci	8	8	5	7	5	6	8	9	5	9	3	9	6	9	—	9	3	—	4	8
Streptococci (not Group D)	7	—	—	—	—	—	—	—	—	—	7	—	—	—	—	—	6	—	—	—
Staphylococcus, coagulase positive	—	—	—	—	—	—	—	—	—	—	—	—	—	—	—	—	—	—	—	—
Candida-yeasts	—	9	4	5	—	—	3	—	—	3	5	5	—	5	—	3	3	—	—	—

* Results are expressed as the colony count (1/log no. of bacteria/g of wet faeces).
† Each subject received 1.2 g of clindamycin/day.
‡ Pre-tr, pre-treatment.

Table 7

Effect of clindamycin on anaerobic flora

	Colony count*/g of wet faeces per subject†									
	1		2		3		4		5	
Organisms	‡Pre-tr	11th day	Pre-tr	12th day	Pre-tr	6th day	Pre-tr	8th day	Pre-tr	7th day
Total count	10	9	11	10	10	10	10	10	10	9
Bacteroides fragilis	9	–	9	–	10	–	9	–	10	–
B. melaninogenicus	–	–	–	–	–	–	–	–	5	–
Acidaminococcus	–	–	–	–	–	–	–	–	8	–
Megasphaera	–	–	–	–	–	–	–	–	8	–
Peptococcus	–	–	10	–	–	–	–	–	9	–
Peptostreptococcus	–	–	–	–	8	–	9	–	–	–
Lactobacillus	–	–	–	–	4	–	9	–	–	–
Bifidobacterium	–	–	–	–	8	–	9	–	10	6
Eubacterium	9	7	11	10	8	9	8	8	6	9
Clostridium perfringens	5	4	4	3	5	–	5	–	4	–
Cl. ramosum	7	–	–	–	5	8	–	9	8	9
Clostridium spp.	8	9	6	10	9	9	6	8	9	–

* Results are expressed as the colony count (1/log number of bacteria/g of wet faeces).
† Each subject received 1.2 g of clindamycin/day.
‡ Pre-tr, pre-treatment.

clindamycin (Wilkins & Thiel, 1973). Additionally, *Cl. ramosum* has been reported to be resistant to several antimicrobial agents usually considered useful in treating anaerobic infections (Tally *et al.,* 1973). Endogenous infection from this kind of intestinal population could pose serious problems in selection of appropriate antimicrobial therapy.

Studies of faecal flora such as these are subject to a number of limitations. A source of error is that anaerobic bacteria were cultivated and enumerated on the open bench and therefore the organisms recovered represent only a percentage of the bacteria present. Studies done with oxygen-free media in either roll-tubes or an anaerobic chamber indicate that the number of bacteria isolated is *c.* 100-fold greater than those found by the use of conventional methods in which oxygen can gain access to the cells during manipulation (Drasar, 1967). Therefore, the effect of cephalexin, cyclacillin and clindamycin on the oxygen sensitive organisms or on those which require a low oxidation-reduction potential is unknown.

The interpretation of the data in terms of what changes have occurred due to antibiotic effect as opposed to what might be expected as normal variation in the bacterial populations of human faeces is based on the somewhat arbitrary criterion that difference in numbers of a particular bacterium of 4 log cycles or greater represents a change due to therapy. Repetitive studies on individuals have indicated that organisms such as *Bacteroides* and *E. coli* remain relatively constant but that organisms usually present in smaller numbers vary from one to 3 log cycles from sample to sample (Levison & Kaye, 1969). A study of quantitative recovery of *Pseudomonas aeruginosa* from human faeces showed that, in many instances, variation was not more than 3 log cycles but in 4 of 19 individuals counts decreased from 10^4/g or 10^5/g to undetectable levels when the method used was sensitive enough to detect 10^2/g (Sutter et al., 1967).

Media for selection or differentiation of specific micro-organisms vary in their effectiveness. Cetrimide agar examined with UV light, for example, allows detection of fluorescent pseudomonads in numbers as low as 10^2/g of faeces. On the other hand, the rifampin agar used in this study to select some *Clostridium* and *Eubacterium* spp. is not selective enough to allow detection of $<10^5$ of the bacteria/g of faeces.

These limitations must be kept in mind in interpreting the data. For example, when an organism was 'acquired', we know only that it was not present at detectable levels in the pre-treatment specimen but was detectable in the specimen taken during therapy. Despite the limitations and reservations, studies such as these should be continued as new antibiotics are introduced to alert clinicians with regard to possible problems of superinfection following the use of antimicrobial agents.

5. Acknowledgements

These studies were supported, in part, by grants from Eli Lilly & Co., Wyeth Laboratories, and the Upjohn Co. We greatly appreciate the excellent technical assistance of W. T. Carter and P. T. Sugihara.

6. References

BAILEY, W. R. & SCOTT, E. G. (Eds) (1966). *Diagnostic Microbiology*, 2nd ed. St Louis: C. V. Mosby Co.

BROWN, V. I. & LOWBURY, E. J. L. (1965). Use of an improved cetrimide agar medium and other culture methods for *Pseudomonas aeruginosa*. *J. clin. Path.* **18**, 752.

DRASAR, B. S. (1967). Cultivation of anaerobic intestinal bacteria. *J. Path. Bact.* **94**, 417.

FINEGOLD, S. M., DAVIS, A. & MILLER, L. G. (1967). Comparative effect of broad-spectrum antibiotics on non-sporeforming anaerobes and normal bowel flora. *Ann. N.Y. Acad. Sci.* **145**, 268.

FINEGOLD, S. M., HARADA, N. E. & MILLER, L. G. (1966). Lincomycin: activity against anaerobes and effect on normal human fecal flora. In *Antimicrobial Agents and Chemotherapy* (1965). Ann Arbor, Michigan: Amer. Soc. for Microbiology.

FINEGOLD, S. M. & SWEENEY, E. E. (1961). New selective and differential mediums for coagulase-positive staphylococci allowing rapid growth and strain differentiation. *J. Bact.* **81**, 636.

HOLDEMAN, L. V. & MOORE, W. E. C. (1972). *Anaerobe Laboratory Manual.* Blacksburg, Va.: V.P.I. Anaerobe Laboratory, Virginia Polytechnic Institute & State University.

LEVISON, M. E. & KAYE, D. (1969). Fecal flora in man: effect of cathartic. *J. infect. Dis.* **119**, 591.

MACLAREN, J. A. & ARMEN, D. (1958). Pigmentation of *Candida albicans* by molybdenum. *Am. J. clin. Path.* **30**, 411.

OHTANI, F. (1970). Selective isolation media for strictly anaerobic non-sporulating Gram-negative rods. *Jap. J. Bact.* **25**, 222.

SUTTER, V. L., ATTEBERY, H. R., ROSENBLATT, J. E., BRICKNELL, K. & FINEGOLD, S. M. (1972). *Anaerobic Bacteriology Manual.* Publ. Univ. of California, Los Angeles Extension Division.

SUTTER, V. L., HURST, V. & LANE, C. W. (1967). Quantification of *Pseudomonas aeruginosa* in feces of healthy human adults. *Health Lab. Sci.* **4**, 245.

TALLY, F. P., ARMFIELD, A. Y. & FINEGOLD, S. M. (1973). *In vitro* susceptibility of *Clostridium ramosum* to eleven antimicrobial agents. *Abst. M 104 annual meeting, American Society for Microbiology.*

WILKINS, T. D. & THIEL, T. (1973). Resistance of some species of *Clostridium* to Clindamycin. *Antimicrob. Ag. Chemother.* **3**, 136.

The Effects of Tetracycline on the Establishment of *Escherichia coli* of Animal Origin, and *in vivo* Transfer of Antibiotic Resistance, in the Intestinal Tract of Man

GLENNA C. BURTON, D. C. HIRSH, D. C. BLENDEN AND JUDY L. ZEIGLER

Department of Veterinary Microbiology, School of Veterinary Medicine, University of Missouri, Columbia, Missouri 65201, U.S.A.

CONTENTS

1. Introduction 241
2. Materials and Methods 243
   (a) Volunteers 243
   (b) Properties of the ingested organism 243
   (c) Antibiotic and X-314 administration 244
   (d) Handling of faecal specimens 245
   (e) Statistical tests 245
3. Results 245
4. Discussion 249
5. Acknowledgements 252
6. References 252

1. Introduction

SINCE THE DISCOVERY of infectious drug resistance in the late 1950's much attention has been focused on this phenomenon. Enterobacteria have been increasingly studied in this respect since the major means by which these bacteria become resistant to antimicrobials is by acquisition of resistance transfer factors (R-factors) (Anderson, 1968). Recent surveys have indicated that enterobacteria resistant to one or more antimicrobials are frequently found in the intestinal tract of both animals and man (Datta, 1965; Loken, Wagner & Henke, 1971; Linton *et al.*, 1972). The proportion of the enteric flora resistant to at least one antimicrobial is thought to be related to the antibiotic exposure of the individuals involved (Mercer *et al.*, 1971). Thus, the level of antibiotic resistant bacteria in livestock raised for food products has been found to be relatively high. In the United States at least, animals such as cattle, swine, turkeys and chickens commonly receive antibiotics in levels from 20 to 500 p/m as feed supplements. It has been our experience that both cattle and swine purchased in the state of Missouri for research purposes possess high levels of antibiotic resistant *Escherichia coli*. All of 69 cattle used in our recent research possessed *E. coli* with multiple antibiotic resistance. Fifty to 100% of the *E. coli* in these animals were multiple resistant. The same was true for swine acquired for research purposes. Other reports have indicated similar experiences with antibiotic resistant bacteria in food producing animals (Walton, 1966; Loken *et al.*, 1971; Mercer *et al.*, 1971).

The incidence and degree of antimicrobial resistance of enteric bacteria in man has also been studied. It has been shown that there is a significant number of persons excreting multiple resistant bacteria. In some cases a high proportion (50-100%) of the total aerobic flora is resistant to one or more antimicrobials (Datta *et al.*, 1971; Loken *et al.*, 1971).

It has been suggested that one source of these multiple resistant bacteria in the human intestinal tract may be food producing animals (Van Houweling & Kingma, 1969; Shooter *et al.*, 1970). As has been mentioned, these animals possess large numbers of multiple resistant organisms in their intestinal tracts. Possible pathways to the human consumer have been suggested by the finding of multiple resistant bacteria on the surface of carcasses (Walton, 1971), in milk (Jones, 1971), sausages (Moorhouse, O'Grady & O'Connor, 1969) and ready-to-cook broiler chickens (Kim & Stephens, 1972).

The importance of these findings lies in the possibility that bacteria of animal origin may serve as sources of infectious drug resistance factors (R-factors) for the human population. Although resistance transfer has been relatively easy to demonstrate *in vitro*, it is more difficult to achieve, at least experimentally, *in vivo* (Salzman & Klemm, 1968). There is enough evidence to consider *in vivo* transfer to multiple antimicrobial resistance a likely occurrence.

Transfer of resistance factors *in vivo* has been demonstrated in experimental animals by several authors. This transfer has usually been attained by modifying the normal flora of the alimentary tract in some manner. Kasuya (1964) and Guinee (1965) brought about *in vivo* transfer in mice whose normal flora had been reduced by the administration of antibiotics. Guinee (1970) studied rats, some of whom were given drinking water containing 20 or 100 p/m of tetracycline hydrochloride. He found that the incidence of R-factor transfer was considerably higher among animals fed tetracycline. Treatment with increasing levels of antibiotics resulted in an increased percentage of animals exhibiting *in vivo* transfer.

Several workers have demonstrated R-factor transfer in the intestinal tract of mice maintained under germ-free conditions (Kasuya, 1964; Reed, Sieckmann & Georgi, 1969). There are very few reports of R-factor transfer in man. Wiedemann *et al.* (1970) failed to demonstrate *in vivo* transmission of drug resistance from an *E. coli* 087 R-factor carrying strain to other enterobacteria in the intestinal tracts of 4 human subjects. Smith (1969) conducted an experiment in which large doses of *E. coli* of animal and human origin were taken by mouth. He concluded that R-factors could be transferred to resident *E. coli* in the enteric tract of man but that the amount of transfer was small and the resident organisms which had acquired resistance did not persist in the intestinal tract for long in competition with the antibiotic sensitive resident strain.

Despite these findings the results of surveys indicate that under natural conditions, large scale transfer of R-factors has occurred among some kinds of bacteria (Anderson, 1968). A good example is the almost epidemic spread of R-factors that occurred in the 1960's among *Salmonella typhimurium* of phage type 29 present in calves in Great Britain (Anderson, 1965; Anderson & Lewis, 1965). Bohus (1971) reported an R-factor transfer from commensal *E. coli* to a *Shigella sonnei* strain during a dysentery outbreak in Czechoslovakia.

Thus it becomes important to determine whether animal strains of enteric bacteria, especially multiple resistant ones can survive in the intestinal tract of man. Smith (1969) has shown that colonization depends upon the particular strain and host to which it has adapted, and dose of ingested bacteria. Cooke *et al.* (1972) reported that colonization depended, in some cases, on the colicinogenic properties of the ingested bacteria.

We have extended the work cited above by using oral administration of an antibiotic to which the ingested organism was resistant. The purpose of our study was two-fold. First we wanted to determine the effect of varying levels of tetracycline upon the establishment of tetracycline resistant *E. coli* of bovine origin in the intestinal tract of man. Secondly, we wanted to know if this R-factor carrying *E. coli* would transfer its resistance markers to resident bacteria and if so, whether tetracycline had an influence on this transfer. Tetracycline was chosen because it is a commonly used antimicrobial both in human and animal medicine.

2. Materials and Methods

(a) *Volunteers*

Sixty-three subjects (43 males and 20 females) aged 20 to 53 were used in this study. They were enlisted on an informed consent basis. Allocation of subjects into experimental groups was done on a double blind basis (Table 1). Six groups of subjects were involved. One group received no tetracycline, another group received 50 mg of tetracycline/day and third group took 1000 mg tetracycline/day. Approximately one half of the persons in each of these groups was given x-314, the experimental organism.

(b) *Properties of the ingested organism*

The ingested organism, designated as *E. coli* x-314, was kindly supplied by H. Williams Smith. The characteristics of this organism were: (1) bovine origin; (2) non-pathogenic as judged by negative keratoconjuctivitis test (Sereney, 1955), non-invasive by the Hela cell assay (LaBrec *et al.*, 1964), negative in the infant mouse test (Dean *et al.*, 1972), negative in the bovine and rabbit ligated gut loop test (Smith & Halls, 1967); (3) non-pathogenic for humans (Smith, H. W., pers.

Table 1

Experimental grouping of subjects with respect to the amount of tetracycline and ingestion of Escherichia coli

Number of subjects	Dose of tetracycline (mg)	Ingestion of E. coli
11	0	–
13	0	+
8	50	–
9	50	+
9	1000	–
13	1000	+

–, no *E. coli* ingested. +, ingestion of *E. coli*.

comm.); (4) serogroup 051; (5) resistant to tetracycline, streptomycin, neomycin, kanamycin, ampicillin, chloramphenicol and nalidixic acid. All these resistances except for nalidixic acid were of a transmissible nature. The minimal inhibitory and the minimal bactericidal concentration for tetracycline was 800 μg/ml.

(c) *Antibiotic and x-314 administration*

The experimental time sequence is shown in Fig. 1. As was previously mentioned, subjects were divided into 6 groups. Faecal samples were collected using 2 different protocols. In the first, faecal specimens were collected 17, 10 and 3 days before subjects received tetracycline. A 4th sample was taken after the subjects had been taking the tetracycline for 3 days and another, 1 day prior to ingestion of x-314. Additional faecal specimens were submitted approximately 1, 4, 6, 11, 18 and 25 days after ingestion of x-314. In the second protocol, samples were obtained daily beginning 7 to 14 days prior to, and continuing for 12-18 days after, x-314 was taken until at least 3 consecutive faecal specimens were negative for the presence of x-314. The purpose of daily sampling was to decrease the likelihood of omitting a faecal specimen containing x-314.

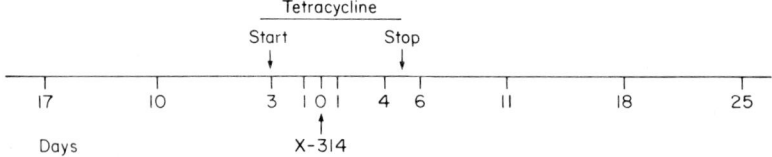

Fig. 1. Time sequence of faecal collection, tetracycline administration and ingestion of *E. coli* x-314. Time is relative to ingestion of x-314.

Subjects were given either 1000 mg (250 mg × 4 times a day), 50 mg (12.5 mg × 4 times a day) or placebo (lactose 4 times a day), orally for 9 days. On the 4th day after tetracycline administration was begun, subjects were given $1\text{-}2 \times 10^6$ viable x-314 organisms in 0.1 ml of tryptose broth on a sugar cube. Control groups received a sugar cube with 0.1 ml of sterile tryptose broth.

(d) *Handling of faecal specimens*

As soon as faecal specimens were obtained, *c.* 1-2 g were taken, weighed, homogenized and diluted serially 10-fold in phosphate buffered saline (0.01 M, pH 7.0). One-tenth ml of each dilution was plated on MacConkey agar, containing 25 µg/ml of tetracycline and MacConkey agar containing 100 µg/ml of nalidixic acid. Plate counts were determined 18-24 h later and the proportion of nalidixic acid and/or tetracycline resistant *E. coli* determined. Five colonies growing on the plate containing nalidixic acid were serotyped in order to confirm the presence of x-314.

Five to ten colonies from the tetracycline plate were identified as *E. coli* using the IMViC series of biochemical tests. The bacteria were then tested for antibiotic resistance patterns according to the method of Bauer *et al.* (1966) using high concentration discs of the following antibiotics: tetracycline, sulfathiazole, streptomycin, kanamycin, neomycin, ampicillin, cephalothin, chloramphenicol, colymicin, nitrofuratoin, gentamicin and nalidixic acid.

Twenty organisms were picked from the plain MacConkey plate and identified as *E. coli*. These organisms were further characterized as to phage type using a set of 24 phages. The method used was a modification of that described by Parisi *et al.* (1969). These phages were isolated from both human and animal sources. Suspensions of the 24 phages were dropped on a plate of trypticase-soy agar which had just been streaked with a log phase culture of the organism to be tested. Phage types were read after 4 h incubation at 37°. Organisms from the tetracycline plate which were shown to have the x-314 pattern of antibiotic resistances minus nalidixic acid were phage typed and serotyped to ascertain whether or not they were x-314.

Minimal inhibitory concentrations for the antibiotics to which x-314 was resistant were determined on isolates suspected of receiving the x-314 R-factor.

(e) *Statistical tests*

Data on colonization and tetracycline resistance were analysed using the Mann-Whitney U-Test. P-values were determined using a one-tailed test.

3. Results

Faeces from subjects not receiving x-314, and from subjects prior to ingestion of x-314, did not contain coliform organisms with the characteristics of x-314.

The effect of the amount of ingested tetracycline upon the proportion of the total coliforms found to be x-314 and the length of time x-314 was found in the faeces is shown in Fig. 2. Since, in most cases, daily samples were not taken, it was assumed that if x-314 was found in 2 consecutive samples, it would also be found on the intervening days. If the proportion of x-314 was different in a later sample, the proportion depicted in Fig. 2 for the intervening days was the lesser of the two.

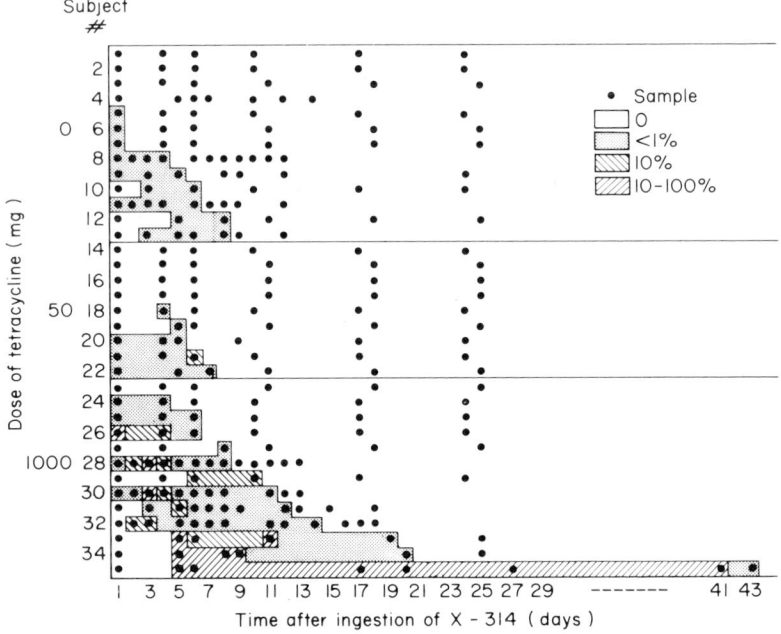

Fig. 2. The effect of the amount of tetracycline upon the proportion of the total coliforms found to be x-314 and the length of time x-314 was detected in the faeces. Each number on the vertical scale represents a volunteer. Numbers on the horizontal scale represent days since ingestion of x-314.

The number of days x-314 was found in the faeces was significantly longer ($P<0.005$) in those individuals who ingested 1000 mg of tetracycline compared to those receiving no tetracycline and significantly longer ($P<0.005$) than those receiving 50 mg of tetracycline. There was no difference ($P>0.05$) in the length of time x-314 was shed in the faeces of subjects taking 50 mg of tetracycline or none.

The maximum proportion of the total number of coliforms found to be x-314 was significantly higher ($P<0.005$) in those subjects receiving 1000 mg of tetracycline compared with those not receiving tetracycline, and significantly higher ($P<0.01$) than those receiving 50 mg of tetracycline. There was no difference ($P>0.05$) between those subjects receiving 50 mg of tetracycline or none.

The effect of tetracycline upon the proportion of total coliforms resistant to tetracycline is shown in Figs 3, 4 and 5. As for the colonization chart, if tetracycline resistant bacteria were found in 2 consecutive samples, it was assumed it would be found on intervening days. Likewise, if the proportion was different in a later sample, the proportion depicted in Figs 3, 4 and 5 for the intervening days was the lesser of the two.

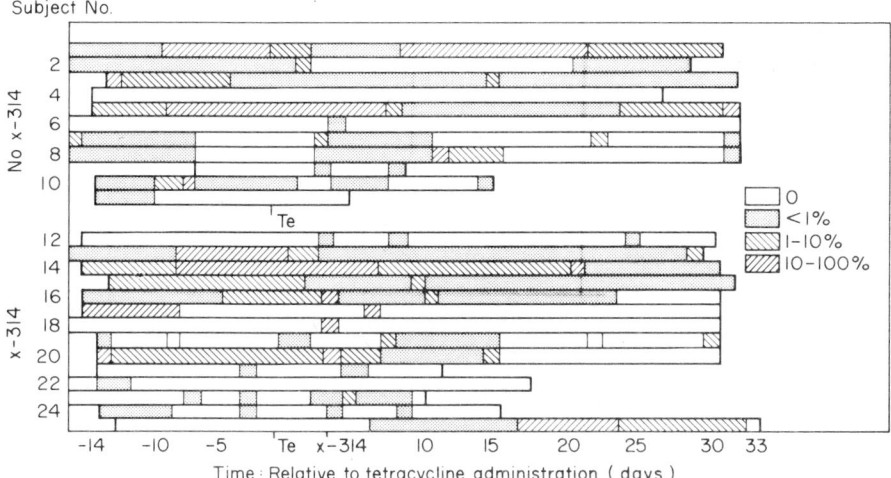

Fig. 3. Percentage of tetracycline resistant *E. coli* in faeces of subjects receiving no antibiotic. Each number on the vertical scale represents a volunteer. Numbers on the horizontal scale represent days prior to or after ingestion of x-314. The shaded areas indicate the percentage of faecal *E. coli* resistant to tetracycline.

Fifty-five of 63 people used in this study excreted tetracycline resistant bacteria at some time prior to tetracycline administration. Twenty-four people had detectable levels of tetracycline resistant bacteria in every faecal specimen submitted. In five of these 24 individuals tetracycline resistant bacteria comprised 10-100% of the flora in all samples. Only 2 of 63 persons never excreted tetracycline resistant coliforms. One of these was in the no tetracycline control group and the other was given 50 mg of tetracycline.

The data on the control groups receiving no tetracycline indicated wide variation both in the number of individuals excreting tetracycline resistant bacteria and the ratio of tetracycline resistant to total coliform bacteria.

Statistical tests showed that ingestion of x-314 in the no tetracycline group had no significant effect on the proportion of tetracycline resistant coliforms in the faeces.

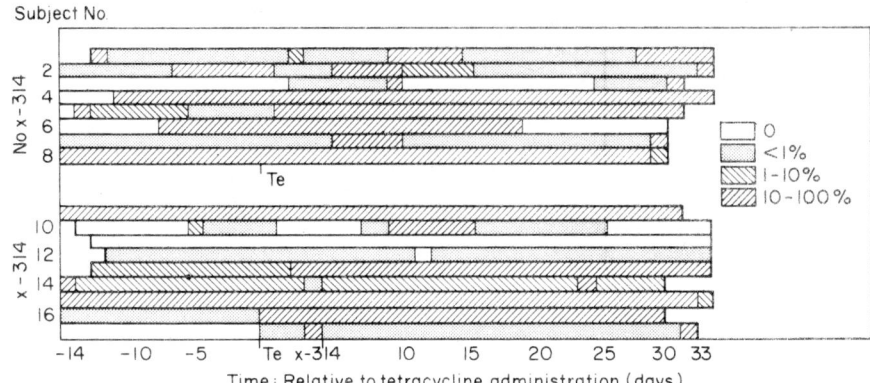

Fig. 4. Percentage of tetracycline resistant *E. coli* in faeces of subjects receiving 50 mg/day of tetracycline. Each number on the vertical scale represents a volunteer. Numbers on the horizontal scale represent days prior to or after ingestion of x-314. The shaded areas indicate percentage of faecal *E. coli* resistant to tetracycline.

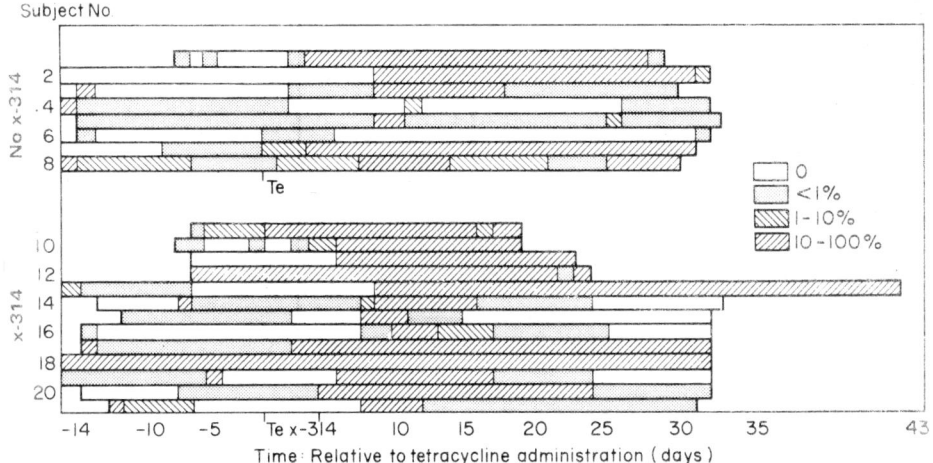

Fig. 5. Percentage of tetracycline resistant *E. coli* in faeces of subjects receiving 1000 mg/day tetracycline. Each number on the vertical scale represents a volunteer Numbers on the horizontal scale represent days prior to or after ingestion of x-314. The shaded areas indicate percentage of faecal *E. coli* resistant to tetracycline.

Twelve of 25 people in the control group excreted tetracycline resistant coliforms as a high proportion (10-100%) of the total coliforms at some time during the study.

In the 50 mg group, 2 people did not excrete tetracycline resistant bacteria until the antibiotic was administered. Statistical tests showed that 50 mg of tetracycline did not have a significant effect on the occurrence of tetracycline resistant bacteria in the faeces.

Giving 1000 mg of tetracycline/day for 9 days had the most pronounced effect on tetracycline resistance in faecal specimens. Nineteen of 21 persons in this group excreted tetracycline resistant bacteria as 10-100% of the total coliform flora after tetracycline administration. The duration of this high level shedding varied from 2 to 43 days. The other 2 subjects excreted tetracycline resistant organisms but in lesser amounts. Statistical tests indicated that 1000 mg of tetracycline significantly influenced the proportion of tetracycline resistant coliforms shed in faeces.

As was mentioned previously, organisms isolated showing the x-314 pattern of transferable resistance were phage typed, serotyped and subjected to MIC determination to ascertain whether *in vivo* transfer of resistance had actually taken place. In faecal samples collected 41 and 43 days after x-314 was ingested, one subject in the 1000 mg tetracycline group excreted organisms with the x-314 pattern of transmissible resistance. These organisms were isolated from tetracycline media at levels of 10^8 organisms/g of faecal material. The proportion of the resident *E. coli* flora which acquired this R-factor varied from 46 to 65% in the 2 faecal specimens in which it was found.

These organisms were found to be of O serogroup 25, H14; x-314 was 0-51, H20. The phage type of these organisms differed from x-314 also. Recipient strains were lysed by the following phages: E, M B4, B10, B21 and B24 but x-314 was not lysed by any of them.

Determination of minimal inhibitory concentrations for 8 antibiotics further confirmed that *in vivo* transfer of resistance had occurred. Minimal inhibitory concentrations of tetracycline (800 μg/ml), streptomycin (25 μg/ml), ampicillin (400 μg/ml), kanamycin (400 μg/ml), neomycin (400 μg/ml), chloramphenicol (100 μg/ml) and nitrofuratoin (100 μg/ml), were identical to those for x-314. The above tests were considered sufficient evidence to indicate *in vivo* transfer of antibiotic resistance from x-314 to resident enteric flora.

In a specimen collected 6 months later, organisms of the O-25, H-14 group were still present at levels of 10^3/g of faecal material. These *E. coli* were resistant to tetracycline and chloramphenicol. These 2 resistance markers (tetracycline and chloramphenicol) were able to be transferred *in vitro*. The MIC for these 2 antibiotics was the same as for x-314. *In vitro* transfer experiments showed that the pattern of resistance to tetracycline and chloramphenicol was transferred frequently from x-314 to a K-12 recipient.

4. Discussion

The criteria used in this study to determine the effect of tetracycline on the establishment of an *E. coli* of bovine origin in the intestinal tract of human volunteers were: the proportion of the total number of coliforms found to be the ingested animal strain and the length of time the organism was found in the faeces compared to subjects not receiving the antibiotic.

That 1000 mg of tetracycline/day potentiated the establishment of a tetracycline resistant *E. coli* of bovine origin is clearly indicated from our data. X-314 in these subjects was shed for a significantly longer period of time and comprised a significantly higher proportion of the total coliforms than subjects not ingesting tetracycline.

On the other hand, 50 mg of tetracycline/day affected neither the proportion of the total coliforms found to be x-314 nor the length of time it was shed when compared to subjects not receiving the antibiotic.

Surprisingly, the dose of 50 mg of tetracycline/day did not change significantly the number nor the length of time x-314 was excreted. Only at the therapeutic level of 1000 mg/day did potentiation occur. A simple selective mechanism could account for these observations. That is, 100 mg of tetracycline may have reduced the intestinal flora (both anaerobic and aerobic) to a greater degree than did the smaller dose because there would undoubtedly be more bacteria sensitive to the levels of tetracycline reached by ingestion of 1000 mg/day than by 50 mg/day. X-314 introduced into this milieu would have an advantage.

Our data suggest, therefore, that if establishment of an animal strain of *E. coli* is to occur in the intestinal tract of man, a possible way to potentiate this establishment is to ingest concomitantly high levels of an antimicrobial agent to which the ingested organism is resistant.

The original outline for these experiments specified the use of subjects excreting no tetracycline resistant coliforms. It soon became apparent that individuals entirely and consistently free of tetracycline resistant enterics were rare. Therefore, the experimental design had to be modified to include persons excreting resistant bacteria. The occurrence and frequency of tetracycline resistant *E. coli* in faeces was not a surprising finding. Linton *et al.* (1972) found that 18-47% of adults tested, depending on occupation, shed tetracycline resistant coliform bacteria. Likewise Datta (1965) reported that 52% of patients prior to hospital admission were shedding tetracycline resistant *E. coli*. In our study we found that 87.5% of persons excreted tetracycline resistant bacteria at some time prior to tetracycline administration.

Several factors may account for the high incidence of tetracycline resistant coliforms seen here. There may have been direct selection for tetracycline resistant *E. coli* by minute quantities of this antibiotic in the food chain. Or the reason may be that there exists a vast environmental pool of resistant *E. coli* to which man is constantly exposed. This could account for the variability seen in this study.

Data obtained by Mercer *et al.* (1971) and by Loken *et al.* (1971) indicated that as many as 85% of animals tested possessed *E. coli* with R-factor mediated multiple antibiotic resistance. Colonization of man with animal types of *E. coli* containing R-factors or *in vivo* transfer of R-factors of animal origin could account for the results seen here. The obvious source for these bacteria is the food man consumes because antibiotic resistant bacteria have been found in numerous food products.

In our studies tetracycline seemed to have a more lasting effect on presence and percentage of tetracycline resistant coliforms shed in faeces than has been reported by other workers. Seventeen of 21 (81%) individuals receiving 1000 mg of tetracycline were still shedding tetracycline resistant bacteria at the termination of the experiment 30 days later. One individual in this group consistently shed tetracycline resistant bacteria as a high proportion (10-100) of the total coliforms for >6 weeks. In the 50 mg group, 14 of 17 (82%) individuals excreted tetracycline resistant *E. coli* at the termination of sampling.

Previous investigators (Datta *et al.*, 1971; Linton *et al.*, 1972) found that, immediately after the withdrawal of tetracycline, a decline in the number of individuals with tetracycline resistant *E. coli* and number of resistant *E. coli*/sample was noted.

As have other workers (Guinee, 1970) we found that tetracycline influenced *in vivo* transfer of antibiotic resistance. However, in our study the effect was more of an indirect one. Guinee (1970) observed enhancement of *in vivo* transfer in rats while they were consuming drinking water containing 20 or 100 p/m of tetracycline. No resistance transfer was evident in our studies until 36 days after tetracycline administration had terminated. The effect of tetracycline was indirect in that 1000 mg of tetracycline selected for resistant organisms, altering the flora of this person to a predominantly resistant coliform population. The 1000 mg dose apparently prolonged colonization by the tetracycline resistant x-314. This allowed more opportunity for resistance transfer to occur.

It should be noted that when transfer did occur, it involved a large percentage of the resident *E. coli*. At the time transfer was first apparent, 46% of the total coliform flora had acquired the x-314 pattern of resistance.

Six months later this person was still excreting *E. coli* with 2 of the resistance markers of the ingested organism. It has been reported by Smith (1969) that resident organisms which acquired resistance did not persist long in the intestinal tract. In our study, resistant resident *E. coli* were able to compete successfully with sensitive strains for long periods. It should be noted that persistence occurred in the absence of antibiotics.

It is interesting to speculate why transfer took so long to occur and why when it did happen, such a large percentage of the coliform flora was involved. Perhaps x-314 did not have a suitable recipient previous to that time. Most of the *E. coli* subsequent to tetracycline administration were multiple drug resistant by virtue of R-factors. Perhaps these R-factors were of the same compatibility groups as the x-314 R-factor and transfer could not occur (Vapnok, Lipman & Rupp, 1971; Grindley, Humphreys & Anderson, 1973). Transfer may not have occurred until a compatible R-factor-carrying bacterium appeared or an antibiotic sensitive *E. coli* was introduced. Unfortunately, serotype and phage type data on sensitive *E. coli* prior to transfer were not available to determine if this was the case.

In conclusion, the oral administration of tetracycline was found to enhance the colonization of the intestinal tract of man by an *E. coli* of animal origin. Tetracycline also influenced *in vivo* transfer of antibiotic resistance from the ingested organism to resident *E. coli*.

5. Acknowledgements

The work upon which this publication is based was performed pursuant to contract FDA-71-306 with the Public Health Service, Food and Drug Administration, Department of Health, Education and Welfare.

We wish to thank Dr. H. Williams Smith for providing *E. coli* strain x-314, Dr. Peter Gemski for performing the rabbit ligated gut loop assay. We are indebted to Joan Ahart, Ronald Macedo and Kaaren Sloan for expert technical assistance and to Robert Tsutakawa and Russ Yang for statistical computations.

6. References

ANDERSON, E. S. (1965). Origin of transferable drug resistance in the Enterobacteria. *Brit. med. J.* **2**, 1289.

ANDERSON, E. S. & LEWIS, M. J. (1965). Drug resistance and its transfer in *Salmonella typhimurium. Nature, Lond.* **206**, 579.

ANDERSON, E. S. (1968). The ecology of transferable drug resistance in the enterobacteria. *Ann. Rev. Microbiol.* **22**, 131.

BAUER, A. W., KIRBY, W. M. M., SHERRIS, J. C. & TURCK, M. (1966). Antibiotic susceptibility testing by a standardized single disc method. *Am. J. clin. Path.* **45**, 493, 496.

BOHUS, J. (1971). Acceptance of R-factor by a *Shigella sonnei* strain from commensal *Escherichia coli* strains during dysentery. *J. Hyg. Epidemiol. Microbiol.* **15**, 225.

COOKE, E. M., HETTIARATCHY, I. G. T. & BUCK, A. C. (1972). Fate of ingested *Escherichia coli* in normal persons. *J. med. Microbiol.* **5**, 361.

DATTA, N. (1965). Drug resistance and R-factors in the bowel bacteria of London patients before and after admission to hospital. *Brit. med. J.* **2**, 407.

DATTA, N., FAIERS, M. C., REEVES, D. S., BRUMFITT, W., ORSKOV, F. & OESKOV, I. (1971). R-factors in *Escherichia coli* in faeces after oral chemotherapy in general practice. *Lancet* **1**, 312.

DEAN, A. G., CHING, Y., WILLIAMS, R. G. & HARDEN, L. B. (1972). Test for *Eschericnia coli* enterotoxin using infant mice: application in study of diarrhea in children in Honolulu. *J. infect. Dis.* **125**, 407.

GRINDLEY, N. D. F., HUMPHREYS, G. O. & ANDERSON, E. S. (1973). Molecular studies of R factor compatibility groups. *J. Bact.* **115**, 387.

GUINEE, P. A. M. (1965). Transfer of multiple drug resistance from *Escherichia coli* to *Salmonella typhimurium* in the mouse intestine. *Antonie van Leeuwenhoek* **31**, 314.
GUINEE, P. A. M. (1970). Resistance transfer to the resident intestinal *Escherichia coli* of rats. *J. Bact.* **102**, 291.
JONES, A. M. (1971). *Escherichia coli* in retail samples of milk and their resistance to antibiotics. *Lancet* **I**, 347.
KASUYA, M. (1964). Transfer of drug resistance between enteric bacteria induced in the mouse intestine. *J. Bact.* **88**, 322.
KIM, T. K. & STEPHENS, J. F. (1972). Drug resistance and transferable drug resistance of *Escherichia coli* isolated from 'ready-to-cook' broilers. *Poultry Sci.* **51**, 1165.
LaBREC, E. H., SCHNEIDER, H., MAGNAI, T. J. & FORMAL, S. B. (1964). Epithelial cell penetration as an essential step in the pathogenesis of bacillary dysentery. *J. Bact.* **88**, 1503.
LINTON, K. B., LEE, P. A., RICHMOND, M. H., GILLESPIE, W. A., ROWLAND, A. J. & BAKER, V. N. (1972). Antibiotic resistance and transmissible R-factors in the intestinal coliform flora of healthy adults and children in an urban and rural community. *J. Hyg., Camb.* **70**, 90-104.
LOKEN, K. I., WAGNER, L. W. & HENKE, C. L. (1971). Transmissible drug resistance in Enterobacteriaceae isolated from calves given antibiotics. *Amer. J. vet. Res.* **32**, 1207.
MERCER, H. D., PERCURULL, D., GAINES, S., WILSON, S. & BENNETT, J. Y. (1971). Characteristics of antimicrobial resistance of *Escherichia coli* from animals: relationship to veterinary and management uses of antimicrobial agents. *Appl. Microbiol.* **22**, 700.
MOORHOUSE, E. C., O'GRADY, M. & O'CONNOR, H. (1969). Isolation from sausages of antibiotic resistant *Escherichia coli* with R-factors. *Lancet* **II**, 50.
PARISI, J. T., RUSSELL, J. C. & MERLO, R. J. (1969). Bacteriophage typing as an epidemiological tool for urinary *Escherichia coli*. *Appl. Microbiol.* **17**, 721.
REED, N. G., SIECKMANN, D. G. & GEORGI, C. E. (1969). Transfer of infectious drug resistance in microbially defined mice. *J. Bact.* **100**, 22.
SALZMAN, T. & KLEMM, L. (1968). Transfer of antibiotic resistance (R-factor) in the mouse intestine. *Proc. Soc. exp. Biol.* **128**, 392.
SERENEY, B. (1955). Experimental Shigella keratoconjunctivitis. *Acta microbiol. Acad. Sci. Hung.* **2**, 293.
SHOOTER, R. A., COOKE, E. M., ROUSSEAU, S. A. & BREADEN, A. L. (1970). Animal sources of common serotypes of *Escherichia coli* in the food of hospital patients. *Lancet* **II**, 226.
SMITH, H. W. & HALLS, S. (1967). Observations by the ligated intestinal segment and oral inoculation methods on *Escherichia coli* infections in pigs, calves, lambs, and rabbits. *J. Path. Bact.* **93**, 499.
SMITH, H. W. (1969). Transfer of antibiotic resistance from animal and human strains of *Escherichia coli* to resident *E. coli* in the alimentary tract of man. *Lancet* **I**, 1174.
VAN HOUWELING, C. D. & KINGMA, F. J. (1969). The use of drugs in animals raised for food. *J.A.V.M.A.* **155**, 2197.
VAPNOK, D., LIPMAN, Muriel B. & DEAN RUPP, W. (1971). Physical properties and mechanism of transfer of R-factors in *Escherichia coli*. *J. Bact.* **108**, 508.
WALTON, J. R. (1966). Infectious drug resistance in *Escherichia coli* isolated from healthy farm animals. *Lancet* **II**, 1300.
WALTON, J. R. (1971). The public health implications of drug resistant bacteria in farm animals. *Ann. N.Y. Acad. Sci.* **182**, 358.
WIEDEMANN, B., KNOTHE, H. & DOLL, E. (1970). Übertragung von R-Faktoren in der Darmflora des Menschen. *Zentbl. Bakt. ParasitKde* Abt. I, Orig. **213**, 183.

Viruses Associated with the Healthy Individual

SYLVIA E. REED AND D. A. J. TYRRELL

Clinical Research Centre,
Watford Road, Harrow, Middlesex HA1 3UJ, England

CONTENTS

1. Introduction 255
2. Viruses of healthy children 255
3. References 257

1. Introduction

THE NORMAL BACTERIAL flora consists of micro-organisms which live in a stable relationship with a host on a surface or in a body cavity. The organism receives nutriment and a home from the host, and it may be that the host receives something in return but it does not eliminate the organism by an immune response or in some other way. A normal viral flora with this type of relationship has not been described in man, but many viruses may be found in healthy subjects, early in life especially; these will be discussed first.

2. Viruses of Healthy Children

The picornaviruses are small and contain an RNA genome within a protein coat constructed with icosahedral symmetry, they are subdivided into rhinoviruses and enteroviruses. The latter include the large group of echoviruses which were so named (Enteric Cytopathic Human Orphan viruses) because they were found in the faeces of healthy infants and were 'viruses in search of a disease'. Subsequent research has indicated that certain serotypes really do not cause disease, though more often — as in the case of the polioviruses — it turns out that they cause disease at times but that infections are usually asymptomatic. Nevertheless, the viruses are shed from the throat for only a few days and in faeces for only a few weeks, then they are eliminated and the subject is resistant to reinfection, probably because he now possesses specific antibodies. Such antibodies can usually be found in the circulation but also, with appropriate techniques, in the intestinal or other secretions. Since there are many serotypes of enteroviruses, children may be found to be excreting one of this group of viruses for a significant proportion of the summer and autumn. Nevertheless after some years they will have acquired immunity to most of the current serotypes and thereafter viruses are rarely to be found in the faeces. Until that time, however, enteroviruses are often found by coincidence in children suffering from a variety of infectious diseases, and this must be borne in mind by all those trying to establish the significance of such isolations or trying to discover the aetiology of such diseases.

Adenoviruses were found first in the tonsils and adenoids removed from children by surgery. Although these children were not healthy and the tonsils, one assumes, were usually enlarged or otherwise abnormal, there was no evidence of an acute infection and it is clear that adenoviruses can not only be shed from the throat of children but can persist in the tissues for a long time. However, if a simple saline extract of the tissue is made and tested, no virus is found, probably because the tissue contains also antiviral antibody. If, instead, the tissue is cultivated it grows at first and then degenerates due to the multiplication of the virus which can then be passaged easily by inoculating it into cultures of any virus-susceptible cell.

There are more striking examples of the persistence of viruses in tissues in members of the herpes virus group. The Epstein-Barr (EB) virus is one which has recently been recognized. It infects a large proportion of the population in childhood and adolescence and is now known to cause glandular fever (infectious mononucleosis) in a substantial proportion of infections from teenage onwards. The virus was first recognized as particles in the cells cultured from the tumours of children with Burkitt's lymphoma. The virus was then found to be present in the cells obtained by culturing the white cells of normal subjects. Furthermore, the cells from persons who have not had this infection cannot be persuaded to grow persistently in the laboratory. It therefore looks as though this virus can persist in at least some cells of the patient, apparently for the whole of his life. No virus proteins or particles can be detected until the cells are cultured. Here, then, is a virus which is normally carried by almost every healthy person. *Herpes simplex* viruses persist in a somewhat similar but more familiar way. The infection is acquired in early life; then it often causes a sore throat. It then disappears and may cause no disease at all, although antibodies against it persist, and then, in an attack of pneumonia, it may multiply freely and cause disease, and virus particles may easily be found in the vesicles. We do not know how the virus is being held in check or how infections, sunshine, or even emotional or hormonal changes can stir it into life, but there is now a good deal of evidence, particularly for EB virus, that the nucleic acid, DNA, becomes closely associated with that of the cell and presumably replicates with it, very much as the DNA of the temperate phages found in lysogenic bacteria. This type of thing certainly seems to happen with certain of the viruses which cause tumourous changes in cells in the laboratory. In the last few years we have some evidence that *herpes simplex* viruses may cause some tumours, in particular, that the herpes type 2 virus which characteristically infects the genital tract, may be involved in the causation of carcinoma of the cervix.

It is suspected, because antibodies which appear after an infection persist for the rest of life, that yellow fever and measles viruses can also continue to survive throughout life after the acute infection. The measles virus sometimes causes a progressive and fatal infection of the brain (subacute sclerosing panencephalitis, or SSPE) some years after the acute infection. The virus is not found free in the brain cell but often virus antigen is found there and, if the cells are cultured, virus can be obtained from them, though it may be necessary to cultivate them with virus-susceptible cells (permissive cells) in order for virus to be produced in a fully infectious form. Looked at from the point of view of the virus, the strange type of multiplication may be an adaptation to growth in an environment surrounded with antibody. There is perhaps less chance that it will be shed into the surroundings and be inactivated and it may be an alternative manoeuvre for an RNA virus which cannot replicate its nucleic acid with the DNA of the host cells. Something rather similar can happen when the virus is grown in tissue cultures containing antibody. Most cells produce antigen, but complete virus particles are rare and the cells themselves continue to multiply. Nevertheless, usually when a virus multiplies and the person infected remains well, the reason is that although perhaps millions of cells are infected and damaged the majority are normal and it is only when a high proportion of cells, perhaps of the mucous membranes or in the CNS, become involved that recognizable illness occurs.

3. References

GINSBERG, H. S. & DINGLE, J. H. (1965). The adenovirus group. In *Viral and Rickettsial Infections of Man,* 4th ed. Eds F. L. Horsfall & I. Tamm. Philadelphia and Montreal: Lippincott.

MELNICK, J. L. (1965). Echoviruses. In *Viral and Rickettsial Infections of Man,* 4th ed. Eds F. L. Horsfall & I. Tamm. Philadelphia and Montreal: Lippincott.

RAPP, F. (1973). Do herpesviruses cause cancer? Of course they do. *J. nat. Cancer Inst.* **50,** 825.

TERMEULEN, V., KATZ, M. & MÜLLER, D. (1972). Subacute sclerosing panencephalitis: a review. *Current Topics in Microbiology and Immunology* No. 57.

Subject Index

Acetone, reduction in self-sterilizing power of skin caused by, 24
Acquired pellicle, 113, 114
Actinomyces spp., in calculus formation, 90
 in dental plaque, 119
 in infants' mouths, 49
 on tooth surfaces, 54, 56, 57, 58, 88
Act. israeli, in mouth, 71
Act. naeslunde, in mouth, 71
Act. odontolyticus, in mouth, 71
Act. viscosus, in mouth, 71
Actinomycetes, in plaque, 115
Adenoviruses, in children, 256
Aerobacter, inhibition of, 123
Agar masking technique, 177
Anaerobic chamber, 62
Anthrax bacillus, inhibition of growth of, 122
Antibiotic resistant bacteria, in livestock, 241
Antibiotics, effect of on plaque, 102, 103
Apples, for controlling dental plaque, 87
Arachnia, 72
Aspergillus awamori, in scalp, 14
Asp. fumigatus, 14
Axilla, apocrine gland secretion from, 21
 effects of demethylchlortetracycline on flora of, 36
 effect of hexachlorophene on, 40
 pH value of, 21
Bacillus spp., inhibition of, 26
 in the mouth, 71
 in the scalp, 14
B. melaninogenicus, 63
 in microbial interrelationship, 116
B. oralis, 63
B. prodigiosus, saliva contaminated with, 144
Bacteriocins, activity of, 26, 27, 28
 definition of, 25
 formation of, 3
 inhibitory effect of, 92
Bacterionema matruchotii, isolation of, 71
Bacteroides spp., as an energy source, 61
 Clostridium perfringens grown with, 208, 209
 culturing of, 4
 in dental plaque, 54, 55, 58, 72, 101
 in faecal specimens, 239
 in infants' mouths, 49
 in intestine, 190, 191
 in intestine of children, 198

Bacteroides spp.,–*cont.*
 in the vagina, 161, 181
 polysaccharide synthesis by, 94
 resistance of to aminoglycoside antibiotics, 210
 taxonomy of, 72, 73
Bact. fragilis, 63, 190, 191, 232, 234, 236
Bact. matruchotii, in batch cultures, 92
Bact. melaninogenicus, in gingival crevice, 60, 87
 on tongue, 51
 prevalence of at different ages, 50
 taxonomy of, 73
Bifidobacterium, 72, 234, 236
 culturing of, 4
Bifid. adolecentis, in Japanese faeces samples, 191
Bile salts, effect of on bacterial survival, 188, 190
Bowel, bacterial colonization of, 229
 Clostridium perfringens in, 216
Bowel surgery, administration of antibiotics prior to, 210
Burkitt's lymphoma, 256
Butyrivibrio fibrisolvens, 63
Calculus, production of, 87, 90, 91
Callus, effects of hydration on bacterial growth in, 19
Candida spp., in faecal specimens, 230, 232, 243
 in infants' mouths, 49
 in the skin, 8
 in the vagina, 159, 161, 165, 171, 179
Cand. albicans, and use of medicated soaps, 40
 effect of steroid-antibiotic mixture on, 44
 in the mouth, 73
 in the vagina, 164, 165, 171, 173, 179, 180, 181
 on the tongue, 103
 rash caused by, 43
 survival of on skin, 24
Cand. krusei, in the vagina, 164
Cand. norvegensis, in the vagina, 164
Carbohydrate, in production of caries, 112
Carcases, multiple resistant bacteria in, 242
Caries, dental, 111, 112, 121
 aetiology of, 101
 causes of, 47
 control of, 108
 effect of fluoride on, 95

INDEX

Caries—*cont.*
  fatty acids as causative factor in, 123, 124, 125
  relation to oral organisms, 94, 95, 120, 126
Cattle, multiple resistant *E. coli* in, 241
Cephalexin, 230, 232, 236, 238
Cervicitis, during pregnancy, 171
Cervix, carcinoma of, 256
Chickens, multiple resistant bacteria in, 242
Children, viruses of, 255, 256
Chlamydia, in the vagina, 181, 182
Chlam. trachomatis, in the vagina, 182
Chloramphenicol, treatment of typhoid patients with, 225
Chlorhexidine, effect of on oral flora, 103, 104, 105, 106, 107, 108
  site of action of, 106
Cholera, epidemics of, 221, 222
Clindamycin, effects of, 37, 234, 236, 238
Clostridia, in faeces of food poisoning sufferers, 212
  in the intestine, 192
Clostridium spp., 232, 234, 236, 239
Cl. bifermentans, in the intestine, 191
Cl. difficile, 234
Cl. innocuum, 234
Cl. novyi type A, 63
Cl. oroticum, 234
Cl. paraputrificum, 192
Cl. perfringens, in the intestine, 191
Cl. ramosum, 234, 238
Coliforms, in the mouth, 72
Colony count, of mixed cultures, 4
Corynebacteria, in the vagina, 161
Corynebacterium spp., in infants' mouths, 49
  in plaque, 115
  in the vagina, 177
  isolation, 71
  on teeth, 188
Coryn. acnes, effect of skin surface lipids on, 24
  inhibition of, 26
  in sebaceous ducts, 17
  recovery of after shampooing scalp, 39
Coryn. diphtheriae, 25, 26
Coryn. minutissimum, 10
Coryn. tenuis, in the hair, 15
Coryn. vaginale, 174
Coryn. xerosis, 26
Crevicular fluid, 92, 93, 96
Cutaneous biocenose, members of, 7, 8
Cyclacillin, 230, 232, 236, 238
Cyclamate-dependent changes in metabolism by flora, 189

Demodex folliculorum, 7
Dense-flora occlusion test, 41
Dental plaque, 53, 54, 55, 56, 57, 58, 59
  bacteria of, 112
  bacterial flora of, 56, 57
  calcification of, 90, 91
  control of, 102, 103, 108
  effect of sugar intake on, 93, 94
  effects of fluoride on, 95, 96
  in primitive peoples, 62
  mechanical removal of, 87, 88
  microscopic count of, 101
  pH range of, 90, 95
Diarrhoea, in infants, 200, 201, 202
  post-gastrectomy, 216
Diphtheroids, anaerobic, 10, 101
  facultative, 101
  in the axilla, 9, 10, 21
  in the mouth, 71
  lipophilic, 16, 17, 23, 24
  non-lipophilic, 17
  on the glabrous skin, 10
  on the skin, 8
  polysaccharide syntheses by, 94
Discharge in pregnancy, 171, 181
Döderlein's bacilli, in the vagina, 158
Droplet nuclei, diameter of, 136
Dysentery, outbreak of, 243
Echoviruses, 255
Entamoeba gingivalis, 73
Enteritis necroticans, 211, 215
Enterobacteria, in axilla, 40
  in occlusion tests, 42
  in S. Indian subjects, 190
Enterobacteriaceae, in the vagina, 161
Enterococci, in specimens from Japan and Uganda, 191
Enteroviruses, 255
Epidermophyton, 14
Epithelium, 112, 113
Epstein-Barr (EB) virus, 256
Escherichia coli, as index of faecal contamination, 207
  colicines in control of, 190
  in faecal specimens, 232, 239
  in the gut, 2
  in the hair, 14
  in the intestine of children, 198, 199, 200, 201, 202
  in the posterior fornix, 162, 179
  inhibition of, 123
  pathogenic strains of in piglets, 211
  R factor transfer to *Shigella sonnei,* 243
  resistance to saturated free fatty acid, 24

Eubacteria, in specimens from India and Japan, 191
Eubacterium spp., 72, 232, 234, 236, 239
Eubact. aerofaciens, in Japanese samples, 191
in faecal specimens, 234
Eubact. lentum, 234
Eubact. ventriosum, 234
Faecal flora, 188, 189, 191, 206, 207, 238
Faeces, of babies, *Clostridium perfringens* in, 206
Fatty acids, antibacterial properties of, 23, 24
Fluoride, effect of, 95, 96
Food chain, tetracycline resistant *E. coli* in, 250
Food poisoning, *Cl. perfringens*, 211, 212, 213, 214, 215, 216
salmonella, 225
Foods, contamination of by *Cl. perfringens*, 208, 212
effects of on caries, 87
Foot and mouth disease, 149
Fornix, posterior, flora of, 173, 179
Fusobacteria, in dental plaque, 56, 57, 101, 115, 116
polysaccharide synthesis by, 94
salivary levels of, 52, 53
Fusobacterium spp., as an energy source, 61
culturing of, 4
in batch cultures, 92
in dental plaque, 52, 56, 57, 72
in infants' mouths, 49
Fuso. nucleatum, 63
Gastrectomy, partial, *Cl. perfringens* in stomach remnants after, 209
Gastric juice, bacterial contents of, 198, 199
Genito-urinary system, diseases of, 2
Gingiva, bacterial flora of, 104, 106
Gingival crevice, bacterial population, 59
effects of IgA and IgM in, 93
Gingival margin, flora at, 103, 106
Gingivitis, prevention of, 103
Glandular fever, 256
Glycogen, in vaginal mucosa, 158
Glycoprotein, salivary, 112, 113, 114
Gonococcus, in the vagina, 179
Gut, large, *Clostridium perfringens* in, 207
Haemophilus, in the mouth, 72
in the vagina, 175, 177
H. influenzae, in the vagina, 175
H. parainfluenzae, in the vagina, 175
H. vaginalis, 174, 175
Herpes simplex viruses, 256
Herpes virus group, 256

Herpesvirus hominis, in the mouth, 73
in the skin, 8
Hexachlorophene, use of in the axilla, 40, 41
Hungate roll-tube method, 62
Hydration, of skin, and release of bacteria, 16, 18
of *stratum corneum*, 20, 22, 29
Hydroxyapatite, 108, 113, 121
IgA, in saliva, 92, 93
in the intestine, 188, 194, 195
IgM, in saliva, 92, 93
Immunoglobulins, in saliva, 92, 93
Infant, development of oral flora in, 49, 85, 86
Intubation technique, 197, 198
Kanamycin, in control of dental plaque, 103
Klebsiella-Enterobacter-Serratia group, colonization of the bowel with, 229, 232, 234, 236
Lachnospira multiparus, 63
Lactase deficiency, 192
Lactate, utilization of and effect on caries, 92
Lactobacilli, in faecal specimens, 234, 236
in plaque, 56, 117
in the intestine, 187, 192, 198
in the mouth, 71, 95
in the vagina, 158, 161, 178
Lactobacillus spp., effect of limited carbohydrate consumption on, 94
in dental plaque, 55
in faecal specimens, 232
in infants' mouths, 49
in the vagina, 177
Lact. acidophilus, symbiotic effect with, 116
Lact. casei, in batch culture, 92
Lauric acid, activity of, 23
Leprosy bacillus, 146
Leptothrix racemosa, in dental plaque, 59
Leptotrichia spp., in calculus formation, 90
in dental plaque, 53, 55, 119
in infants' mouths, 49
Lepto. buccalis, 72
Leucocytes, in saliva, 93
Lipids, skin surface, 22, 23, 24, 25
Listeria, inhibition of, 26
in the vagina, 160
Livestock, antibiotic resistant bacteria in, 241
Lysozyme, in skin, 22, 26
in the intestine, 188
in the saliva, 92
Measles, infection by, 142
virus, 257

Media used to study vaginal flora, 159, 160
Medium, for recovery of anaerobic bacteria, 63
Meningococci, inhibition of, 123
Micrococcaceae, on the skin, 17
Micrococci, in the vagina, 162
Micrococcus spp., 51
Micrococcus M2, 9
Micrococcus M3, 9
M. epidermidis, production of inhibitory substances by, 26
M. mucilagenosus, 51
M. pyogenes, production of 'staphylococcins' by, 26
Microsporum, 14
Mic. audouini, 15
Milk, multiple resistant bacteria in, 242
Mitis-salivarius agar, 70
Morbidity, vaginal, 171
Mycobacterium tuberculosis, dispersal of, 152
Mycoplasma, 160, 161
   T strains, 173, 182
Mycop. hominis, in the vagina, 162, 173
Mycoplasmas, genital, 173
   in the mouth, 73
Neisseria spp., adherence of, 64, 65
   identification of, 71
   in dental plaque, 54, 55, 56, 57, 58
   in infants' mouths, 49, 86
   in microbial interrelationships, 116
   in saliva, 52
   in the gingival crevice, 60
   in the vagina, 161, 162
   suppression of staphylococcal growth by, 123
N. catarrhalis, 71, 162
N. gonorrhoeae, in vagina, 174
N. perflava, in batch cultures, 92
N. pharyngis, 71, 162
Neomycin, cream, effect of on *Staph. aureus*, 44
   preparations, use of before bowel surgery, 210
Nocardia spp., in dental plaque, 55, 119, 188
   in infants' mouths, 49, 86
Nose, contamination levels in, 138, 139, 150, 152
Nutrients, on skin surface, sources of, 20
Occlusion of skin *in vivo*, 41
   for testing antimicrobial agents, 40
Occlusion tests, 41, 42, 43
Odontomyces viscosus, in batch culture, 92
Oral hygiene, 102
   cessation of, 104, 105, 106

Papanicolaou smear, 174
Penicillin, 210
   growth-promoting effect of, 209
Peptostreptococci, 101
Peptostreptococcus, in gingival crevice, 60
Pepto. elsdenii, 63
Periodontal disease, 47, 101, 108, 111, 112, 113, 125, 126
Peritonitis, organisms associated with, 210
Peroxide, inhibitory effects of, 92
pH of plaque, 90, 91, 94
   of saliva, 90, 95
Picornaviruses, 255
Pig-bel, 211, 215
Pityrosporum, recovery of, 38, 39
Pityro. orbiculare, 14
   in the skin, 8
Pityro. ovale, inhibitory action against dermatophyte fungi, 15
   in the skin, 8
   sensitivity to undecylenic acid, 24
Plaques, other types of, 121
Pneumococcal strains, pathogenicity of, 123
Polioviruses, 255
Polymyxin, for control of dental plaque, 102, 103
Polysaccharides, 122
   addition of to pure cultures of plaque *Nocardia*, 119
   extracellular, production of, 66, 113, 124, 125, 126
   in fibrillar matrix, 117
   linked by calcium bridging, 115
Premolar teeth, plaque samples from, 57
Primitive peoples, oral flora of, 61, 62
Propionibacterium, 72
Proteus spp., antibacterial substances lethal to, 123
   colonization of bowel with, 229
Pr. mirabilis, in the vagina, 162
Proteus-Providencia organisms, 232, 234
Protozoa, in the mouth, 73
Pruritus, 180, 181
Pseudomonas spp., and use of medicated soaps, 40, 41
   colonization of bowel with, 229, 232, 234
   in the mouth, 72
   in the vagina, 161
Ps. aeruginosa, from human faeces, 239
   inhibition of, 123
   resistance of to saturated free fatty acid, 24
   survival rate of, 24
R (resistance) factors, transfer of, 242, 243, 249, 251, 252

INDEX

Respiratory viruses, dispersal of, 149, 151
Rhinoviruses, 255
Rothia dentocariosa, isolation of, 71
Saliva, antibacterial factors in, 92, 93, 96
    bacterial composition of, 52, 53, 94, 102, 104, 105, 106, 135, 138, 143, 144, 146
    biological role of, 121
    composition of, 89, 90, 91, 94
    effects of, 87, 89
    particle size distribution of, 136
    sugar content of, 91
Salmonellae, in food poisoning, 226
Salmonella typhi, excretion of, 223, 225, 226
Salm. typhimurium, 226, 243
Salmonellosis, 225, 226
Sampling skin bacteria, methods of, 16
Sarcina spp., 17
Satellitism, in skin microbiology, 29
Sausages, multiple resistant bacteria in, 242
Scalp, effects of shampooing, 38, 39
Selective localization of flora, explanations for, 2, 3
Selective media, for culturing body flora organisms, 4
Selenomonas spp., in the mouth, 73
Sel. ruminatium, 63
Septicaemia, bacterial causes of, 179
Skin contamination with *Cl. perfringens*, 210
Soap, effects of, 40, 41, 43, 45
Spirochaetes, in crevicular crevice, 92
    in gingival crevice, 61, 73, 87
    in plaque, 115, 116
Staphylococcal dispersal by babies, 147, 148
Staphylococci, aerobic, control of in newborn infants, 28
    effect of lipids on, 23
    in infants' skin, 25
    in the vagina, 160
    on skin, 117
Staphylococcins, production of, 26, 27
Staphylococcus spp., in infants' mouths, 49
Staphylococcus SII, 9
Staph. aureus, avirulent, penicillin-sensitive strain of, 28
    dispersal of, 146, 149
    effect of neomycin on, 45
    effect of steroid antibiotic mixture on, 44
    'hospital strain' of, 123
    in the axilla, 36
    in the bowel, 229
    in the hair, 14

Staph. aureus—cont.
    in the nose, 3
    in the vagina, 160
    inhibition of, 23
    moisture requirement of, 19
    on the skin, 9
    pyoderma due to resistant strain of, 43
    survival of on the skin, 24, 26, 27, 28
Staph. epidermidis, 22
    in the vagina, 160, 162, 178
    staphylococcin obtained from, 27
Stratum corneum, structure of, 22
Streptococci, α-haemolytic, dispersal of, 143, 144
    effect of chlorhexidine on, 106, 108
    facultative, 101
    in classroom air, 141
    in infants' mouths, 86
    in plaque, 55, 126
    in the intestine, 186
    in the mouth, 67
    in the vagina, 161, 162, 178
    involvement in dental caries, 120
    polysaccharide synthesis by, 94
    role in controlling pathogens, 124
    salivary, 104, 135, 140, 143, 146, 150, 152
Streptococcus spp., in faecal specimens, 232
    in infants' mouths, 49
Strep. brevis, in saliva, 135
Strep. faecalis, 191
Strep. faecium, 191
Strep. milleri, identification of, 57, 70, 168
Strep. mitior, identification of, 68
Strep. mitis, inhibition of growth of diphtheria bacilli by, 123
Strep. mutans, adherence of, 65
    chlorhexidine resistance of, 107
    distribution of, 87
    growth of, 64, 66
    identification of, 68, 69
    in dental plaque, 101
    in infants' mouths, 50
    in microbial interactions, 92, 116
Strep. pyogenes, in the mouth, 67
    in the pharynx of children, 3
    inhibition of, 23, 26
    survival of on skin, 24
Strep. salivarius, adherence of, 64
    detailed examination of, 51
    distribution of, 87
    growth of, 64, 65
    identification of, 68, 69, 70
    in dental plaque, 57
    in microbial interactions, 92
    in saliva, 52

Strep. salivarius–cont.
  in the mouths of newborn babies, 49
  on the tongue, 51
  transmission from mother to baby, 86
Strep. sanguis, 56, 66, 92
  adherence of, 65
  identification of, 68, 69
  in dental plaque, 106, 107
  in microbial interrelationships, 116
  in the mouths of infants, 50, 64
Strep. viridans, 14
  suppression of staphylococcal growth by, 123
Streptomycin, action of, 225
Succinivibrio dextrinosolvens, 63
Sucrose, effect of on plaque, 94
Sugar, effect of on plaque pH, 91, 92, 93, 94
Swallowing, quantitative study of, 86
Sweat, apocrine, 21
  eccrine, 21
Swine, multiple resistant *E. coli* in, 241
TCID (tissue culture infecting doses) of respiratory viruses, 149, 150, 151
Teeth, effects of on development of oral flora, 49, 50, 86, 87, 88
Tetracycline, for control of plaque, 102, 103
Tetracyclines, effects of, 36
Throat, contamination levels in, 138, 150
Tinea capitis, 15
Tinea pedis, 15
Tinea versicolor, relation to *Pityrosporum orbiculare*, 8
Tongue, cultivable flora of, 51
  flora on surface of, 102, 103
Tooth, brushing, 87, 94, 101
  chlorhexidine absorption by, 108
  composition of enamel, 112
  deposits on surface of, 113, 114, 116, 117
  flora on surface of, 102, 104, 106, 119
  plaque composition in various sites, 57, 88, 120
Torulopsis glabrata, in the skin, 8
  in the vagina, 164, 180, 181
Treponema denticola, 61, 63, 73

Trep. dentium, in microbial interrelationships, 116
Trep. macrodentium, 63, 73
Trep. microdentium, in the gingival crevice, 61
Trep. oralis, 63, 73
Trichomonas, in the mouth, 73
  in the vagina, 165
Tr. vaginalis, 161, 165, 174, 181
Trichomycosis axillaris, 15
Trichophyton, 14
Tubercle bacilli, dispersal of, 144, 145
  bacillus, infecting dose, 151
Typhoid bacillus, discovery of, 222
Typhoid fever, epidemiology of, 222, 223
  prevention of, 225
Umbilicus, colonies of aerobic flora on, 18
Vaginitis, 161, 164, 171, 175, 179, 180, 181
Vancomycin, effect of on plaque, 102, 103
Veillonella spp., adherence of, 64, 65
  in dental plaque, 54, 55, 56, 57, 58, 88, 101, 104, 105, 115, 116
  in infants' mouths, 49, 86
  in the gingival crevice, 60
  in the mouth, 70
  lactate utilization by, 66
  on clean tooth surfaces, 54
Veill. alkalescens, 70, 92
Vibrios, in dental plaque, 101, 115, 116
Vibrio cholera, infecting guinea pigs with, 123
V. fetus, 63
V. parvula, 70
V. sputorum, in the gingival crevice, 61, 63
Vulvovaginitis, mycotic, 180, 181
Washing, effects of on skin flora, 38
Water supply, influence on size of cutaneous populations, 9, 19
Yeasts, in dental plaque, 102, 103
  in the vagina, 161, 162, 164, 165, 179, 180
  inhabiting the skin, 8
  symbiotic effect with, 116
Yellow fever virus, 257
Zymotic disease, epidemic spread of, 135